FREETHINKERS OF MEDIEVAL ISLAM

FREETHINKERS OF MEDIEVAL ISLAM

Ibn al-Rāwandī, Abū Bakr al-Rāzī, and Their Impact on Islamic Thought

BY

SARAH STROUMSA

TUTA SUB AEGIDE PALLAS · 1683

BRILL
LEIDEN | BOSTON

This paperback was originally published in hardback as Volume 35 in the series *Islamic Philosophy, Theology and Science. Texts and Studies.*

Cover illustration: Samarkandskii bazar i ego tipy. Doktor prodaiushchii lekarstva (tabibi daru-furush). (Translation: Samarkand bazaar and its types of vendors. Doctor selling medicine (doctor selling medicine). From Turkestanskiĭ al'bom, chast' ėtnograficheskaia͡..., 1871–1872, part 2, vol. 2, pl. 145. Photo taken between 1865 and 1872. Courtesy of the Library of Congress Prints and Photographs Division.

The Library of Congress has cataloged the hardcover edition as follows:

Library of Congress Cataloging-in-Publication Data

Stroumsa, Sarah.
Freethinkers of medieval Islam : Ibn al-Rāwandī, Abū Bakr al-Rāzī and their impact on Islamic thought / by Sarah Stroumsa.
p. cm — (Islamic philosophy, theology and science, ISSN 0169-8729 ; v. 35)
Includes bibliographical references and index.
ISBN 9004113746 (cloth : alk. paper)
1. Rāzī, Abū Bakr Muḥammad ibn Zakarīyā, 865?-925? 2. Rāwandī, Aḥmad ibn Yaḥyá, 9th cent. 3. Philosophy, Islamic 4. Philosophy, Medieval 5. Free thought. I. Title. II. Series.
B753.R384S77
297.2'092'2—dc21

99-20112
CIP

Want or need Open Access? Brill Open offers you the choice to make your research freely accessible online in exchange for a publication charge. Review your various options on brill.com/brill-open.

Typeface for the Latin, Greek, and Cyrillic scripts: "Brill". See and download: brill.com/brill-typeface.

ISBN 978-90-04-31547-1 (paperback, 2016)
ISBN 978-90-04-11374-9 (hardback, 1999)

This book is printed on acid-free paper and produced in a sustainable manner.

For Guy

CONTENTS

PREFACE TO THE PAPERBACK EDITION

The years and months preceding the publication of this paperback edition have witnessed an appalling rise in the "dissemination of discord" and bloodshed, for which, according to Nāṣir-i-Khusraw, Abū Bakr al-Rāzī blamed the "machinations" of impostors claiming to be divine messengers.[1] The presence of religious violence carried out in the name of Islam in the daily news stands in stark contradiction to the existence of independent, free thought in the world of Islam. The book published here, however, does not deal with these very disturbing contemporary questions. Its concern is limited to the medieval world of Islam, and to one particular kind of free thought in it, namely the expressed denial of prophecy.

My narrow definition of freethinking as the denial of prophetic religions may disappoint those who expect to find behind the title "freethinkers" a comprehensive exposition of enlightened Islam, or a complete list of open-minded humanists and independent thinkers in the world of medieval Islam. A broader use of the term "freethinking," reflecting a more pluralistic view of medieval Islam, may indeed be considered useful today in response to the growing, oppressing tidal waves of benighted Islam as well as to the equally oppressing and benighted waves of Islamophobia. But the narrow definition of "freethinking" adopted in the present book allows us to identify a specific concern of medieval Islam in this period. It also allows us to treat medieval Islamdom as a culturally unified world, in which Jews and Christians shared the theological concerns of Muslims.

In the years that have passed since the first publication of this book in 1999, some of the issues discussed in it have been revisited by other scholars, who have debated or challenged some of the arguments it offers, or added new material deserving of consideration. For some, the thinkers discussed in this book were in fact not freethinkers in this sense at all,[2] while others see the phenomenon of freethinking in medieval Islam as deeper, or as broader and more widespread than

[1] See below, p. 106.

[2] For example Marwan Rashed, "Abū Bakr al-Rāzī et la prophétie," *Mélanges de l'Institut Domincain d'Études Orientales du Caire* 27 (2008): 169–82.

presented here.[3] Some have pointed to the fact that preoccupation with the authentication of prophets can be detected already before the rise of Islam,[4] while others have highlighted the far-reaching implications of medieval Islamic freethinking in later periods.[5] In a series of stimulating workshops held at the Institute for Advanced Study in Princeton the late Patricia Crone invited colleagues to explore various aspects connected to the broader issues of free thinking.[6] Some new studies have appeared, relevant to specific parts of the book, or challenging a specific argument in it.[7] In still other cases, the arguments and the material presented in this book were integrated into studies of deviant Islam, with or without mention of their origin in the present book.[8] Moreover, many new relevant texts pertaining to philosophy, theology and the intellectual history of Islam have been edited in the last few years.

The arguments presented against the book's main thesis did not, on the whole, change my mind: I still believe that within the general flux of intellectual opinions in Abbasid Islam, prophetology and its counter position of freethinking can be seen as specifically Islamic concerns, that played a major part in shaping Islamic thought. Ibn al-Rāwandī

[3] See, for example, Patricia Crone, "The Dahrīs according to al-Jāḥiẓ," *Mélanges de l'Université St Joseph* 63 (2010–11):63–82; Eadem, "Al-Jāḥiẓ on *aṣḥāb al-jahālāt* and the Jahmiyya," in *Medieval Arabic Thought: Essays in Honour of Fritz Zimmermann*, ed. R. Hansberger, M. Afifi al-Akiti and C. Burnett (London and Turin 2012), 27–40; and compare below, pp. 121–30. See also the collection of articles by various authors, published under the title *Dangerous Freethinkers* in *Critical Muslim* 12 (2014).

[4] M. E. Pregill, "Ahab, Bar Kokhba, Muhammad, and the Lying Spirit: Prophetic Discourse before and after the Rise of Islam," in Ph. Townsend and M. Vidas (eds.), *Revelation, Literature and Society in Antiquity* (Tübingen: Mohr Siebeck, 2011), 271–313; compare below, 23–26. Pregill convincingly shows that the interest of Christians in listing the proofs for prophesy is recorded in late antiquity before the rise of Islam. Nevertheless, it is also obvious that it took a new turn, and became much more pronounced, in response to the Muslim challenge

[5] For example, Jonathan I. Israel, *Enlightenment Contested: Philosophy, Modernity, and the Emancipation of Man 1670–1752* (Oxford: Oxford University Press, 2006), 631–39.

[6] *The Materialistic World View from Late Antiquity to Islam* (workshop held in 6–10 May 2007); *Islamic Freethinking and Western Radicalism* (workshop held in 20–24 April 2008, organized by P. Crone and Jonathan Israel).

[7] Thus, for example, Maribel Fierro, "Ibn Ḥazm and the Jewish *zindīq*," in Camilla Adang, Maribel Fierro and Sabine Schmidtke (eds.), *Ibn Ḥazm of Cordoba: The Life and Works of a Controversial Thinker* (Leiden: Brill, 2013), 497–509. Compare, in the same volume, David J. Wasserstein, "Ibn Ḥazm and al-Andalus", 69–85, at 76; and compare below, pp. 198–213.

[8] To these categories one may add the translation of the book to Indonesian, which despite repeated efforts I was unable to see; *Para Bemikir Bebas Islam: Mengenal Pemikiran Teologi Ibn ar-Rawandi dan Abu Bakr ar-Razi*, LKiS, 2006.

and al-Rāzī remain, I believe, the foremost protagonists of this kind of freethinking, and I have not been convinced by the attempts to turn them into good orthodox Muslims. On the other hand, I am also not at all convinced by the efforts to view freethinking as omnipresent in this period, let alone by the presentation of people like Ibn al-Naghrilla as tapping into the freethinkers' wellspring. I have therefore chosen to republish this book in its original form. Now that it is accessible in paperback, and thanks to the new texts that keep surfacing, I hope that it may trigger new research, permitting us to continuously sharpen our understanding of the medieval Islamicate world.

PREFACE

> In our attempts to verify historical traditions, our only possible stratagem is the cautious application of independent reasoning as best we can. It is reprehensible to refrain [from discussing] the things we know because there are [other] things which we do not know. Therefore, let us offer, as an excuse [for our shortcomings], the muddle [in our sources].[1]

The present book revolves around two paradoxes. Its main protagonists, Ibn al-Rāwandī and Rāzī, were, according to the testimony of their contemporaries, among the brightest intellects of their time, and both gave considerable thought to religion. And yet in histories of medieval Islamic religious philosophy they usually occupy a relatively marginal place, and are considered as two eccentrics rather than as major luminaries. The second paradox is related to the first: the phenomenon studied in this book is indeed, from a sociological viewpoint, marginal and short-lived in the world of medieval Islam. And yet the vehement reaction of Muslim thinkers can give the illusion that freethinking was a widespread phenomenon, serious enough to present a real threat to orthodox Islam over a few centuries.

These two paradoxes form the axis of this book. Its two main goals are to study the freethinkers of medieval Islam, and to point to their impact on later Muslim thought.

The enormous difficulties of such a project are obvious: even if all the texts were extant, the missing facts and our own distance from the historical context and cultural milieu give ample room for misreadings and misinterpretations. In the case of the topic at hand, almost all the original texts are lost, and one has to rely on later, often biased sources. The likelihood of misinterpretation becomes here almost a certainty.

---

[1] *Lā ḥīlata lanā fī taṣḥīḥ al-aḫbār illā bi-ghāyat al-ijtihād wa'l-iḥtiyāṭ, wa-qabīḥ tarku mā naʿlamu li-mā lā naʿlamu; falnabsuṭ min al-iḍṭirābi ʿudhranā*; al-Bīrūnī, quoted as epigraph by H. Steiner, *Die Muʿtaziliten oder die Freidenker im Islam*, Leipzig 1865. I was unable to identify the exact source of this citation, although its content is certainly in line with al-Bīrūnī's spirit; see, for instance, his introduction to *Hind*, esp. 19:4.

The epigraph of this book is derived from Bīrūnī, as quoted in H. Steiner's *Die Muʿtaziliten oder die Freidenker im Islam*. Bīrūnī's words offer a model of both humility and perseverance in the pursuit of knowledge based on confused and limited evidence. On the other hand, Steiner's title provides us with a hopeful note: in the century that has passed since he wrote, the publication of new material has substantially added to our knowledge of the Muʿtazila. At some point in the future, we may know a bit more than we do today about the freethinkers of medieval Islam. It is hoped that this book will be a contribution toward this goal.

The scarcity and contradictory nature of the sources offer fertile ground for disagreements. In almost every part of this book, some of my suggestions are at variance with those of a number of scholars; in some cases, the publication of an earlier version of my argument has triggered responses to which I try to respond here. Before plunging into the *mêlée*, however, I would like to acknowledge my debt to these scholars. The magisterial works of Paul Kraus, Shlomo Pines, and Josef van Ess, not only made available a substantial body of texts, but also have offered a profound and inspiring reading of these texts. It is their own work and erudition which makes it possible to suggest other interpretations, including those I offer here.

Although chapters of the book, or significant parts of them, were published earlier as articles, they have all undergone drastic changes. Previous versions of parts of the book appear in the following articles:

"The Signs of Prophecy: The Emergence and Early Development of a Theme in Arabic Theological Literature," *HTR* 78 (1985), 101–14 (revised in Chapter One).

"The Blinding Emerald: Ibn al-Rāwandī's *Kitāb al-Zumurrud*," *JAOS* 114 (1994), 163–85 (revised in Chapter Two).

"The Religion of the Freethinkers of Medieval Islam," in F. Niewöhner, ed., *Atheismus im Mittelalter und in der Renaissance* (Wolfenbüttel, forthcoming), (a revised version of which is included in Chapter Four).

"The Barāhima in Early Kalām," *JSAI* 6 (1985), 229–41 (revised in Chapter Five).

"Ibn al-Rāwandī's *sūʾ adab al-mujādala*: the Role of Bad Manners in Medieval Disputations," in M. R. Cohen, S. H. Griffith, H. Lazarus-Yafeh, and S. Somekh, eds., *The Majlis-Book*, (Studies in Arabic Language and Literature; Wiesbaden: Harassowitz, forthcoming), (a revised version of which is part of Chapter Six).

"From Muslim Heresy to Jewish-Muslim Polemics: Ibn al-Rāwandī's *Kitāb al-Dāmigh*," *JAOS* 107 (1987), 767–72 (revised in Chapter Seven).

"Elisha Ben Abuyah and Muslim Heretics in Maimonides' Writings," *Maimonidean Studies* 3 (1992–93), 173–93 (revised in Chapter Eight).

Over the years, I had the opportunity to present bits and pieces of my research, in writing and orally, and to benefit from the comments and corrections of friends and colleagues. In particular, I would like to acknowledge, thankfully, the help of Meir M. Bar-Asher, Haggai Ben-Shammai, Rémi Brague, Michael A. Cook, Patricia Crone, Yohanan Friedmann, Daniel Gimaret, Sidney H. Griffith, Etan Kohlberg, Shaul Shaked, Frank Stewart, Guy Stroumsa, Dominique Urvoy, Steven Wasserstrom, and David Wasserstein. Their willingness to read drafts of my articles and comment on them had preserved me from many pitfalls.

A lecture at McGill University in 1986 gave me the opportunity to present a first, hesitating form of the idea behind the present book. A fellowship at the Annenberg Institute in Philadelphia in 1991–92 allowed a wonderful year of concentrated work. Philippe Hoffmann's kind invitation to give a series of lectures at the Ecole Normale Supérieure (Paris) in 1993 offered me the opportunity for a preliminary synthesis of the material.

I am grateful to Hans Daiber, the editor of *Islamic Philosophy and Theology*, for having accepted this book in the series, and for his erudite remarks on the manuscript. I also wish to thank Trudy Kamperveen for her efficient and cheerful help during the preparation of the book for publication; Alifa Saadya, who copy-edited it; and Ronit Nikolsky, who prepared the index.

My daughters Rachel and Daphna not only gracefully indulged my preoccupation with the freethinkers, but also actively encouraged it and helped with their perceptive remarks. I dedicate this book to my husband Guy, whose eager curiosity and attentive listening have kept me from despairing of this long project.

INTRODUCTION

FREETHINKING AND FREETHINKERS IN MEDIEVAL ISLAM

1. The Boundaries of the *Umma*

A central issue in early theological discussions in Islam is the question of who is to be counted as a believer (*mu'min*). In asking this question early medieval Muslim thinkers sought to define the essence of Islam and to delineate the borders of the Muslim community.[1] Their views were sometimes formulated in positive terms,[2] but quite often they addressed the issue in a roundabout way, seeking to define who was *not* a believer.

In this context, Muslim writers paid relatively little attention to non-Muslims, who were regarded as outright infidels.[3] Instead, they focused their attention on those who are considered to be heretics, that is to say: people who are nominally Muslims, but who adhere to non-orthodox beliefs. The study of accusations of heresy can thus serve as an important tool in the attempt to understand the development of Muslim self-perception.

[1] The terms "medieval (or: early medieval) Islam" and "early Islam" are used in the present book as near synonyms, to denote the period around the third-fourth/ninth-tenth centuries, a period which has also been called "The Classical Age of Islam"; see Marshall G. S. Hodgson, *The Venture of Islam: Conscience and History in a World Civilization*. I. *The Classical Age of Islam*, Chicago and London 1979. For the periodization suggested by Hodgson, who distinguishes between "High Caliphal Period" (692–945) and "Earlier Middle Period" (945–1258), see pp. 234–35. See also R. Brague, "Elargir le passé, approfondir le présent," *Débats* 72 (1992), 34.

[2] See, for example, Ghazālī, *Fayṣal*, 195: "The principle [doctrines which constitute] faith are three: The belief in God, His Messenger and the Last Judgement. Everything else is derivative." See also below, apud n. 9.

[3] See, for instance, Ghazālī, *Fayṣal*, 134. Ghazālī's initial incentive for writing the book was the accusations leveled against him for professing views that were not strictly Aš'arite. His detractors had claimed that "straying from Aš'arī's positions, even by an inch, is *kufr*" (*Fayṣal*, 127). Nevertheless, although *kāfir* was indeed a fairly frequent term for Muslim opponents, it essentially designated non-Muslims; see art. "Kāfir," *EI*², IV, 407–409 (W. Björkman). For a similar distinction in Judeo-Arabic literature, see below, Ch. 8, n. 25.

Not all heresies were seen as equally deviant. While some opinions, though fiercely disputed, were treated as wrong interpretations of Islam, others were viewed as being definitely outside Islam.[4] One of the most striking yet neglected features of early Islam is the nature of the internal opposition to which it gave rise. Although the ideal of unity is implied in the very concept of *umma* or community, doctrinal disagreements and plurality of opinions have been part of the reality of the Muslim community since its early days. And, being part of the reality, they were also accommodated by theory. When limited in scope, pluralism—notably in legal matters—was regarded in early Islam as benign, legitimate, at times even fruitful. It was this kind of pluralism which was greeted by the *ḥadīth*: "The disagreements of my community are a blessing" (*iḫtilāf ummatī baraka/raḥma*).[5]

Much less welcome were major doctrinal disputes, which often came hand in hand with political dissension and communal split. This negative attitude is also reflected in a *ḥadīth*, which says: "Every community is tested by a predicament: the predicament of my community are the sects."[6] Still more famous is the *ḥadīth* about the seventy-three sects of Islam, which clearly deplores the reality of such disputes. The *ḥadīth* also grimly predicts the future outcome of sectarianism, namely, the perdition of seventy-two sects. This would seem to imply that only the seventy-third, the "saved" sect (*al-firqa al-nāǧiya*), is truly Muslim. The other seventy-two were perceived only as what in Christian parlance would have been called "heresies:" deviations from the right path.[7]

---

[4] See, for instance, Laoust, "Classification," 22–25.

[5] Cf. *Mawsūʿat aṭrāf al-ḥadīth*, I, 183 (which cites several sources, e.g. *Kanz al-ʿummāl* 28,686; the *ḥadīth* is not listed in the canonical collections); Lewis, "Heresy," 53. On the early meaning and later developments of this dictum, see R. Paret, "Innerislamischer Pluralismus," in U. Haarmann and P. Bachmann, eds., *Die islamische Welt zwischen Mittelalter und Neuzeit: Festschrift für Hans Robert Roemer zum 65. Geburtstag*, Beirut 1979, 523–29, esp. 524–25 (I am indebted to Yohanan Friedmann for this reference). A similar notion of benign diversity may be reflected in the *ḥadīth* which compares the difference of opinions to the distinction between the prophet's fingers (. . . *fa-ka-annī anẓuru ilā iḫtilāf aṣābiʿ rasūl allāh*; cf. Wensinck, *Concordance*, II, 68; and see, for instance, Muslim, *Masāǧid*, 26–27; *Ṣalāt*, 146). On the question of the legitimacy of disagreements in legal matters, see É. Chaumont, "'Tout chercheur qualifié dit-il juste?' (*Hal kull muǧtahid muṣīb*): La question controversée du fondement de la légitimité de la controverse en Islam," in Le Boulluec, *Controverse religieuse*, 11–27.

[6] *Inna li-kulli ummatin fitnatan, wa-fitnatu ummatī al-milal*; cf. Wensinck, *Concordance*, I, 93.

[7] For this *ḥadīth*, see Wensinck, *Concordance*, I, 93; I. Goldziher, "Le dénombrement

And yet in Muslim doxographical works—in the works of ʿAbd al-Qāhir al-Baghdādī (d. 429/1037), Ibn Ḥazm (d. 456/1064) or Šahrastānī (d. 568/1153), for instance—this last prophetic tradition often served as a framework for discussing the Muslim community and drawing the boundaries of Islam in its broadest sense.[8] However scandalous a heresy could have seemed to the heresiographer, as long as it could fit into the framework of the seventy-three sects, it remained, for better or for worse, part of the Muslim community. Thus Baghdādī, in an attempt to define the idea underlying the expression "the Muslim community," quoted the Muʿtazilite Abū al-Qāsim al-Kaʿbī (d. 319/931), who had said:

> When one uses the expression *ummat al-islām*, it refers to everyone who affirms the prophetic character of Muḥammad, and the truth of all that he preached, no matter what he asserted after this declaration.[9]

The fact that so many heretical views were considered to belong within the community highlights the existence of some other views, which were considered to be outside the boundaries of Islam properly so called. This fact is perhaps best illustrated by Šahrastānī, whose *Book of Religions and Sects* represents, in many ways, one of the climaxes of medieval Arabic heresiographical literature. Writing in the middle of the sixth Islamic century, Šahrastānī made good use of the abundance of previously composed heresiographies. To the information provided by his sources he added his own insights, and applied to the whole material his remarkable analytical and synthetical talents. Šahrastānī discussed the seventy-three sects of Islam in the first part of his book, the part dedicated to scriptural religions (*al-diyānāt wa'l-milal*).[10] But some Muslims—including philosophers such as Ibn Sīnā—Šahrastānī relegated to the second part of his

---

des sectes mohammétanes," *RHR* 26 (1892), 129–37; idem, *Introduction to Islamic Theology and Law*, trad. A. and R. Hamori, Princeton 1981, 167; Laoust, "Classification," esp. 22–25; cf. also Shahrastani, *Livre des religions*, 108, n. 3. Another version of the same prophetic tradition approves of the seventy-odd sects, condemning for perdition only one; see, for instance, Ghazālī, *Fayṣal*, 206–207.

[8] See, for instance, Baghdādī, *Farq*, 4–8; Ghazālī, *Fayṣal*, 206.

[9] Baghdādī, *Farq*, 8–9 (= Seeyle's translation, 27).

[10] *Milal*, 2, 24. As scriptural religions Šahrastānī counted Manicheism, Zoroastrianism, Judaism, Christianity, and Islam. Šahrastānī's views of the identity of the seventy-third sect, the only one which will be saved, are discussed in G. Monnot, "Les controverses théologiques dans l'œuvre de al-Šahrastānī," in Le Boulluec, *Controverse religieuse*, 281–96; see also D. Gimaret's Introduction to Shahrastani, *Livre des religions*, 11–20.

book, where he discussed the "arbitrary notions" (*al-ahwā' wa'l-niḥal*): Indian religions, Sabeans, religions of the Arab *Jāhiliyya*, and all the philosophers.[11] According to Šahrastānī, the common denominator of these groups was the fact that they relied on traditions which did not originate in a revealed book, and that they resorted exclusively to the use of their personal opinion. Šahrastānī called this attitude *al-istibdād bi'l-ra'y* ("the exclusive, willful use of [personal] opinion"), a term aptly translated by Daniel Gimaret as "libre pensée."[12] The layout of Šahrastānī's book suggests that, unlike the seventy-three sects, the formally Muslim philosophers discussed in the second part were not "heretics" in the proper sense, since they were not accused of offering an unacceptable interpretation of Islam. According to Šahrastānī's understanding, they fell outside the pale of Islam altogether.[13]

For Šahrastānī, the first one to have erred in this way had been Iblīs, who had "relied solely on his own opinion and did not accept the authority of [God's] explicit order; . . . he opted for vain notions, in opposition to the [divine] order . . . because of his vanity." Šahrastānī presents Iblīs as the prototype of the freethinker "the first Cursed one (*la'īn*), for he imposed the government of reason on that which cannot be governed by reason."[14]

---

[11] *Milal*, 2, 25. Cf. also art. "*al-Milal wa'l-Niḥal*", *EI*[2], VII, 55 (D. Gimaret).

[12] Shahrastani, *Livre des religions*, 13, 160; *Milal*, 24–25. In this context, Šahrastānī quotes a prophetic tradition (*ḥadīth*): "No person was ever made unhappy because of consulting [other people], and none was made happy by relying solely on [his own] opinion" (*mā šaqiya imru'un 'an mašūra wa-lā sa'ida bi-istibdād bi-ra'y*; cf. *Mawsū'at aṭrāf al-ḥadīth*, IX, 155, citing *Tafsīr al-Qurṭubī*, IV, 251). In Šahrastānī's use of the term, however, the most blameworthy trait of the *mustabiddūn bi'l-ra'y* is not their refusal to listen to other people's opinion, but rather their rejection of religious (especially prophetic) authority and their exclusive reliance on reasoning (*ra'y*); see further below.

[13] The observation that radical heretics are not counted among the seventy-three sects was already made by Gabrieli, "Zandaqa," 24. In practical terms this could mean, as suggested by Kraemer, that "the 'zanādiqa' were *quasi-ḥarbis*, internal members of the *dār al-ḥarb*, who were treated as enemies whose blood could be spilled with impunity" (Kraemer, "Heresy," 178). In fact, as noted by both Gabrieli ("Zandaqa," 37) and Kraemer ("Heresy," 175), the kind of *zindīqs* discussed in the present book suffered from surprisingly little persecution; see also below, Ch. 3, apud nn. 168–76.

[14] *Milal*, 5. See further below, Ch. 4, apud n. 47. A similar view was expressed by Ibn al-Jawzī, who said that Iblīs deluded the philosophers so that they "relied solely on their personal opinions and intellects. They supported doctrines which were dictated by their own misconceived notions, disregarding the prophets" (*innahum infaradū bi-ārā'ihim wa-'uqūlihim, wa-takallamū bi-muqtaḍā ẓunūnihim, min ghayri iltifāt ilā al-anbiyā'*; cf. *Talbīs*, 45). For Ibn al-Jawzī, the philosophers' sin (their *infirād bi'l-ra'y*, which is equivalent to Šahrastānī's *istibdād bi'l-ra'y*) is not their refusal to listen

In Šahrastānī's view the freethinkers par excellence, in whose case this title requires no further qualification, were those who rejected all revealed religions (*al-mustabiddūna bi'l-ra'y muṭlaqan hum al-munkirūna li'l-nubuwwāt*). People who belonged to this category, like the philosophers, the Sabeans, or the *Barāhima*, did not believe in religious precepts and in given laws (*šarā'i' wa-aḥkām amriyya*). Instead, they invented rational strictures (*ḥudūd 'aqliyya*), to insure the possibility of communal human life.

One could justifiably argue that Šahrastānī's choice of thinkers figuring in this category is in itself an "arbitrary notion." Nevertheless, although one can take exception to Šahrastānī's choice of freethinkers, his concept of freethinking is worth borrowing. In Islamic heresiographical literature, heretics of all shades and colors were labelled rather loosely as *zanādiqa* (sing. *zindīq*), a term which could refer to dualists (especially Manicheans) as well as to philosophers or theologians whose doctrinal audacity irked their *bien pensants* Muslim adversaries.[15] Another term, *malāḥida* (sing. *mulḥid*), was sometimes used more specifically to denote apostates or heretics who radically opposed the contemporaneous idea of religiosity. According to the mu'tazilī scholar Zamaḫšarī (d. 539/1144), for example, a person was labelled a *mulḥid* not when he abandoned one religion for the sake of another, but only when he adopted a system which deviated from all religions.[16] But *mulḥid* was also used more vaguely, to designate other kinds of religious deviations, or as a synonym of *zindīq*.[17]

---

to other people's opinion, but the fact that they disregard the prophets. A demonization of heretics in general and of freethinking heretics in particular is reflected also in the *majālis mu'ayyadiyya*, where an Ismā'īlī *dā'ī* identifies the demons (*ghīlān*) mentioned in the *ḥadīth* as *ahl al-ilḥād wa'l-zanādiqa*; see Kraus, "Ketzergeschichte," 108:3–5.

[15] Cf. van Ess, *TG*, I, 416–18; Mohaghegh, *Bist Guftar*, 191–92; Chokr, *Zandaqa et zindīqs*, esp. 64–68, and see the preface of D. Gimaret, ibid., 7–8; M. I. Fierro Bello, "Accusations of 'Zandaqa' in al-Andalus," *QSQ* 5–6 (1987–88), 251–58. One example may suffice to illustrate the ubiquitous use of this term: Abū Nu'aym (al-Mulā'ī, d. 219/834) testified that in his own experience, any person who adhered to the doctrine of the created Qur'ān and tried to support it in a theological discussion would be labelled a *zindīq*; see 'Aynī, *'Umdat Al-Qārī'*, I, 296; I owe this reference to Vardit Tokatly.

[16] *Kaššāf*, II, 429 (commenting on Q. 16:103): *wa-minhu al-mulḥid, li-annahu amāla madhhabahu 'an al-adyān kullihā, lam yumil 'an dīn ilā dīn*. On other Qur'ānic instances of *ilḥād* and its derivatives, see Q. 7:180; 22:25; 41:40. On *mulḥid* in Aš'arite terminology, see Gimaret, "Bibliographie," 228 and nn. 13, 15; and see below, Ch. 4, n. 16.

[17] See, for instance, the explanations listed in *Lisān al-'arab*, 3, 388–89. Cf. also Gabrieli, "Zandaqa," 24.

The term *mustabidd bi'l-ra'y*, or freethinker, allows us to distinguish what Muslims felt to be a distinct and particularly pernicious category of disbelief. Muslim heresiographers, including Šahrastānī, did not give this category a special name. But their description of its characteristics and the particularly violent language they used when discussing it demonstrate the degree of Muslim sensitivity to its threat.[18]

As mentioned above, a close examination of Šahrastānī's list of *aṣḥāb al-niḥal* discloses surprisingly few people who would truly fit his definition of real freethinkers. Šahrastānī's inflated list allows us to diagnose an important trait of freethinking in medieval Islam, which is precisely its imaginary aspect. Quite often, Muslim authors who launch an attack against freethinking and its dangers do not attempt to illustrate their claims with facts, and their diatribes are accompanied by warnings against unidentifiable shadows. When they do identify the object of their criticism, the individuals they accuse of denying prophecy often turn out to be thinkers who, quite on the contrary, incorporated prophecy in their thought or who seem to have taken the notion of prophecy for granted. The danger presented by freethinkers was felt by Muslims as profound and pressing, out of all proportion to their actual number or real strength.[19] The few people who really deserved to be called freethinkers remained individualists and did not establish any sect. But they embodied the idea that one can wish to establish a personal ethical system, organize a human society, or even worship God without relying on the authority of revealed scriptures. This very idea was perceived by Muslims as an abomination, a radical revolt against Islam. The actual number of intellectuals who consistently adhered to the ideal of freethinking in medieval Islamic society is in all probability negligible. But the very existence of such individuals, however few and eccentric they may have been, triggered a massive reaction among mainstream thinkers, a reaction which sometimes reflected an obsession

[18] See, for example, Ḫayyāṭ, *Intiṣār*, 11; Ibn al-Jawzī, below, Ch. 7, n. 31; and see also Shmu'el ben Ḥofni, below, Ch. 8, n. 26.

[19] This feeling of being besieged by a ridiculously insignificant minority is reflected in the words of 'Abd al-Jabbār, who complained that the books of heretics, written in extreme seclusion, become in a short time widespread and even popular; see *Tathbīt*, 129, 374.

or even a phobia. The fact that as a prototype of the freethinker Šahrastānī chose none other than Iblīs himself is an apt illustration of this reaction, which literally demonized the freethinkers.[20]

2. Freethinking

One may question the legitimacy of borrowing a term from early-modern European intellectual history to describe a phenomenon in the medieval Islamic world. Even in Europe, the term "freethinking" was often used loosely and was applied to various movements and persons, especially in a derogatory sense.[21] Similarly, in modern Islamic scholarship the term "freethinking" (and its equivalents in other European languages) has been frequently used, like *zandaqa*, to cover anything from dualism to libertinism.[22] At other times, it was attached to a specific movement within Islam. Thus the Muʿtazila were called "the freethinkers of Islam"[23] as were also the *falāsifa*, and even the *Ḥašīšiyya*.[24] When used in this indiscriminate way, the term "freethinking" (like the term *zandaqa*) becomes a general label under which one could lump together all those who were, or were thought to be, more rationalistically minded than the orthodoxy. For a term to be historically useful, it must be defined more accurately. In European intellectual history, from which the term "freethinking" is borrowed, it suggested some kind of independent thought. In the various dictionaries and encyclopedias, the main trait of the free-thinkers is their rejection of ecclesiastic authority, a rejection that encompassed the institutions as well as the scriptures from which the former derived their authority.[25]

---

[20] The depiction of heretics as followers of Satan is, of course, a topos of heresiographical literature. For an example of particularly thorough demonization, see Ibn al-Jawzī's *Talbīs*, which identifies Iblīs's fingerprints in practically every trend of thought in Islam. See also above, n. 14.

[21] See Spink, *French Free-Thought*, 3–4 (on the terms "libertins" and "Free-thinking").

[22] Cf., for example, van Ess, *TG*, I, 416; Brion, "Philosophie et révélation," 147, uses it to translate *zandaqa*; so also McCarthy, *Freedom and Fulfillment*, 118, n. 20; de Smet, "Polémique ismaélienne," 85.

[23] See Steiner (above, Preface, n. 1).

[24] Cf. F. Nietzsche, "Zur Genealogie der Moral," *Werke in drei Bänden*, Munich 1955, 889, quoted in: Niewöhner, *Veritas*, 252.

[25] Spink, *French Free-Thought*, 4.

I suggest that we limit the application of the term "freethinkers" in early Islam to those Muslim intellectuals who, in opposition to other heretics, did not adhere to any scriptural religion. Although the position of these thinkers within Muslim society was very different from that of the philosophers in the ancient city, one might get some insight by comparing them to those Greek thinkers accused of atheism in the ancient world, because they denied any truth value to the cultic religion of the polis and proposed alternative systems of religious belief. The animosity of the freethinkers of Islam sometimes turned especially against the religion into which they were born, but in principle they displayed the same aversion to all revealed religions. One cannot really consider them to have been "Muslim heretics," since they did not offer a new, heretical, interpretation of Islam.[26] Rather, it was the very message of Islam, its very foundations—The Qur'ān and the Prophet—which they rejected.

The term "freethinking," in the narrow sense of the rejection of revelation, focuses on a common trait in the thought of intellectuals whose beliefs may otherwise differ considerably. It is argued that the paramount importance of this trait justifies a study that will combine them together, despite their differences.

In modern scholarship these intellectuals are sometimes referred to as "atheists." This, however, is certainly incorrect, since their criticism of religion never included the negation of God's existence.[27] What they did deny was the scriptural religions' idea of God, His epithets, and His interference in the world through revelation. Another phenomenon in modern European thought, deism, seems at first sight to offer a better parallel to the Islamic freethinkers. The deists' rejection of the authority of the church and of the scriptures was perceived by Christian theologians as a radical religious rebellion, despite the deists' belief in God. Thus Pierre Viret, a sixteenth-century Calvinist, admitted that the deists retained some kind of belief in God, "like the Turks and the Jews," but he accused them of regarding the doctrines of the Evangelists and the Apostles as "fables and reveries."[28] In modern scholarship we often see a conflation of the two terms, "deism" and "freethinking." R. L. Emerson, for instance,

[26] Nevertheless, and for lack of a better concise term, in what follows I will often use the word "heretic" to denote these people who, technically speaking, are even beyond heresy, or to translate the term *mulḥid* when it applies to them.

[27] See below, Ch. 4.

[28] P. Viret, *Instruction Chrestienne* (1564), quoted by Emerson, "Deism," 648a.

began the bibliographical part of his article on deism with two eighteenth-century German works on the *Freidenker*,[29] who, he said, are still useful as guides to the background on which deism grew. The thinkers who are the subject of the present book did indeed have some traits in common with the deists; like them, they believed that natural reason was sufficient to attain truth; like them, they believed in the natural laws and science which should be the only guide to human conduct; and like them, they regarded with suspicion all established religions. But other central components of deism, and primarily the belief in the goodness of divine providence, were missing in the thought of some of the freethinkers of Islam.[30] In the search for a term that will identify the common traits of these people, "deists" thus seems unsuitable.

For the same considerations it is also unsuitable to call them "materialists." If we understand materialism in the widest sense, as a system which "excludes the existence of entities that are radically different in kind from, and in some sense superior to, the matter of our ordinary experience,"[31] then the thinkers to be discussed below were far from agreeing with each other on that score.[32]

I have therefore opted for the term "freethinkers," which I use to denote the advocates of autonomous reflection on the major metaphysical and human issues, with no commitment to the monotheist tradition.[33] Because of the connotation of the term in modern history, I would like to stress the obvious, namely, that the use of this term here carries absolutely no judgement of value. These freethinkers were not necessarily more rationalists than the religious *ahl al-raʾy*. They were also not always more humanistic or more tolerant in their approach; in fact, one may sometimes detect a definite misanthropic streak in their expressions. The term "freethinking" is used here solely to denote the rejection of the authority of both revelation and of the revealed religions.

[29] J. A. Trinius, *Freidenkers Lexicon*, Leipzig and Bernburg 1759; U. G. Thorschmidt, *Versuch einer Vollständiger Engelländischen Freidenker-Bibliothek*, Halle 1765–66; cf. Emerson, "Deism," 652.

[30] Rāzī could, according to some interpretations, be called a deist. See below, Ch. 4, apud n. 44.

[31] Cf. E. McCullin, "Materialism," *Encyclopedia of Religion*, ed. M. Eliade, New York and London 1987, 9, 279–80.

[32] See below, Ch. 4.

[33] See also J. M. Robertson, *A History of Freethought in the Ninteenth Century*, New York 1930, esp. xxxiv–xxxvi.

The insistence on the precise meaning of the term is essential if we wish to use it for historical research in any meaningful way. A book published recently by Dominique Urvoy draws a wide and fascinating picture of the rationally inclined intellectuals in the medieval Islamic world. The book, entitled *Les penseurs libres dans l'Islam classique*, offers a series of portraits of thinkers belonging to different religions, all of whom manifested remarkable intellectual rigor and originality. According to Urvoy, the purpose of his book is "to draw attention to the existence of . . . a critical approach within the religious domain."[34] The term chosen by Urvoy, "penseurs libres," may therefore best be translated as "independent thinkers." But this newly-coined appellation closely resembles the established term "libres penseurs." Consequently, the book's title, which faithfully reflects the book's approach, is, in a way, misleading. It gives the impression that an intellectual who adopted rationalistic methods and who hailed the role of the intellect was, ipso facto, "free." Urvoy's book overlooks the fundamental difference between the vast majority of critically-minded intellectuals, who accepted the presuppositions of the monotheistic religions, and between the handful of freethinkers, who did not. In the world of medieval Islam, where beliefs and intellectual attitudes were part of the "religion in a religious world,"[35] the acceptance or rejection of the notion of revelation was not a minor, negligible detail. Admittedly, the religious thinkers discussed by Urvoy were independent, daring and sometimes original. But they were not all "free" in the same sense. A strict definition of freethinking is thus a necessary prerequisite for the study of the extent and nature of free thought in medieval Islam.[36]

3. The Freethinkers of Islam

In the introdction to his *History of Muslim Philosophy* Majid Fakhry wrote:

---

[34] Urvoy, *Penseurs libres*, 8.

[35] It was S. D. Goitein who saw the importance of spelling out this seemingly self-evident fact; see the volume which he edited, *Religion in a Religious Age*, Los Angeles 1974.

[36] Cf. my review of Urvoy in *Bulletin critique des annales islamologiques* 14 (1998), 97–100, and see below, apud n. 48.

> Some of the most fascinating Muslim thinkers, such as al-Naẓẓām (d. 845), al-Rāzī (d. 925) and al-Maʿarrī (d. 1057), fall outside the mainstream of thought in Islam. Their dissident voices lend a discordant note to an otherwise monotonous symphony. The difficulty of expounding their thought with any degree of completeness is bound up with its very nonconformist character. Islam did generate such dissentient and solitary souls, but it could not tolerate or accept them in the end.[37]

We must, first of all, take issue with a presentation that puts al-Naẓẓām and Rāzī on the same footing. Non-conformist as al-Naẓẓām may have been, he saw himself, and was seen by others, as a Muslim, whereas Rāzī was neither.[38] Fakhry's presentation seems to reflect the assumption that the "dissentient voices" were only the expression of particularly strange—and strong—personalities. Indeed, if we were to take eccentricity as a main criterion by which to classify dissent, we would be justified in considering together people with equally strong but very different ideas. H. S. Nyberg, for example, could thus consider together ʿAmr b. ʿUbayd and Ibn al-Rāwandī as two outcasts ("deux réprouvés").[39] Lumping these thinkers together, however, blurs the fundamental difference between an ʿAmr b. ʿUbayd and an Ibn al-Rāwandī, between an innovative nonconformist—at worst a heretic—and a radical freethinker.

Equally misleading is the approach that confuses the rejection of some specific Muslim dogma with the total rejection of Islam. Muḥammad b. Karrām (d. 869), for example, is said to have formulated some hard questions on theodicy. As a result, Muslim authors attacked him and compared him to Ibn al-Rāwandī, thus defaming him by association with a figure considered to be the arch-heretic of Islam. But we would be mistaken to conclude, as Joseph van Ess had suggested, that the nature of Ibn al-Rāwandī's criticism of Islam can be equated with the critical thought of Ibn Karrām. Whereas Ibn Karrām can still legitimately be said to "simply start from different theological assumptions," Ibn al-Rāwandī's rebellion goes far beyond simple theological differences.[40]

In this context, it is illuminating to note the distinction which al-Ḫayyāṭ made between Shiʿites and "those who claim to belong to

---

[37] Fakhry, *Islamic Philosophy*, xxiv.

[38] On al-Naẓẓām, see *EI*², VII, 1057–58 (J. van Ess); van Ess, *TG*, III, 296ff.

[39] Nyberg, "Deux réprouvés;" on ʿAmr, see van Ess, *TG*, II, 280ff; S. Stroumsa, "The Beginnings of the Muʿtazila Reconsidered," *JSAI* 13 (1990), 265–93.

[40] Compare van Ess, "Image," 24–25; *TG*, IV, 300.

the [Muslim] nation" on the one hand, and Ibn al-Rāwandī's brand of disbelief on the other. Turning to Ibn al-Rāwandī, al-Ḫayyāṭ said:

> I do not blame you for [holding] the views of the *rāfiḍa*, nor for [belonging to any] of the schools of those who claim to belong to the *umma*; rather, I blame you because of your own doctrine, that is to say, your belief in the eternity of the world . . . and for composing the *K. al-Zumurrud*.[41]

The approach of Majid Fakhry and of Josef van Ess, who regard freethinking as only a form of eccentricity or of non-conformism, is tantamount to the denial of the very existence of freethinking in medieval Islam. Some modern scholars (foremost among whom is, in fact, van Ess) indeed reject the heretical image of the freethinkers and raise the suspicion that this image, concocted by their jealous opponents, does not stand on any firm ground. We shall return below to examine this view, which I do not share. For our present purpose suffice it to say that in most medieval reports the radical heretical character of the freethinkers is stated as common knowledge. Their rehabilitation to Islam thus requires an arbitrary re-reading of the sources, as well as an equally arbitrary dicision to reject the most prevalent reports.

Perhaps the best example of a freethinker would have been Abū'l-'Abbās al-Īrānšahrī, who flourished in the ninth century C.E. Unfortunately, we know very little about this intriguing personality.[42] Bīrūnī (d. ca. 442/1050) tells us that Īrānšahrī's accounts of the Jews, the Christians, the dualists, and their respective scriptures were exceptionally objective, because he had no religious prejudices to keep at bay. In fact, says Bīrūnī,

> he himself did not believe in any of the then existing religions, but was the sole believer in a religion invented by himself, which he tried to propagate.[43]

---

[41] *Intiṣār*, 123.

[42] On him, see Pines, *Atomism*, 41–42.

[43] "*Lam yakun min jamī' al-adyān fī šay', bal munfaridan bi-muḫtara' lahu, yad'ū ilayhi*;" Bīrūnī, *Hind*, 4 (= *Alberuni's India*, I, 6). According to Nāṣir-i Ḫusraw, however, Rāzī plagiarized Īrānšahrī's ideas and distorted them, turning them into heresy; cf. Pines, *Atomism*, 42, 65–67; Nāṣir-i Ḫusraw, *Zād al-Musāfirīn*, 98, 102 (quoted by Kraus, *Rasā'il*, 255–56, 258–59); cf. German translation in Meier, "Der 'Urknall'," 12–13 (I am indebted to Hans Daiber for calling my attention to this article).

Unfortunately, however, our knowledge concerning Īrānšahrī or his religion is too scant to allow even speculations.[44]

The two most notorious freethinkers, Ibn al-Rāwandī and Rāzī, on whom we shall focus the discussion in what follows, exemplify the fragmentary, often second-hand state of our knowledge. In spite of their very different intellectual backgrounds and tastes, Ibn al-Rāwandī and Rāzī had in common a career of equally vociferous rejection of all religions based on revelation. They founded this rejection on the assumption that the human intellect should be, and indeed is, a sufficient source for all knowledge. Because of this common contention the two are often mentioned together, by Muslim writers like Tawḥīdī as well as by modern scholars like Shlomo Pines and Mehdi Moḥaghegh. Usually, the comparison between the two is limited to pointing out a common trait or a curiously recurring phenomenon. At times this recurring phenomenon is interpreted in terms of *Zeitgeist.* According to Joel Kraemer, for example, political and economic uncertainties in the tenth century encouraged skeptical attitudes. At the same time, the belief that only the Imām possesses the authoritative knowledge—a belief widespread in this "Shiʿite century"—encouraged critical and relativistic approaches to any other kind of knowledge.[45] A similar view has been suggested by van Ess, who pointed out the skepticism (*taškīk*) fostered by the Shiʿites. According to him, this skepticism played down reason and stressed the dead-end to which it lead: "This sort of skepticism could become . . . some sort of vital feeling in the fourth century. It was no radical skepticism, . . . it was skepticism for belief's sake. . . . One approached [the esoteric doctrine] leaving reason behind: one had only to open oneself for the *taʿlīm*, the instruction of the Imām and his emissaries."[46]

Although freethinking in its radical form appeared in or around the fourth/tenth century, treating this phenomenon as if it were only an expression of the *Zeitgeist*, a pure product of the period, is as misleading as the view that regards freethinkers as a few individual eccentrics. Freethinking in its radical form was a typically Islamic phenomenon, a heresy whose particular character developed in response to the centrality of the concept of prophecy in Islam. The

---

[44] See below, Ch. 4, apud. n. 65.

[45] Cf. J. L. Kraemer, *Humanism in the Renaissance of Islam: The Cultural Revival During the Buyid Age*, Leiden 1986, pp. vi, 189–90.

[46] Van Ess, "Skepticism," 91–93.

circumstances of the period probably encouraged the full-fleged manifestations of this phenomenon, but they were not the main cause for its appearance. The study of freethinking in medieval Islam is therefore conceived here as part of the attempt to shed light on the development of the Islamic creed and on its manifestations. This most radical form of disbelief is, paradoxically, an integral part of Islamic religiosity.

Furthermore, if we identify freethinking with skepticism, and consider it as resulting from a malaise of the times, we may cast our net too wide, and end up finding many more "freethinkers" than there actually were. As mentioned above, a similar tendency to equate any manifestation of skepticism or of rationalism with radical apostasy was common among Muslim authors. Thus the ḥanbalite Ibn ʿAqīl (d. 513/1119) could include Tawḥīdī in the same group with Ibn al-Rāwandī and Rāzī, because he suspected this tenth-century man of letters of harboring some skeptic notions (mainly *taʿṭīl*). We would also have to take into account cases like the muʿtazilite leader Thumāma b. al-Ašras, to whom Baghdādī attributed deprecating remarks about the Prophet.[47]

In his aforementioned study of the independent thinkers of classical Islam, Urvoy discussed several Muslim personalities: Ibn al-Muqaffaʿ, Abū ʿĪsā al-Warrāq, Ibn al-Rāwandī, Abū Bakr al-Rāzī and Abū'l-ʿAlāʾ al-Maʿarrī. He also included in his discussion some non-Muslims: Ḥunayn ibn Isḥāq and Ḥīwī al-Balḫī, Ramon Lull, and Ibn Kammūna. Obviously, the inclusion of Ramon Lull as an example of independent thinking in classical Islam is hard to justify. But the difficulty in explaining Urvoy's choice of people, both Muslims and others, extends beyond Ramon Lull: Ḥunayn ibn Isḥāq, an open-minded, liberal, and deeply Christian intellectual, can hardly be seen as a freethinker. Neither can Ḥīwī al-Balḫī and al-Warrāq, whose loyalty to dualist beliefs transpires through their respective criticism of Judaism and Islam.[48] While both liberal-minded intellectuals and heretics often employed Scriptural criticism of the kind

[47] Seeing men hastening to the Friday prayer, Thumāma labeled them "donkeys and cows." "Then he said: 'What has that Arab made out of men,' meaning the Prophet of Allāh"; *Farq*, 158 (Seelye's translation, 179). Although Thumāma must have expressed quite unorthodox opinions, less stern people than Baghdādī may not have been so appalled by them. This is perhaps reflected in the fact that al-Jāḥiẓ included them in his *Book of Jest*; cf. Baghdādī, Ibid., 158–59.

[48] See below, Ch. 2 and Ch. 8; and above, n. 36.

propagated by the freethinkers, this in itself does not make them freethinkers themselves. As much as possible, this name should be given to those who are not part-time freethinkers, i.e., those who are not adherents of any revealed religion.

The cases of Ibn al-Muqaffaʿ (d. ca. 139/756) and Abū al-ʿAlāʾ al-Maʿarrī (d. 449/1058) are more complicated. On various scales, these two qualify as authentic freethinkers: their criticism of religion is both systematic and severe, and their loyalty to established Islam is highly doubtful. But Ibn al-Muqaffaʿ sometimes appeared to be more of a *zindīq* (crypto-Zoroastrian, or someone who was loyal to the Zoroastrian heritage) than a freethinker, whereas Maʿarrī represented not only the fruit of rational speculation, but also of the skeptical tradition prevalent among poets on the one hand and Ṣūfī ascetics on the other.[49]

In the present study, I have therefore chosen to focus on the ninth-century theologian Ibn al-Rāwandī and the tenth-century philosopher Abū Bakr al-Rāzī. These two figures appear in most lists of freethinkers (whether under this heading or under another). They can thus be regarded as representing the hard core of freethinking in medieval Islam. Most of our sources present them as methodically and constantly denying the validity of prophecy. Their apostasy was not mitigated in any way: they did not substitute Islam with any other monotheistic religion, nor did they embrace a dualist religion. The consistent rejection of the notion of prophecy was the most striking trait of their treatment of religion. In the world in which they lived, the repeated, outspoken denial of prophecy was remarkable enough to set them apart from other thinkers, who did not share this view. This distinct religious attitude justifies, I believe, their combined study. Although, as mentioned above, both medieval and modern scholars often mention them together, no separate study was dedicated to the one aspect of their thought to which they owe their notoriety: their freethinking. Such a separate study is necessary not only in order to gain a fuller understanding of the thought of

[49] On Ibn al-Muqaffaʿ, see Guidi, *La Lotta*; S. Shaked, "From Iran to Islam; Notes on Some Themes in Transition," *JSAI* 4 (1984), 31–67; Urvoy, *Penseurs Libres*, 29–66; art. "Ibn al-Muḳaffaʿ," *EI*², III, 883–85 (F. Gabrieli) and the bibliography there. On Maʿarrī, see A. von Kremer, "Ein Freidenker des Islam," *ZDMG* 29 (1875), 304–12; I. Goldziher, "Abū-l'-ʿAlā al-Maʿarrī als Freidenker," *ZDMG* 29 (1875), 637–41; Urvoy, *Penseurs Libres*, 163–76; art. "al-Maʿarrī," *EI*², V, 927–35 (P. Smoor) and the bibliography there.

these two individuals. It is also a first step towards a better appreciation of the role played by freethinking in the development of medieval Islamic thought.

Of course, a full appreciation of a religious phenomenon like freethinking cannot be separated from the study of the social and political circumstances of the period. The tensions between Arabs and *mawālī*, between rulers and the people dominated by them, were channeled into religious dissent and translated into religious terms. Radical heresy should thus be studied in its political and social context. In the following study, however, this context will be, more often than not, ignored. This is so not because I do not appreciate its importance, but because in the search for a clearer understanding of the phenomenon, as in any laboratory study, one must isolate the factors before trying to evaluate their combined effect. This book sets as its main purpose, therefore, the analysis of the data concerning the religious rebellion of Ibn al-Rāwandī and Rāzī and the evaluation of its impact on Islamic religious thought.

I should perhaps clarify that although I believe the number of true freethinkers to have been extremely small, I do by no means argue that there were only two of them. I hope, however, that the concentration on the two most notorious freethinkers may allow us a better understanding of the phenomenon of radical freethinking, its real and imagined existence, and its repercussions on various aspects of medieval Islamic thought.

4. The Method

The study of the freethinkers, and in particular that of Ibn al-Rāwandī, can be seen as falling within the scope of what Daniel Gimaret has described as an excessive preoccupation with "the pre-Socratic" phase of *kalām*.[50] The loss of some of the most interesting compositions of both Ibn al-Rāwandī and Rāzī often renders the search for their true, historical personalities highly speculative. This situation is aggravated by the fact that many of the writings of their contemporaries are also lost, and thus it is difficult to achieve even a trustworthy image of what the freethinkers were then perceived to be. Van Ess has criticized some of my previous works on Ibn al-

[50] Cf. D. Gimaret, "Pour un rééquilibrage des études de théologie musulmane," *Arabica* 38 (1991), 11–18, esp. 12.

Rāwandī, dismissing it as relying on material which is merely "Aussagen über Aussagen . . ., nicht Aussagen über die Wirklichkeit."[51] That my attempts to interpret the highly fragmentary evidence remain speculative is, perforce, quite true. But it is, unfortunately, as true for practically everything written on ninth-century Islamic theology, including van Ess's own magisterial work. Quite often we are forced to rely on third- and fourth-hand material, and the extant manuscripts depend on the vicissitudes of fate. Even then, the material actually available to the modern reader is probably only a partial reflection of what the treasures of existing manuscripts could disclose. Many later *kalām* compilations still remain unpublished, and I have not tried to unearth those manuscripts. But even if such compilations were to surprise us with additional, unknown material, their information could also be dismissed as secondary or prejudiced, "Aussagen über Aussagen." Fully aware of the problematic nature of our sources, we are still bound to make do with what they offer.

The working method chosen here results from the fragmentary state of our sources. It is of course possible that future manuscript discoveries will add substantially to our knowledge and change the picture drastically. But at the present, the information given by our sources is both scant and contradictory. The present study attempts to collate the various pieces of information in order to find in them some consistency. If we were to put these pieces, as it were, one on top of the other, like transparent paper-silhouettes, the end-result would be a pile in which one could distinguish a recurrent pattern. Besides faintly drawn arms and legs sticking to all sides, a darker figure may emerge, the outcome of the points where all the silhouettes (or most of them) overlap. The ensuing portrait of the freethinkers, based on the points where all our sources agree, is not necessarily the "true one" in the sense that this is how things "really happened." But it is necessarily the portrait that most faithfully reflects the impact of our topic on medieval Muslim thought.

Had the sole goal of this book been to exhume the historical figures of Ibn al-Rāwandī and Rāzī, the endeavor would have indeed been doomed to be frustrating. Their historical personalities, however, constitute only a part of the investigation. Their image in the eyes of their contemporaries as transmitted to later generations, and the impact they left on them, is as significant as the question of the

[51] *TG*, IV, 320.

extent to which this image corresponds to the historical reality.

In this book I try to make a case for the argument that the disparate pieces of information concerning Ibn al-Rāwandī and Rāzī add up to a fairly consistent picture, which justifies the claim that they were indeed freethinkers. An unbending skeptic in the value of our sources could still argue that Ibn al-Rāwandī and Rāzī may have been two good Muslims who were thoroughly and consistently wronged by the press of their times. True; but as far as the history of Islamic thought is concerned, these two good Muslims do not exist. The significance of these two thinkers for the development of Islamic thought lies only in their radical image, as it transpires from the texts.

It is their image as freethinkers which accounts for the fact that their impact on posterior generations far exceeds the insignificant number of people involved. Whether rightly or wrongly, the idea of freethinking was blown to huge and menacing dimensions among intellectuals of various trends in the Islamic world, out of all proportion to the actual size of the phenomenon. Orthodox theologians regarded freethinking as an imminent danger that lurked behind every rationalistic enterprise. Philosophically-minded thinkers, both Aristotelian *falāsifa* and Neoplatonizing Ismāʿīlīs, saw the freethinkers as dangerously close to their own rationalism, and they struggled to distance themselves from their contaminating heresy.

This book is divided into eight chapters, which fall into two parts; The first five chapters present the case for freethinking in medieval Islam, the last three its repercussions in Islamic thought.

I will first present the Muslim dogma which is the backdrop to freethinking in Islam, by following the development of prophethood as a polemical theme. The topic of the proofs of prophecy highlights the paramount importance of prophecy in medieval Islamic discourse. It allows us to appreciate what medieval authors felt to be the enormity of the rejection of prophecy by the freethinkers. It also introduces the main arguments the freethinkers used in order to reject it (Chapter One).

Focusing on the extant fragments of Ibn al-Rāwandī's *K. al-Zumurrud* and of Rāzī's *K. Maḫārīq al-anbiyā'*, I will then endeavor to establish the authenticity of these books and their true heretical character. These two books, read in the context of their authors' other writings, testify to the paramount importance of the rejection of prophecy in the religious thought of Ibn al-Rāwandī and Rāzī (Chapters Two and Three).

Chapter Four examines the image of the freethinker in Islam and analyzes its main religious characteristics, as reflected in the thought of Ibn al-Rāwandī and Rāzī. As I will try to show, the examination of the disparate information about Ibn al-Rāwandī and Rāzī allows us to see the fundamental differences between them regarding major religious issues, such as theodicy and the temporal creation of the world. The rejection of prophecy, which is the core of their freethinking, emerges as the main common component of their religious thought, a component which nevertheless reflected on all aspects their thought, especially in the eyes of their fellow Muslims.

The associations of the freethinkers with non-Muslim religions are investigated in Chapter Five: the possible links of continuity between the freethinkers' criticism of the Scriptures and between dualist and Hellenic Bible criticism, the association of Ibn al-Rāwandī with the *Barāhima* and of Rāzī with the Sabeans.

The last three chapters deal with the impact of the freethinkers on later generations. I will first discuss the theological and philosophical literature, and the possible influence of the freethinkers on the development of atomistic theories, on the standards of polemical and logical argumentation (*jadal*), and on the development of Islamic political philosophy (Chapter Six).

I will then study the avatars of the freethinkers in inter-religious polemical literature, focusing in particular on two case-studies: the anti-Muslim epistle of the Christian ʿAbd al-Masīḥ al-Kindī, and the anti-Qurʾānic treatise attributed to Ibn al-Naghrilla by Ibn Ḥazm. These two compositions show striking similarities to Ibn al-Rāwandī's books. It is argued that they testify to the depth of the penetration of the freethinkers' ideas into the intellectual discourse of their times (Chapter Seven).

As far as I can see, freethinking in the strict sense of the term did not appear in the Middle Ages among the Arabic-speaking Christian and Jewish communities. To the extent that it did, this should be seen as an incidental spillover of Muslim freethinking, rather than a fully assimilated influence. The last chapter of the book investigates possible reflections of this Islamic heresy outside Islam, in Byzantine and Western Christianity and in medieval Judaism. As this book endeavors to show, however, freethinking in its radical form remained a typical Islamic phenomenon.

CHAPTER ONE

THE SIGNS OF PROPHECY: THE TOUCHSTONE OF MUSLIM PROPHETOLOGY

The basic Islamic position towards other religions is reflected in the two "testimonies" (*al-šahadatayn*) which constitute the Muslim declaration of faith: the belief in the unity of God and the belief in the prophecy of Muḥammad. These two testimonies shaped the theory and practice of the Muslim attitude to religion: the first expresses the Islamic unequivocal rejection of polytheism, the second is marked by the awareness of both continuity with and departure from earlier revelatory religions. Indeed, the Qur'ānic insistence that every nation in history has had a prophet sent to it is the touchstone of Islamic comparative religion.[1]

The central importance of prophecy in Islam is evident in many areas of Muslim thought.[2] The present chapter will examine its manifestation in one aspect of early theological literature, which is particularly relevant to the subject of freethinking.

In their effort to discredit the notion of prophecy, the freethinkers proceeded in several directions. On the theoretical front, they attempted to show the implausibility of the very notion of prophecy.[3] On the polemical front, they endeavored to show that throughout history, people who claimed to be prophets were in fact manipulative tricksters and impostors. In this context, the analysis of each prophet and his credentials became a major issue. The evidentiary miracles related in the Scriptures thus became a pivotal concern of the freethinkers and of their opponents. The following chapter is an attempt to reconstruct the evolution of the topos of the signs of prophecy in early Arabic literature.

---

[1] See Q. 10:47 (*li-kulli ummatin rasūlun*); 13:7 (*li-kulli qawmin hādin*), as well as 16:36, 84, 89; 22:34, 67; 28:75; 35:24.

[2] See, for example, Rahman, *Prophecy in Islam*; Y. Friedmann, *Prophecy Continuous: Aspects of Aḥmadī Religious Thought and its Medieval Background*, esp. 49–82.

[3] See below, Ch. 5.2.

1. The Emergence of the Theme in Islam

Among the standard components of comprehensive *kalām* works one finds the discussion of prophecy, which figures as a prominent part of any *summa theologica*.[4] Moreover, many entire works were dedicated solely and specifically to this question, usually bearing such titles as "The Signs of Prophecy" (*Aʿlām al-nubuwwa*), "Establishing the Proof of Prophecy" (*Ithbāt al-nubuwwa*), and the like. What normally comes under these titles is twofold: an attempt to prove the human need for a prophet, and a list of arguments meant to support the claim of one particular prophet, and his superiority over other prophets.[5] Contributions to this theme in Arabic literature were made by authors from various religious affiliations. Among Muslims we find Muʿtazilite,[6] Ašʿarite[7] and Māturīdite theologians,[8] and Sunnīs as well as Zaydīs and Ismāʿīlīs.[9] Among the Christians, Melchites,[10] Jacobites,[11] and Nestorians[12] dedicated works to this subject; and Jewish theologians, both Rabbanites and Karaites, dealt with it in their writings.[13]

In comparison to the abundant Arabic literature dealing with this question, the meager part allotted to it in theological literatures before the rise of Islam is indeed striking. Although the book of Deuteronomy explicitly defines the criteria by which a true prophet is to be distinguished from a false one,[14] these criteria had not elicited more

---

[4] ʿAbd al-Jabbār dedicated to this complex no less than three volumes of his *Mughnī* (vols. XV, XVI, and also XVII in the Cairo edition).

[5] See, e.g., ʿAbd al-Jabbār's *Mughnī*, XV, 7–8.

[6] From al-Jāḥiẓ (*Ḥujaj al-nubuwwa*) to ʿAbd al-Jabbār (*Tathbīt*). The crucial importance of this theme for the Muʿtazila is reflected in the fact that Muʿtazilites gladly dwell on it in other contexts. Thus, for example, in a *Maqāla fī al-radd ʿalā al-mujbira* attributed to ʿAbd al-Jabbār, the author dedicates a lengthy discussion to prove that God will not accord miracles to impostors. (I am indebted to Hans Daiber for the reference to this unpublished treatise and for kindly sending me a copy of the manuscript).

[7] Like Mawardī (*Aʿlām al-nubuwwa*).

[8] Cf. Māturīdī, *Tawḥīd*, pp. 176–210.

[9] See, for example, Qāsim b. Ibrāhīm, *Rādd ʿalā al-mulḥid*, 50–54; Sijistānī, *Ithbāt al-nubuʾāt*.

[10] Like Abū Qurra; see Qustantīn Bāsha, *"Maymar fī ṣiḥḥat al-dīn al-masīḥī li'l-ab ... Abī Qurra usquf Ḥarrān," al-Machriq* 6 (1903), 633–43, 693–707, 800–809.

[11] Abū Rāʾiṭa, *Rasāʾil*, 131ff.

[12] ʿAmmār al-Baṣrī, *Burhān*, 32ff; idem, *Masāʾil*, 136ff.

[13] See, for example, Saʿadya, *Amānāt*, Chapter 3, 117ff., as well as his Arabic Introduction to *Sefer Ha-Galuy*, in A. Harkavy, *Zikkaron la-Rishonim*, 150–51, 162–63; Qirqisānī, *Anwār*, 576–86.

[14] Deut. 12:2–7; 18:15–22.

than a few occasional comments among either Jewish or Christian thinkers before the Islamic period. When Philo, for example, refers to the relevant biblical passage, he sums up its content in a concise, matter-of-fact way, without any elaboration.[15] For Rabbinic Judaism as well as for mainstream Christianity, the era of prophecy has come to an end. There was thus no pressing need to ponder the criteria by which a true prophet can be recognized, and the study of these criteria became largely obsolete.[16]

Another biblical passage, concerning the miracles performed by Moses in Egypt and in the desert, later served Arabic writers as a favorite reference for their "signs of prophecy."[17] But where a Muʿtazilite would analyze the characteristics of real miracles as opposed to those contrived by sleight of hand, Philo prefers to draw a colorful and vivid picture of the various afflictions that befell the Egyptians.[18] Philo certainly had a strong feeling for the difference between a sophist or trickster and a prophet.[19] One can occasionally recognize in his writings elements that later writers would include in their "signs" (like the duration of a miracle, or the moral qualities and the intentions of the performer). But Philo does not underline these elements, let alone isolate or regroup them systematically.[20]

In the same biblical passage, describing the plagues in Egypt, the Christian Gregory of Nyssa saw only an allegorical representation of the virtuous soul's striving for perfection.[21]

---

[15] Philo, *The Special Laws* (*De Specialibus legibus*), 1.315–18, tr. F. H. Colson, LCL 7; London 1984, 283–84.

[16] See Urbach, *The Sages*, 267–73; idem, "When Did Prophecy End in Judaism?" (in Hebrew) *Tarbiz* 17 (1946), 1–11; D. E. Aune, *Prophecy in Early Christianity and the Ancient Mediterranean World*, Grand Rapids 1983. The conspicuous absence of a discussion of the signs of prophecy in these studies faithfully reflects the topic's marginality in the literature they discuss.

[17] See, e.g., Abū-Qurra, 639–40; Abū Rā'iṭa, *Rasā'il*, 137:12–18; Bāqillānī, *Bayān*, 59–60; Saʿadya, *Amānāt*, 127–28. This passage also serves the author of *Qiṣṣat mujādalat al-usquf* as an example of real miracles, see *Qiṣṣa*, section 21, and see below, apud nn. 33, 78.

[18] Philo, *Moses* (*De Vita Mosis*), 1.96, tr. F. H. Colson, LCL 6; London 1984, 325ff.

[19] See esp. *Spec. leg.* 6. 50–51 (LCL 7, 39) and *Vit. Mos.* 2. 185ff. (LCL 6, 541ff.).

[20] On Philo's view of prophecy, see Wolfson, "Veracity of Scripture," 604; H. A. A. Kennedy, *Philo's Contribution to Religion*, London 1919, 226–30; E. Bréhier, *Les idées philosophiques et religieuses de Philon d'Alexandrie*, Paris 1950, 180ff. One should, however, note the tendency of these works to articulate and systematize Philo's remarks more than he might have cared for.

[21] Gregory of Nyssa, *La vie de Moïse*, trans. J. Daniélou, Paris 1955, 48ff.

When Arabic-speaking theologians, both Jews and Muslims, set out to establish the legitimacy of prophecy in general, the polemical edge of their writings is directed against the *Barāhima*,[22] or else against Hellenizing philosophers.[23] Pre-Islamic Christian writers do not seem to have so much as heard of "Brahmans" in this context. References to Indian wisdom is, of course, a *topos* in Greek literature since Hellenic times.[24] But the Indian sages of this literature are not concerned with the denial of prophecy and do not advocate the sufficiency of human intellect, as the *Barāhima* do. Now the absence of these *Barāhima* in pre-Islamic Christian literature is easy enough to explain.[25] But, unlike the "Brahmans," representatives of Greek philosophy were readily available to the Church Fathers, and yet the fight against pagan philosophy hardly touches upon prophetology.[26] To cite just one example: in his polemical treatise against Celsus, Origen does mention the role of miracles, and argues that, unlike the trickster, the true prophet is not lost in ecstasy.[27] But this is clearly a marginal issue in his defense of Christianity against pagan philosophy.

[22] See, e.g., Bāqillānī, *Bayān*, 26:9–11. On the *Barāhima*, see below, Ch. 5.

[23] Sometimes referred to as the *dahriyya*; see Jāḥiẓ, *Ḥujaj al-nubuwwa*, 119:7; also Māturīdī, *Tawḥīd*, 187:19 (cf. Qirqisānī, *Anwār*, 578:6–7); Kraus, "Raziana," 369:22–25; Pines, *Atomism*, 162–64 (= *Beiträge*, 90, n. 2).

[24] On the image of India in the west, see W. Halbfass, *India and Europe: An Essay in Understanding*, New York 1988.

[25] It could be interpreted as another proof for the late, fictitious nature of the *Barāhima*. See Kraus, "Ketzergeschichte"; but it can also be seen as further circumstantial evidence for the argument that the *Barāhima* theme reflects the impact on Islam of real encounters with Indian religions, see below, Ch. 5. On the other hand, the absence of the *Barāhima* from later, post-Islamic Christian (and, in particular, Christian Arabic) texts, remains puzzling. As a tentative explanation I would suggest that, although often incorporated in Muslim prophetology, the *Barāhima* theme was still perceived as part of Muslim heresiography. Generally speaking, Christian writers showed little interest in Muslim heresiography; they turned to it mainly when it served their own polemical anti-Muslim purposes. The *Barāhima* were of no use for the Christians in their debate with Islam, and since the Christians already had their own traditional pattern for proving the idea of prophecy, they did not feel the need to change it by adding the *Barāhima*.

[26] And vice versa: the fight of the pagan philosophers against Christianity gives relatively little place to the issue of prophecy. To the extent that the freethinkers of Islam inherit themes from the pagan reaction, as suggested by Kraemer, "Heresy," 168, 174, the topic of prophecy, which constitutes the core of their revolt, allows us to see in them a fresh start.

[27] Origen, *Contra Celsum* 7.3 (H. Chadwick, tr., *Origen, Contra Celsum*, Cambridge 1953, 397). The same argument was brought up against the Montanists by Eusebius (*Historia Ecclesiastica* 5.16.7–9, LCL 1, 474–77, referred to in the *Theological Dictionary of the New Testament*, ed. G. Kittel, s.v. "*prophètès*," vol. 6, p. 861). On other early

A plausible explanation of this phenomenon seems to be that from an early date in mainstream Christian theology, Jesus was not considered to be a prophet. Those who spoke of Jesus as a "righteous man" and a prophet (the Ebionites and other Jewish Christians) were definitely in the minority, and their views were pushed aside by the predominant christology. Prophets were thus usually given a secondary role; they are those who announce the coming of Jesus;[28] who are, in some respects, his *figurae*;[29] or even those who propagate his gospel.[30] In any case, the signs that mark a true prophet had become the problem of bygone generations.

This Christian attitude had its effect on the Jewish approach. The Christians readily admitted the prophecy of Moses, although they saw it as an outdated issue. Therefore, for contemporary Jews the encounter with Christianity did not require that they should dwell on the proofs of prophecy.[31]

With the rise of Islam the scene changed drastically. Since Muḥammad had claimed to be a prophet, and since this claim had been rejected by both Jews and Christians, the traits that distinguish a true prophet from an impostor at once became a key issue. And although by the ninth century we find this topic elaborated by Muslims, Christians, and Jews alike, it is a safe assumption that in the first round it was Islam which had to come up with "proofs of prophecy," in response to Christian and Jewish incredulity. Already John of Damascus (d. 724) challenges the Muslims to substantiate Muḥammad's claim to revelation, calling him a "false prophet."[32]

An interesting early example is the discussion of the miracles proffered by Moses. All three religions believe in the prophecy of Moses and accept as true the miracles recorded in the Bible. A detailed list of Moses' miracles (based on the Qur'ān as well as on

---

Christian references to true prophecy, or true religion, see Wolfson, "Veracity of Scripture," 604–605.

[28] See, e.g., Origen, *Contra Celsum* 1.51ff. (Chadwick, 47ff.); see also *Dictionaire de Théologie Catholique*, s.v. "prophécie," 711.

[29] See the classic work of J. Daniélou, *Sacramentum Futuri: Etudes sur les origines de l'apologie biblique*, Paris 1950.

[30] Cf. *TDNT*, vol. 6, p. 861. See also the intriguing passage in Abū Qurra's defense of orthodoxy, 669–70, esp. 670:22.

[31] The rabbinic interest in miracles versus magic was rarely linked to prophetology. See Urbach, *The Sages*, 81–102, 502–13; and see above, n. 16.

[32] John of Damascus, *De haeresibus*, 101; Cf. Saint John of Damascus, *Writings*, tr. F. H. Chase, Jr.; FC, Washington 1981, 153; D. J. Sahas, *John of Damascus on Islam: The "Heresy of the Ishmaelites"*, Leiden 1972, 79ff.

the Jewish midrashic literature) is incorporated, for example, in Ṭabarī's *History*.[33] One could thus expect these miracles to be a neutral subject. But if fact, we find this subject at the forefront of interreligious polemics. In particular it appears in the anti-Christian debate, where the miracles performed by Jesus are compared unfavorably to those performed by Moses (as well as by other prophets). This topos appears in early (ninth century) Jewish and Muslim works. The christological parts of the same works demonstrate the authors' awareness of the fact that the Christians do not consider Jesus to be a prophet, and yet, unlike pre-Islamic polemics, the polemicists dwell on Jesus' miracles as if they were his prophetic signs. The similarities of these Jewish and Muslim anti-Christian discussions suggest a common source to all of them, and the internal evidence favors the assumption that the Muslim sources are dependent on a Jewish text.[34] But it is noteworthy that such an anti-Christian Jewish discussion surfaces under Islamic rule. In other words: the topic of the signs of prophecy gained prominence in the Jewish anti-Christian context only after the appearance of Islam on the polemical scene.

The earliest "signs of prophecy" works written by Muslim *mutakallimūn* are, unfortunately, lost.[35] What we find in the ninth century is already a well-shaped prophetology. In the discussion, old "biblical" arguments were adduced together with new ones, which were shaped and reshaped in the course of the interreligious debate.

The general frame of the discussion seems, at first sight, to be one and the same for the three religions, as do the particular "signs". However, a closer look reveals specific components characteristic of each religion. I shall not draw up here a list of the "signs" that were alluded to.[36] Rather, in what follows I shall try to point out some of the above mentioned components and to analyze them.

---

[33] *Ta'rīḫ al-rusul wa'l-mulūk*, Leiden 1964², I, 513–15, 633–40.

[34] See D. Sourdel, "Un pamphlet musulman anonyme d'époque 'abbāside contre les chrétiens," *REI* 34 (1966), 1–33; "*Risālat Abī Rabī' Muḥammad ibn al-Layth 'llatī katabahā ilā Qusṭanṭīn malik al-rūm*," *Jamharat rasā'il al-'arab*, III, ed. Aḥmad Zākī Ṣafwāt, Cairo 1937, 252–324; *Qiṣṣa*, II, 29–33 (Judeo-Arabic text), I, 54–56 (translation). On the chronology of these texts, see *Qiṣṣa*, I, 22, 140–42.

[35] For a few examples, see Ibn al-Nadīm, *Fihrist*, who lists works by Bišr b. al-Mu'tamir (p. 162:15), by Abū Sahl al-Nawbaḫtī (p. 177:3), and by Kindī (p. 259:15), all from the ninth century. Of earlier books dedicated to this subject we have no knowledge.

[36] For examples of various criteria of true prophecy and their place in the discussion, see Vajda, "Prophétologie."

2. Early Developments of the Signs of Prophecy

We shall begin at the beginning, which in this case means the Muslims. Among the earliest preserved works wholly dedicated to the "signs of prophecy" is Jāḥiẓ's *Proofs of Prophecy* (*Ḥujaj al-nubuwwa*).[37] In this text, Jāḥiẓ enumerates two categories of proofs: those belonging to sense experience (*ʿiyān ẓāhir*) and those based on indisputable tradition (*ḫabar qāhir*).[38] Jāḥiẓ regards both these proofs as dependent upon the intellect (*ʿaql*).[39] The "sense experience" in this context is, first of all, the miracle performed by the prophet, which the intended people can observe with their own eyes.[40] The "tradition" was understood as referring to the historical traditions about the prophet and his miracles and to the transmission of the revealed Book.[41] Now for Muslims, of course, the revealed Book was also, in itself, a miracle. In fact, Jāḥiẓ goes to great lengths to show that the miraculous beauty of the Qur'ān, which humans are unable to imitate (*iʿjāz*) is as good a miracle as any other.[42] The apologetic, defensive vein of his words is unmistakable.

It has been suggested that the doctrine of *iʿjāz al-qur'ān* may be seen as an eventual outcome of the importance of Arabic grammar in early Islam.[43] Admittedly, this doctrine could not have developed without the importance attached to the Qur'ān and to linguistics. A similar development of Scripture-centered linguistic interest can be seen around the same period in Judaism, where Saʿadya laid down the foundations for the study of Hebrew as part of his sustained theological and exegetical effort. And yet the importance Saʿadya accorded to biblical Hebrew did not lead him to invest it with miraculous beauty.[44] The Muslim transformation of linguistic interest into dogma

---

[37] See n. 6 above.

[38] Jāḥiẓ, *Ḥujaj al-nubuwwa*, 118:16.

[39] Ibid., 118:17–22.

[40] Ibid., 139:19ff., 143:5.

[41] In this context, tradition means "knowledge based upon revelation as recorded in Scripture" (to use Wolfson's remark concerning Saʿadya, *Philo*, 62–63).

[42] Jāḥiẓ, *Ḥujaj al-nubuwwa*, 136:1–7, 11–14; 138:14–17; 141:20ff.; 143:4f. On the doctrine of the miraculous nature of the Qur'ān, see art. "Iʿdjāz," *EI*², III, 1018–1020 (G. E. von Grunebaum); and see below, Ch. 2, n. 182

[43] Griffith, "Comparative Religion," 80.

[44] This is all the more striking, considering Saʿadya's undeniable indebtedness to Muslim theologians as well as to Arab grammarians. On Saʿadya's linguistic theory and its place in his thought, see, for instance, R. Drory, *The Emergence of Jewish-Arabic Literary Contacts at the Beginning of the Tenth Century* (in Hebrew), Tel-Aviv 1988,

thus cannot be explained solely as a natural outcome of an inner development, the result of a purely scholarly interest. In the reconstruction of the shift in Muslim thought it seems necessary to assume an additional, external factor: the impact of the encounter with both Christians and Jews. Relying on the Deuteronomic criteria of true prophecy, Jews and Christians had challenged the Muslims to adduce a miracle of some sort to buttress their claims, thus driving the latter to ascribe to the Qur'ān a role "comparable to the role of the theory of evidentiary miracles in Christian apologetics."[45]

Moreover, another distinctly Islamic science must also be taken into consideration in this context. Besides the miraculous beauty of the Qur'ān, Jāḥiẓ emphasizes the miraculous transmission of all traditions about Muḥammad's prophecy. He sets out to prove that such unanimous transmission cannot be mistaken, or result from purely natural circumstances.[46] In this argument, Jāḥiẓ relies heavily on the criteria developed in *'ilm al-ḥadīth*.[47] Elsewhere, Jāḥiẓ quotes Wāṣil b. 'Aṭā' (d. 131/748), who had listed four ways through which truth could be reached: (1) a revealed book (*kitāb nāṭiq*);[48] (2) a widely-accepted tradition (*ḫabar mujtama' 'alayhi*); (3) intellectual argument (*ḥujjat 'aql*); (4) the consensus of the people (*ijmā'* or *ijmā' min al-umma*).[49] This schema is, in fact, "the quaternion Koran, *sunna*, consensus, and *qiyās*, which comprises the recognized sources or principles (*uṣūl*) of law in the classical theory" of Islamic jurisprudence.[50] But

esp. 156–84; A. Dotan, *The Dawn of Hebrew Linguistics: The Book of Elegance of the Language of the Hebrews by Saadya Gaon* (in Hebrew), Jerusalem 1997, esp. I, 17–40, 94–110.

[45] Cf. Griffith, "Comparative Religion," 80; idem, "'Ammār al-Baṣrī's *Kitāb al-Burhān*: Christian Kalam in the First Abbasid Century," *Le Muséon* 96 (1983), 165. In fact, Jāḥiẓ's epistle, which addresses itself at first broadly to many opposing religious beliefs (see *Ḥujaj al-nubuwwa*, 119), occasionally narrows its scope and concentrates on the vindication of Islam against Christianity (ibid., esp. 125:23–126; 128:19–23; 131:15ff.).

[46] Jāḥiẓ, *Ḥujaj al-nubuwwa*, 129:22–130:5; 136.

[47] For example, ibid., 132:3–4: "*mā naqalathu al-jamā'āt 'an al-jamā'āt*"; cf. Schacht, *Origins*, pp, 42, 51. See below, n. 50.

[48] The idea conveyed by this term is expressed more clearly by Māturīdī, *Tawḥīd*, 209:4: "*wajada kitāb Allāh nāṭiqan bi-iẓhār dīnihi.*" Cf. also Ṭabarī, *al-Dīn wa-l'dawla*, 18 (*nubuwwa nāṭiqa fī kitābihi*).

[49] See 'Askarī, *Awā'il*, II, 134:6–7; Pines, *Atomism*, 149, n. 31.

[50] See Schacht, *Origins*, 134–35. The supposition that this list had its origin in a traditionalist setting, from which it was passed over to *kalām* (rather than the other way around) is established by the very nature of the criteria in the list. Three of the four criteria are of a non-intellectual character. It is, therefore, more plausible to see the origin of these three criteria in a traditionalist setting, than in the intel-

it is also easy to see the similarity between this list and Jāḥiẓ's two categories of proofs and what they stand for.[51]

Josef van Ess has argued that the adoption of this schema by Mu'tazilite theologians is a late phenomenon (appearing around the mid-ninth century), the result of the *mutakallimūn*'s succumbing to the stronger orthodoxy, and giving up their former rejection of traditions. Van Ess sees here a "shift in the polemical accent," in which "the Mu'tazilites, instead of attacking the *muḥaddithūn*, turn now . . . against non-Muslims."[52] But it is perhaps more plausible to see here two facets, very possibly contemporary, of the same theological phenomenon, rather than two stages of a theological development. Again, evidence from the seventh and eighth centuries is too scarce, and usually comes from secondary sources. But we do know that the intellectual religious encounter with Christianity had started not in the ninth century, but much earlier. The spokesmen for Christianity did not wait politely until the Mu'tazilites and orthodox Muslims had settled their disagreements. Early Mu'tazilites could not afford the luxury of dealing only with *muḥaddithūn* while ignoring *ahl al-kitāb*, leaving to their heirs the arduous task of refuting them. Thus the same early *mutakallimūn* may at one and the same time have opposed the proliferation of *ḥadīth*, and yet applied criteria, originally developed to check the authenticity of *ḥadīth*, to the scrutiny of religions.[53] It seems to me quite untenable to claim that the "truth" (*al-ḥaqq*) which Wāṣil tries to achieve in his fourfold schema referred only to the veracity of *ḥadīth*.[54] Already for Wāṣil, in all probability,

lectually-oriented early *kalām*. This supposition is strengthened by another such list, in which the intellectual argument is absent altogether. Al-Dārimī (*al-Radd 'alā al-Jahmiyya*, ed. G. Vitestam, Lund and Leiden 1960, 93) mentions only a revealed book (*kitāb nāṭiq*), tradition (*athar*), and consensus (*ijmā'*).

[51] See above, nn. 38–40.

[52] Van Ess, "L'autorité de la tradition prophétique," 220–22. On the Mu'tazilite rejection of tradition, see Schacht, *Origins*, 41–42, and 258; Cook, "'Anan and Islam," esp. 166–69. For all we know, however, what the Mu'tazilites rejected was not the very idea of tradition (for they did accept *ijmā'*, a concept which implies some non-scriptural source of knowledge). Their objections seem to have been limited to "traditions" in the sense of *ḥadīth*, separate traditions about the *sunna* of the prophet as a basis for legal decisions.

[53] The phenomenon is not uncommon; Mu'tazilites also use in their discussions with the *mujbira* arguments which according to their own testimony were hurled at the *mu'tazila* by Manicheans. The same practice is found also among Christians. See, e.g., Vollers, "Religionsgespräch," 65:7–13, 69:10–14.

[54] Cf. van Ess, "L'autorité de la tradition prophétique," 213. Other quotations of Wāṣil by al-'Askarī, which do refer specifically to *ḥadīth*, do so explicitly; see *Awā'il*, 134:8ff.

the "truth" this list endeavors to establish ultimately refers to the truth of one religion, over against the others.

This analysis, and in particular the affinity of the "proofs of prophecy" with *ʿilm al-ḥadīth*, corroborates the suggestion[55] that it was the Muslims who had developed the disparate Jewish and Christian criteria of true prophecy into a crystallized system.

3. The Christian "Negative Signs" of the True Religion

Nevertheless, the Christians were quick to pick up the challenge. Already in the ninth century one finds a Christian version of this "Signs of Prophecy" theme, with a peculiar Christian twist. Like their Muslim contemporaries, the Christians built their religious *apologiae* on the evidence of scriptures, traditions, and intellectual reflection, referring often to miracles and to consensus. But in its Christian garb, this theme is not formulated as referring to the prophet and to the way in which his authenticity can be evaluated in his lifetime. Rather, Christian writers in Arabic usually assemble the criteria by which an unbiased observer should be able to judge whether the success and propagation of an established religion is, in itself, a proof of its authenticity.[56] One can discern here the old Christian reluctance to overemphasize the role of the prophet,[57] as well as a defensive note: the success of Islam results from this-worldly circumstances (*al-asbāb al-arḍiyya, asbāb al-dunyā*).[58] On the other hand, the spread of Christianity is depicted as defying any normal expectations;[59] it is,

[55] See above, apud n. 32.

[56] For a detailed description and analysis of this theme see Samir Khalil, "Liberté religieuse et propagation de la foi chez les théologiens arabes chrétiens du IXe siècle et en Islam," *Witness of Faith in Life and Worship; Tantur Yearbook, 1980–81*, Jerusalem 1982, 93–160; Griffith, "Comparative Religion," 63–87.

[57] Sometimes the differences between Jesus and the prophets are clearly spelled out. See Vollers, "Religionsgespräch," 45–46, 61, and 64:20 (where, for argument's sake, the monk includes Muḥammad with all the prophets; cf. also Marcuzzo, *Dialogue*, 387–89, 405); I.-A. Khalifé and W. Kutsch, "*Ar-Radd ʿalā-n-naṣārā* de ʿAlī-at-Ṭabarī," *MFOB* 36 (1959), 139:7–8, where the juxtaposition of Jesus and the prophets, stressing the prophets' weakness (*ʿajaz*), might be a pun intended to play down their miracles (*muʿjizāt*); see also Abū-Rāʾiṭa, *Rasāʾil*, p. 38.

[58] The fear of the sword, ethnic or familial ties, and the hope for worldly gains (wealth or sexual pleasure, both in this world and the next) are regularly mentioned in this context. See, e.g., ʿAmmār al-Baṣrī, *Burhān*, 32; idem, *Masāʾil*, 136–37.

[59] Because of Jesus's humble birth, and because of his teaching which rejects accumulation of wealth and imposes stringent restrictions on sexual behavior, the

therefore, a divinely-sent miraculous proof of the authenticity of the Christian religion.

Instead of positive "ways" to truth, it is thus on "negative attributes" that the Christian Arabs rely in the search for the true religion.[60] In this line of argumentation, the events that take place after the death of the prophet gain major importance.

These events (i.e., the success of the religious campaign and the means by which this success is achieved) become the staple proof put forward by Christian Arabs.[61] The Nestorian ʿAmmār al-Baṣrī even claims that a continuation of miracles in all generations would have constituted a compulsion (*ijbār*) restricting the free choice of humanity.[62] What is implied in ʿAmmār's statement is that evidentiary miracles, although necessary in the establishment of a new religion, have a theoretical drawback. For him, intellectual reflection on history in later generations can provide a religion with a proof of higher value than miracles. For Muslims, on the contrary, the recourse to traditions from the prophet, after his death, is usually seen as a necessary technical device to bridge the time-gap between the actual miracle and later generations. Thus, the superior value of miracles is generally admitted on the Muslim side.[63] At best, the post-prophetic propagation of traditions is granted the role of an auxiliary miracle.[64] Muslim (and Jewish) authors often state that, as a source of knowledge, an authentic tradition ranks as high as sense-perception.[65] But this very

growth of Christianity was surprising. See, e.g., Vollers, "Religionsgespräch," 62–63; Marcuzzo, *Dialogue*, 399–401; ʿAmmār al-Baṣrī, *Burhān*, 36–38.

60 See the references in nn. 10–12 above.

61 See, for instance, the Christian apologist Kindī, whose list of conditions for the success of a religion includes the absence of military force (*lā bi'l-ghalba wa'l-qahr*), of group loyalty (*ḥamiyya wa-ʿaṣabiyya*), of fame and noble decent (*al-šaraf fī al-ḥasab wa'l-nasab*), the lack of the support by a strong or big family, the absence of temptations such as monetary gains, easy laws, and permissive mores, and the absence of the ruler's coercion; cf. Kindī, *Risāla*, 99. On this anti-Muslim Christian text, see below, Ch. 7.

62 ʿAmmār al-Baṣrī, *Burhān*, 27.

63 See, e.g., ʿAbd al-Jabbār, *Mughnī*, XV, 148:3: "Nothing but miracles is a proper proof for the sending of [divine] messengers"; the traditions are accepted as proofs for past events only (*māḍiyāt*), *faute de mieux*. The same view, which technically relies only on traditions, but ultimately gives a theoretical preference to miracles, was probably held by ʿAbbād b. Sulayman (*Mughnī*, XV, 257–58, 317–38; XVI, 242:8–10. Cf. van Ess, "L'autorité de la tradition prophétique," 220).

64 This sometimes appears to be Jāḥiẓ's view; see above, n. 46.

65 See, for instance, M. Zucker, *Saadya's Commentary on Genesis* (in Hebrew), New York 1984, 13–14, and n. 9. For a similar Muslim position, Zucker quotes al-Āmidī's *K. al-Aḥkām fī'l-iḥkām*, I, 9.

statement reveals the a priori theoretical precedence they give to actual miracle, performed in front of the community.

It is thus the Christians who introduced into the discussion this set of "negative attributes" of the true religion. But this original contribution[66] was not a totally new invention; rather, it was the Christian version of—and response to—the Muslims' "signs of prophecy."[67] In this Christian version, one can discern a few adaptations of the theme for Christian use. Against the Muslim emphasis on the marvelous nature of the Qur'ān's language, Christians repeatedly emphasize the multilingual spread of the Gospels.[68] As far as I know, this linguistic argument is not regularly adduced in Patristic literature. It is thus possible that it was developed especially by Christians in the Muslim world, in response to the Muslim doctrine of *iʿjāz al-qur'ān*. This was for them an additional proof of the divinely-sponsored propagation of Christianity, as well as an argument against the possible fabrication through conspiracy (*tawāṭu'*) or falsification (*taḥrīf*) of the Bible.

Another important Christian development concerns the role of the intellect. ʿAmmār al-Baṣrī opens his discussion by paying tribute to the twin proofs of miracles (*āyāt*) and an intellectually-tested proof (*dalālat ʿaql*).[69] Immediately thereafter he sets out to strip this "intellect" of its important role. For him, intellectual judgment refers not to reflection upon the content of religious teaching, but, again, to its spreading.[70] ʿAmmār is anxious to emphasize that the logical content, however appealing, of a given religion (*istiḥsān*) has no bearing on the question of its authenticity.[71] On the contrary, only a religion which is widely accepted in spite of its apparently illogical claims can be the true one.[72] On several occasions ʿAmmār argues that knowledge of God does not come naturally to human beings. It is

[66] See Griffith, "Comparative Religion," 75.

[67] The development of the Christian version out of the Muslim one can be reconstructed in al-Muqammaṣ's version of this theme; see below, apud nn. 80–83.

[68] See ʿAmmār al-Baṣrī, *Burhān*, 32, 41, 72; idem, *Masā'il*, 141; Abū Qurra, "*Maymar fī taḥqīq al-Injīl*," *Mayāmir*, 74; I. Dick, "Deux écrits inédits de Théodore Abū Qurra," *Le Muséon* 72 (1959), 64:11ff.; Abū Rā'ita, *Rasā'il*, 136. Al-Bāqillānī sees this argument as typical of Jews and Christians; see *K. al-Tamhīd*, ed. R. J. McCarthy, Beirut 1957, 160ff. and esp. 161:14–17; 173:1–6; 385:13–20.

[69] ʿAmmār al-Baṣrī, *Burhān*, 26:10–13. Cf. Jāḥiẓ's proof, above, nn. 38–39.

[70] Ibid., 31–32.

[71] The proof thereof is the logical pretense of Manicheism, which may seem congenial to the human intellect, but (as all ʿAmmār's readers are supposed to know) is Falsehood Incarnate; cf. ibid., 31, 36; idem, *Masā'il*, 136.

[72] Ibid., 136, 138.

not inherent to their constitution, it is not part of their sense experience, only revelation can give them the right knowledge of God.[73]

If we consider the long Christian tradition of "natural knowledge of God,"[74] the prevailing contemporary Muʿtazilite belief in inborn religiosity (*fiṭra*),[75] and also ʿAmmār's overall intellectual inclination, then his statement is somewhat surprising. It would seem that this position was meant to enable him to respond to Muslim and Jewish attacks on what was regarded as the illogical Christian doctrine of the Divinity.[76]

4. The Polemical Exchange of Signs

A Jewish contemporary of ʿAmmār and of Jāḥiẓ, Dāwūd b. Marwān al-Raqqī al-Muqammaṣ, allows us a glimpse into the intricacy of the debate between religions. His *Twenty Chapters* (*ʿIšrūn maqāla*), the earliest surviving work of Jewish theology in Arabic, is in fact also the earliest surviving *kalām Summa* of any religious affiliation.[77] Chapter fourteen of the *ʿIšrūn maqāla* deals with the "signs of prophecy," referring constantly to the biblical passages concerning Moses' miracles.[78] Technically, however, this chapter is presented as concerned with the ways of verifying a tradition (*ḫabar*) about a prophet after his death. As we have seen, this preference for miracles, complemented by a

[73] ʿAmmār al-Baṣrī, *Burhān*, 25:19–20; 64–65.

[74] See H. de Lubac, "Les religions humaines chez les Pères de l'Eglise," in his *Paradoxe et Mystère de l'Eglise*, Paris 1967, 120–40.

[75] See Gardet-Anawati, *Introduction*, 350; "*fiṭra*," *EI*², II, 932 (D. B. MacDonald); and cf. Griffith, "Comparative Religion," 80.

[76] This doctrine is understood to contain both the trinitarian and the christological aspects of Christian theology. See. e.g., al-Jāḥiẓ, *Ḥujaj al-nubuwwa*, 131:19–132:7. We have no indication that ʿAmmār was already aware of the "intellectual," "anti-prophecy" *Barāhima*. However, if he was, this position would have the additional benefit of avoiding the "*Barāhima*-catch": for, if it is openly stated that the human intellect is not the proper vehicle for religious knowledge, the *ilzām* used by the *Barāhima* becomes void.

[77] For the Judeo-Arabic text (with an English translation) of the *Išrūn Maqāla*, see Muqammaṣ, *Twenty Chapters*. For biographical outline, see Vajda, "Finalité"; S. Stroumsa, in the Introduction to Muqammaṣ, *Twenty Chapters*, 15–23; B. Chiesa, "Dāwūd al-Muqammiṣ e la sua opera," *Henoch* 18 (1996), 121–55; On al-Muqammaṣ's sobriquet and its meaning, see S. Stroumsa, "Al-Muqammas, Daud," *Routledge Encyclopedia of Philosophy*, London and New York, 1998, VI, 604.

[78] Muqammaṣ, *Twenty Chapters*, 263–71. The first lines of this chapter are, unfortunately, missing in the only extant manuscript. A French summary and partial paraphrase of Chapter 14 was published by Vajda "Prophétologie," 231–34.

resigned reliance on traditions in later generations, has its equivalent—and probable origin—in Muslim argumentation.[79] Another component of al-Muqammaṣ's prophetology, similar to Muslim conceptions, is the demand for logical content (*al-ṣaḥīḥ ʿinda al-qiyās*) of the prophecy. Al-Muqammaṣ explicitly interprets this condition as excluding Christianity, together with other false religions.

However, some of the other "signs of prophecy" listed by al-Muqammaṣ point to a different direction. He insists that the tradition about the prophet's first victory over his enemies should recount a victory miraculously achieved, rather than a military one ("*lā bi-ḥarb wa-lā bi-sayf*"). This demand is clearly intended to play down Islam and its success. He also asks that the tradition (which, in this case, is identical with the revealed book) be transmitted by several nations in many languages ("*wa-lā min umma wāḥida wa-lisān wāḥid*"), a typical Christian requirement.[80] Al-Muqammaṣ conveniently ignores the reference to tribal or familial links (*ʿaṣabiyya*) which Christians directed also against Judaism.[81]

Altogether, al-Muqammaṣ lists ten characteristics of the authentic tradition about the true prophet, all of them presented in the form of quotations from some unidentified book ("and as to his saying" . . . "*wa-ammā qawluhu*"). Now we know that al-Muqammaṣ had received his education from Christians, and was familiar with their writings.[82] We can be fairly confident that it is some early Christian list of the "signs of prophecy" that he quotes.[83] It is noteworthy that this early

---

79 See above, nn. 63–65.

80 See n. 68 above.

81 See, e.g., ʿAmmār al-Baṣrī, *Burhān*, 31, 36. For later, more sensitive adaptations of the Christian argument in Jewish texts, see Stroumsa, "Jewish Polemics Against Islam and Christianity In the Light of Judaeo-Arabic Texts," *Judaeo-Arabic Studies; Proceedings of the Founding Conference of the Society for Judaeo-Arabic Studies*, ed. N. Golb (Studies in Muslim-Jewish Relations 3; Amsterdam 1997, 241–50, esp. 248–49).

82 Cf. Qirqisānī, *Anwār*, 44:10–16. According to Qirqisānī, al-Muqammaṣ had actually converted to Christianity. His *ʿIšrūn maqāla*, however, was clearly written after his return to Judaism. On al-Muqammaṣ's Christian background, see the references cited above, n. 77.

83 A Muslim source is unlikely because of the above mentioned anti-military remark. The possibility of a Jewish source, earlier than al-Muqammaṣ and yet so well structured, is improbable. The reference to the transmission of scriptures by "many nations" renders the possibility of a Jewish source even more unlikely. The quoted source enumerates only the conditions; it is clearly al-Muqammaṣ who applies the conditions to Moses, and thus adapts the general frame provided by his source to Jewish theology. Moreover, the Christian identity of al-Muqammaṣ's source is apparent also in his use of the term "covenant" (*ʿahd*) to designate the revealed book. See *Twenty Chapters*, 267; "*ʿAhd*," *EI*², I, 255 (J. Schacht).

Christian work also incorporated some elements of Muslim prophetology. Like its Muslim models, this Christian list is specifically intended to evaluate a prophet, not a religion ("and the tradition about [him] should relate . . ." "*wa-an yakūna fī ḫabarihi* . . ."). However, it also includes some of the negative characteristics of the true religion, typical of Christian texts. Unexpected as it may seem, it is thus a Jewish text which enables us actually to see how the Christians turned the Muslim "signs of prophecy" theme into their own "negative description" of the true religion.

The attempt made above to isolate typical Christian or Muslim arguments is an attempt at reconstructing the earliest encounters, for which we have no reliable records. All we can do is to discern the arguments favored by each religion, or those that fit best into the framework of each prophetology. But we should also bear in mind that the arguments and "signs" introduced by Muslims directly influenced those used by Christians; and vice versa.

Thus, the eleventh-century Muʿtazilite *Qāḍī* ʿAbd al-Jabbār goes out of his way to demonstrate that Muḥammad's victories were indeed miraculous, and not at all military; that Muḥammad came from a humble family and that, consequently, it was not for earthly gain that his followers joined him; that his rejection by his tribe invalidates the accusation of *tašāʿub*; and that Muḥammad's message spread among many nations.[84] Together, these contentions put forward an uncommonly Christian, "Jesus-like" portrait of Muḥammad's prophecy.

It appears that here again we can capture one moment in the process of the interreligious debate in which each side helped to shape and remodel the theological edifice of the other. Muḥammad's claim for prophecy triggered a Jewish and Christian attack on his prophetic qualifications, which in its turn forced the Muslims to establish a system of vindication of Muḥammad's prophecy. The existence of this system obliged the Christians to respond with "the (negative) signs of a true religion," their own version of the "signs of prophecy"; this response in its turn influenced later Muslim depictions of Muḥammad and of early Islamic history.

[84] ʿAbd al-Jabbār, *Tathbīt*, 8:16–17, 24; idem, *Mughnī*, XVI, 27–38 (based on Abū Hāšim's refutation of Ibn al-Rāwandī's *K. al-Farīd*). Some of these elements appear already in Jāḥiẓ's *Ḥujaj al-nubuwwa*, as well as in Ṭabarī, *al-Dīn wa-l'dawla* (cf. esp. 64). But the Qāḍī stands out in his consistent and lengthy depiction of the meekness of the Prophet.

The analysis of the "signs of prophecy theme" highlights the particularly important place of prophecy in Islam. Unlike Christianity, and much more than Judaism, Islam defines itself around the concept of prophecy. Its early theology was shaped as it struggled to define the particular role and the specific nature of the Prophet and his role in history, and this topic became a shibboleth that distinguished the Muslim from other monotheists. The appearance of the freethinkers, and of the very idea of freethinking, i.e., the rejection of prophecy, in early Islam, is a reflection of the centrality of prophecy in Islamic thought. For most early Christian heretics and for the *minim* (i.e., heretics) of Talmudic Judaism, the notion of prophecy and its validity does not seem to have been *the* major concern, and their rebellion against the authority of the Church or of the Rabbis took other venues. In Islam, the Qur'ānic declaration that every single nation, throughout history, has had a prophet sent to it, also became the basis of the heretical comparative analysis of religion: every nation claims to have had a prophet, but the contradicting messages of these prophets show that they are all frauds or impostors. The development of freethinking as defined in the introduction, with prophets and prophecy as its main target, is thus a new phenomenon, connected to the rise of Islam and typical of it. We can say that the centrality of prophecy in Islam engendered a radically new kind of heresy in its mirrored image, which aimed at nothing less than a total rejection of prophetic religions.

CHAPTER TWO

IBN AL-RĀWANDĪ AND HIS BAFFLING "BOOK OF THE EMERALD"

The question whether or not there were any true freethinkers in medieval Islam revolves mainly around the figure of Ibn al-Rāwandī. Although in Muslim sources he is depicted as an arch-heretic, this image has been questioned in modern scholarship. The main reason for this skeptical attitude is Ibn al-Rāwandī's *K. al-Zumurrud*, concerning which we have particularly baffling and contradictory information. The following chapter is an attempt to re-examine Ibn al-Rāwandī's image, in order to establish the fact of his freethinking. I will first briefly review the relevant part of his biography, then present and analyze the data concerning the *K. al-Zumurrud*.

1. Biography

In view of his notoriety, one would have expected to know something concrete about Ibn al-Rāwandī. Paradoxically, however, this notoriety is almost the only certain fact known about him. Concerning all aspects of his life and thought we possess a plethora of contradictory information.

If one were to write a biography of Ibn al-Rāwandī which contained only the information on which there is relative consensus, it would be quite short.[1] Abū al-Ḥusayn Aḥmad b. Yaḥyā b. Isḥāq al-Rāwandī was born in Khorāsān[2] around the year 205/815. He joined

---

[1] The various versions of Ibn al-Rāwandī's biography have been summarized and analyzed several times: cf. Kraus, "Ketzergeschichte," 371–79; Nyberg's introduction to his edition of al-Ḫayyāṭ's *K. al-Intiṣār*; van Ess, "Image," 8, n. 2; Vajda, "Ibn al-Rāwandī"; Aʿsam, *History*, passim; *Bist Guftar*, 27–28; Sezgin *GAS*, I, 620–21. For a thorough compilation of sources concerning Ibn al-Rāwandī, see Aʿsam, *History*, and Aʿsam, *References*. For recent studies, see van Ess, *TG*, IV, 295–349; VI, 433–90; Urvoy, *Penseurs libres*, 117–33.

[2] His birthplace was either in the vicinity of Marwarrūdh (cf. *Fihrist*, 216), or near Nīšāpūr, or else near Iṣfahān (*TG*, IV, 296).

the Muʿtazila of Baghdād, and gained prominence among them. But when he approached the [topical] age of forty he became estranged from his fellow Muʿtazilites, and formed close alliances with non-Muʿtazilites, both Muslims (Shiʿites) and non-Muslims (Manicheans, Jews and perhaps also Christians). He wrote against the Muʿtazila, and they responded in kind.

Our sources contain much material beyond this short biography. But for every detail other than those mentioned above there are at least two contradictory versions. His name has been variously spelled as Ibn al-Rāwandī/al-Rawandī/al-Rīwandī.[3] Some sources suggest that Ibn al-Rāwandī may have died around 245/860,[4] others that he lived to the year 298/912 or slightly later.[5] The exact dating of Ibn al-Rāwandī has important implications for the evaluation of some aspects of his thought. If we adopt the early dating, it would be easier to depict him as an original thinker, the first to introduce into the Islamic world ideas that would later be used by others, but that often remain connected to his name. For example, the question of whether or not he is to be credited with the invention of the rationalist, anti-prophecy *Barāhima* could be more neatly solved if we knew the exact date of his death. But the basic questions regarding Ibn al-Rāwandī's religious identification are not really affected by his precise dating. This observation does not mean to say that in this last domain—Ibn al-Rāwandī's religious identification—we have a clear picture, for, unfortunately, the fact is quite the opposite: while most of the sources describe him as an outspoken and dangerous heretic, some appear to present him in a neutral or even positive light.

---

[3] Aʿsam, *History*, 7, 9, opts for the last reading, assuming apparently that the name must reflect the area of Ibn al-Rāwandī's birth. But, of course, the name could just as well reflect the area of the family's distant origin, as names often do; see van Ess, *TG*, IV, 296–97. I have adopted the reading Rāwandī, which seems, on the face of it, to be more current in the Arabic sources.

[4] For example Yāfiʿī, Masʿūdī, and Ibn Ḫallikān; cf. Kraus, "Ketzergeschichte," 375; Aʿsam, *History*, 7; Abrahamov, "Barāhima," n. 3. This date is preferred by Vajda ("Ibn al-Rāwandī"). However, Vajda's main evidence for it (the section on prophecy in al-Muqammaṣ's *Twenty Chapters*) may actually be a proof for the contrary position. See below, Ch. 5, n. 40; and see *Twenty Chapters*, 31, 254–63; Stroumsa, "*Barāhima*," 234–40.

[5] For example Ibn al-Murtaḍā and Ibn al-Jawzī. Other dates are listed in Aʿsam, *History*, 9. See Nyberg's introduction to his edition of *K. al-Intiṣār*, 40–43; van Ess, "Image," 8–9; idem, *TG*, IV, 344–45; Schacht, "New Sources," 42; Madelung, "Abū ʿĪsā," 311; Dhanani, *Physical Theory*, 8.

These differing perceptions of Ibn al-Rāwandī are reflected in Ibn al-Nadīm's *Fihrist* (compiled ca. 988), which offers two lists of Ibn al-Rāwandī's books, the one listing respectable scholarly publications, the other an impressive collection of heretical works: in support of the eternity of the world, against the idea that God is wise, against the Qur'ān, against the prophet Muḥammad, against all prophets. This dichotomy is often explained in the sources as the result of the development of Ibn al-Rāwandī's thought. Thus al-Balḫī (d. 931), cited by Ibn al-Nadīm, can speak of books written by Ibn al-Rāwandī in his righteous period (*ayyām ṣalāḥihi*), as opposed to those written after he had gone astray (*ayyām fasādihi*).[6]

The turning point between these two periods is described in our sources as connected on the one hand to Ibn al-Rāwandī's estrangement from his fellow Muʿtazilīs, and on the other to the tightening of his friendship with al-Warrāq.

The Muʿtazilite Connection

Ibn al-Rāwandī attacked several tenets of Muʿtazilī cosmology. He is reported to have criticized doctrines relating to Atomism.[7] Muʿtazilī sources are unanimous in their animosity toward him. The earliest such source, al-Ḫayyāṭ (d. ca. 300/913), dedicated his *K. al-Intiṣār* to a methodical refutation of Ibn al-Rāwandī. The book, composed around 279 H., is, strictly speaking, an *apologia* of the Muʿtazila, that is to say, a response to Ibn al-Rāwandī's attack on the Muʿtazila in general and on al-Naẓẓām in particular.[8] Significantly, al-Ḫayyāṭ does not regard Ibn al-Rāwandī's book as directed only against the Muʿtazila, but rather as reflecting Ibn al-Rāwandī's general attitude. He thus says:

> I realised that this is the book of a person who is resentful of the people of religion, wrathful against the Muslims.[9]

According to al-Ḫayyāṭ, Ibn al-Rāwandī's wrote several books, whose subject was "the establishment of irreligious ideas (*fī tathbīt al-ilḥād*), the negation of God's unity, the denial of [the notion of] prophecy

[6] See Fück, 73; Kraus, "Ketzergeschichte," 378; Aʿsam, *History*, 39.
[7] See below, Ch. 6, apud n. 10.
[8] On Ibn al-Rāwandī's *K. Faḍīḥat al-Muʿtazila*, see van Ess, *TG*, IV, 300–301.
[9] *Intiṣār*, 11; Aʿsam, *History*, 18.

and the vilification of the prophets and the *imāms*." As far as al-Ḫayyāṭ knew, these books of Ibn al-Rāwandī's were "famous and well known" (*mašhūra maʿrūfa*).[10]

Indeed, according to al-Ḫayyāṭ, Ibn al-Rāwandī made no secret of his heresy. Al-Ḫayyāṭ reports that Ibn al-Rāwandī was "constantly declaring concerning himself: Know that I am an apostate (*mulḥid*)."[11] Nevertheless, there were apparently times where Ibn al-Rāwandī had to be more cautious. At least one of his "righteous books," the *K. al-Tawḥīd*, was written already after Ibn al-Rāwandī had changed colors. In this book, says al-Ḫayyāṭ, Ibn al-Rāwandī, fearing for his life (apparently because of the heretical opinions he had already had the opportunity to express), was trying to ingratiate himself to the Muʿtazila.[12]

Al-Warrāq

Closely linked to Ibn al-Rāwandī's estrangement from the Muʿtazila, and perhaps simultaneous with it, was the tightening of his association with al-Warrāq. Abū ʿĪsā Muḥammad b. Hārūn b. Muḥammad al-Warrāq flourished in the first half of the third/ninth century.[13] In Muslim heresiographical sources he is remembered either as a scholar or as the source of other scholars (since most of his own books are no longer extant). He had a keen heresiographical interest, which, typically for the period, was not expressed in anthropological but rather in theological analysis of his subjects. He wrote a detailed description (*cum* refutation) of Christianity, as well as of both Judaism and dualist religions. His *K. al-Maqālāt*, one of the earliest doxographies written in Arabic, unfortunately did not survive. It is, however, quoted extensively by later authors, such as Ašʿarī, Malāḥimī and ʿAbd al-Jabbār, particularly in their accounts concerning the dualists.[14] His thorough study and refutation of Christianity allows us to

---

[10] *Intiṣār*, 11; Aʿsam, *History*, 18–19.

[11] *Intiṣār*, 38; Aʿsam, *History*, 23. On the etymology of this term, see above, Introduction, n. 16.

[12] *Intiṣār*, 19; Aʿsam, *History*, 20–21.

[13] On the dates of his *floruit*, see Thomas, *Anti-Christian Polemic*, 17–18; on al-Warrāq, cf. Sezgin, *GAS*, I, 620; Massignon, *Recueil*, 182–83; art. "Abū ʿĪsā al-Warrāḳ," *EI*², I, 130 (S. M. Stern); Urvoy, *Penseurs libres*, 102–17. For a list of his works see van Ess, *TG*, VI, 430–33.

[14] See Monnot, *Penseurs musulmans*, 60–61, 104; Thomas, *Anti-Christian Polemic*, 12–16; Dhanani, *Physical Theory*, 182–84; Madelung, "Abū ʿĪsā"; van Ess, *TG*, I, 429, 431.

see the scope of his knowledge and the precision of his records.[15] Even his opponents regarded him as a reliable authority on non-Muslim religions, and the marks of his research in later works show him to have been "an astonishing intellectual genius."[16]

Apart from this academic role, al-Warrāq expressed religious ideas of his own. These unorthodox ideas earned him a place in heresiographical literature not only as a source, but also as the object of study and of criticism. He is accused of affiliation with the Shiʿa as well as with Manicheism. Some modern scholars have raised doubts concerning these accusations, and in particular concerning al-Warrāq's identification as a Manichean. David Thomas, for example, has suggested that al-Warrāq's precision and objectivity in his presentation of other religions was mistaken by his fellow Muslims for identification with the subject of his books.[17] It is, of course, quite possible that al-Warrāq's interest in other religions could seem suspicious to his fellow Muslims. But such suspicion does not in itself preclude the possibility that al-Warrāq held some unorthodox views, or even that his unorthodox views were a reflection of a specific religion, other than Islam, to which he adhered. Norman Calder, on his part, has attempted to offer an interpretation that would, on the one hand, take into account al-Warrāq's unorthodox image, and on the other hand keep him clear of Manicheism. According to Calder, al-Warrāq was really what he calls an "Abrahamite."[18] This ingenious idea cannot, however, be seriously considered until we establish the existence of a distinctly identifiable Abrahamite religion in the ninth century; at present, such a religion remains, at best, what Calder himself would call a literary construct.

Much in our sources points to some Shiʿī sympathies, at least in some part of al-Warrāq's life;[19] and yet he is said not only to have spoken derisively of the Prophet Muḥammad, but also to have expressed particular animosity to ʿAlī because of the blood he had spilled.[20] Muslim sources are quite unambiguous in labelling al-Warrāq

---

[15] Thomas, *Anti-Christian polemic*, 12.

[16] Calder, "Literary Construct," 49. See also Thomas, *Anti-Christian Polemic*, 9, 12; and see, for example, Masʿūdī, *Murūj*, V, 23.

[17] Cf. Thomas, *Anti-Christian Polemic*, 11; cf. also van Ess, *TG*, I, 424, n. 2.

[18] Calder, "Literary Construct," 49.

[19] See, for instance Masʿūdī, *Murūj*, IV, 77, where he is counted among *mutakallimī al-šīʿa*; cf. Thomas, *Anti-Christian Polemic*, 21–22. According to Thomas, Shiʿism remains al-Warrāq's more likely form of religious affiliation.

[20] See Murtaḍā, *Šāfī*, 98 (quoting ʿAbd al-Jabbār); *Intiṣār*, 111–12; Tawḥīdī, *Imtāʿ*, III, 192; Thomas, *Anti-Christian Polemic*, 11–13.

a Manichean, "not just *zindīq*, or *thanawī*, or *rāfiḍī*," as Carsten Colpe emphasized.[21] It is of course true that people were sometimes wantonly accused of Manicheism, but such slanderous accusations were usually expressed in a formulaic way, and the accusers did not go out of their way to substantiate their claims. The bits of information that surface, almost unconsciously, in our sources concerning al-Warrāq, ring differently from such flat formulaic accusations. They suggest that al-Warrāq was believed to be affiliated with a real, identifiable religion. It thus seems difficult to see him as "another one of those famous 'heretics' who without sufficient cause became anathema to the later Sunni tradition."[22] In various sources we find him criticized for advocating views that are typically Manichean, formulated in typical Manichean vocabulary. Māturīdī's *K. al-Tawḥīd*, for example, quoted Ibn al-Rāwandī as expressing his astonishment from al-Warrāq, who rejected well-established traditions about the prophets, but advocated Manicheism. In this context, Ibn al-Rāwandī attacked "the follies of these people (i.e., the Manicheans), such as [the belief that] the heavens were made from the skins of the archons (*šayaṭīn*), and that the earth was corrupted when it became swarming with snakes and scorpions, and . . . their accounts of the deeds of Light and of Darkness."[23] What Ibn al-Rāwandī attributed here to al-Warrāq was not just "unerring reports of Manichean myths," as this passage is read by Thomas,[24] although, considering al-Warrāq's heresiographical expertise, such unerring report would hardly be surprising. But Ibn al-Rāwandī explicitly accused him of advocating these myths himself.

---

[21] "Anpassung des Manichäismus an den Islam (Abū 'Īsā al-Warrāq)," *ZDMG* 109 (1959), 86. Contrary to Vajda's opinion ("Zindiqs," 197 and 221), al-Ḫayyāṭ's information is consistent in seeing al-Warrāq as a dualist, not a Shi'ite. Al-Ḫayyāṭ, *Intiṣār*, 108:10–14, should be translated and understood as a hypothetical clause, meaning that the opinions expressed by Faḍl al-Ḥadhdhā' and Ibn Ḥā'iṭ should not be attributed to the Mu'tazila, just as the Shi'a should not be held responsible for the dualism of Abū 'Īsā al-Warrāq and his companions, merely on the ground that the latter had feigned to be Shi'ites.

[22] Van Ess, Introduction to *Bist Guftar*, 9.

[23] Māturīdī, *Tawḥīd*, 199:17–20. On this Manichean myth, see, for instance, F. Cumont, *Recherches sur le Manichéisme, I: La cosmogonie manichéenne d'après Théodore bar Khoni*, Brussels 1908, 25–26. The Manichean myth described there is much too precise to be explained away by seeing in al-Warrāq "plutôt . . . un sceptique, un critique indépendant, . . . qu'un partisan d'une doctrine déterminée" (Vajda, "Zindiqs," 196–97); and see below, Ch. 3, apud n. 97.

[24] *Anti-Christian polemic*, 21.

Furthermore, both ʿAbd al-Jabbār's *Tathbīt dalāʾil al-nubuwwa* and Tawḥīdī's *K. al-Imtāʿ waʾl-muʾānasa* mention al-Warrāq's book, "which he imputed to the eastern stranger" (*alladhī nasabahu ilā al-gharīb al-mašriqī*). Van Ess has dismissed this title as a literary device which al-Warrāq used to distance himself from dangerous ideas, by putting them in the mouth of some distant and obscure orientals.[25] But the "stranger," and especially "the Eastern Stranger" or the stranger who departed to the east, carried a definite meaning in Manichean mythology, at least as recorded in the most important of all Arabic sources on Manichean theology. According to Ibn al-Nadīm, Šāthil, the pure seed of Eve and Adam, was recognized by the archon Ṣindīd as "a stranger," (*gharīb*) that is to say, not of his own progeny. As the archon manifested his animosity to Adam and his children, Šāthil decided to go east "towards God's light and His wisdom."[26] Šāthil's description as a stranger bespeaks the fact that he was not under the sway of the archons, and his going east manifests his choice to live where the archons do not rule.[27] If we accept the sources' claim that al-Warrāq was a Manichean, the title of his book would thus declare his sense of alienation from the world created by the demiurge, and his desire to liberate himself from it.[28]

Indeed, the few lines quoted by Tawḥīdī from this book testify to a preoccupation with God's injustice, and seem to corroborate the book's Manichean tendency. Tawḥīdī attributed to al-Warrāq the usage of typical Manichean arguments, in typical Manichean terminology, such as the epithet "fool" (*safīh*) attached to the unjust God.[29] Al-Warrāq presented his argument as a general rule: "He who orders his slave to do things that he knows him to be incapable of doing, then punishes him, is a fool." The formulation itself, and the fact that al-Warrāq did not explicitly mention God, were not likely to fool his audience. The context as well as the familiarity with this kind of argument made his purpose of denouncing God's governance

---

[25] *TG*, IV, 289. See further below, n. 50.

[26] "*Halluma tanṭaliqu ilā al-mašriq, ilā nūr allāh wa-ḥikmatihi*;" Cf. *Fihrist*, 331:26–332:4 (Dodge, 785–86).

[27] Cf. Flügel's Commentary, *Fihrist*, Leipzig 1862, 270–71.

[28] Cf. also G. Stroumsa, *Another Seed: Studies in Gnostic Mythology*, Leiden 1984, 145–52, who discusses the Manichean notion of being "from another seed" (*allogenes*) and its place in Manichean mythology.

[29] See *Imtāʿ*, III, 192 (German translation in van Ess, *TG*, VI, 432); H.-C. Puech, "Le Prince des Ténèbres en son Royaume," *Sur le Manichéisme*, Paris 1979, 104–51, and especially 145–47.

perfectly clear.[30] It is also very unlikely that this argumentation, which was so common in the discussion of theodicy as to sound almost banal, could have been put forward as "sardonic humor," let alone that such humor would be expected to amuse the audience.[31]

Often mentioned in proximity to the *K. al-Gharīb al-mašriqī* is another book of al-Warrāq's, the *Book of Lament over Animals* (*K. al-Nauḥ ʿalā al-bahāʾim*) or the *Book of Animals' Lament* (*K. Ḥanīn al-bahāʾim*).[32] This book, which probably dealt with the question of the animals' sufferings, would also tally with al-Warrāq's Manicheism.

The association of such mythological and theological elements with al-Warrāq sets him apart from other rationalists, who were accused of heresy or of skepticism. Al-Warrāq mastered the dialecticians' method,[33] and often used it to discredit theologians of various religions; this was a widely-used technique among Manichean propagandists, as Augustine already witnessed. The mythological elements which crop up in association with al-Warrāq, however, testify to the fact that the epithet "rationalist" suits him only to a degree.[34]

Al-Warrāq is frequently mentioned as Ibn al-Rāwandī's colleague and mentor. Ibn al-Rāwandī's name is often associated with other heretics too.[35] But the consistent association of al-Warrāq with Ibn al-Rāwandī as well as the words chosen to describe their relations (al-Ḫayyāṭ, when he addressed Ibn al-Rāwandī, spoke of "your brother al-Warrāq") indicate clearly that Ibn al-Rāwandī's friendship with him was significantly closer than with others.[36] This close friendship accounts, to a large extent, for the confusion and uncertainty concerning Ibn al-Rāwandī's heresy.

---

[30] See below, Ch. 4, n. 50; and compare Thomas, *Anti-Christian Polemic*, 21, according to whom al-Warrāq "conspicuously avoids commenting on the morality of God's action."

[31] Compare Thomas, Ibid.

[32] Cf. Murtaḍā, *Šāfī*, 13, 16ff.; van Ess, *TG*, VI, 432.

[33] Tawḥīdī, *Imtāʿ*, III, 192, describes him as "*min ḥudhdhāq al-mutakallimīn.*"

[34] Compare van Ess, *TG*, IV, 289.

[35] As noted by van Ess ("Image," 22: *Īcī*, 223; *Bist Guftar*, 11), ʿAbd al-Jabbār and al-Ḫayyāṭ mention al-Ḥaddād, al-Ḥuṣrī, Abū Šākir al-Dayṣānī and Ibn Ṭalūt together with al-Warrāq and Ibn al-Rāwandī.

[36] See, for instance, *Intiṣār*, 111: "Your teacher and evil ancestor, the one who introduced you to heresy, Abū ʿĪsā al-Warrāq, who took you out of the honorable Muʿtazila to the degradation of heresy and infidelity;" cf. also *Intiṣār*, 73. Abū ʿAlī al-Jubbāʾī is said to have recounted that "the Sultan searched for Abū ʿĪsā al-Warrāq and for Ibn al-Rāwandī," and that the two had to run away from Baghdād (Ibn al-Jawzī, *Muntaẓam*, 112:7–9 [= Ritter, 5]; *Maʿāhid*, 158). Abū ʿAlī's statement suggests that the hunt for al-Warrāq was somehow connected with the hunt for Ibn

As mentioned above, the Muslim sources do not offer a clear picture concerning Ibn al-Rāwandī's religious integrity. On this topic, modern scholars remain as divided as the medieval authorities. Most scholars assume that Ibn al-Rāwandī was indeed a heretic, but there is no agreement between them as to the nature of his heresy. Some look for the roots of his heresy in his connections with Shiʿism,[37] and depict him as a Muʿtazilite gone wild.[38] Some regard him as an Aristotelian philosopher, mainly because he is often associated with the *dahriyya*—that is, those who assert the eternity of the world.[39] Others see him as a radical atheist,[40] while some stress the political challenge he presented to the Islamic polity.[41] Most modern scholars describe his move from the Muʿtazila as a gradual process, in which he passed through several religious affiliations, including, perhaps, a Shiʿite period.[42] At the same time, they try to account for the more positive view of Ibn al-Rāwandī in some Muslim sources.[43] In particular, Josef van Ess has offered an original interpretation that aims at accommodating all the contradictory information. Van Ess has noted that the sources which portray Ibn al-Rāwandī as a heretic are predominantly Muʿtazilite and stem from Iraq, whereas in eastern texts "the fact that he is seen in positive light jumps to the eye."[44] As an explanation for this difference, van Ess suggests "a collision of two different intellectual traditions," i.e., the tradition from Iran and that from Iraq. He further suggests that Ibn al-Rāwandī's notoriety was the result of the fact that after Ibn al-Rāwandī left Baghdād, "his colleagues in the metropolis . . . profiting from his absence . . . could create a black legend."[45] In other words, van Ess believes that

al-Rāwandī, and this is how the statement is understood by Ibn Kathīr, who says: "Abū ʿĪsā al-Warrāq befriended Ibn al-Rāwandī . . .; thus, when people learnt of their deeds, the Sultan searched for Abū ʿĪsā . . . and as to Ibn al-Rāwandī, he fled . . ." (*Bidāya*, 113).

[37] Cf. Nyberg, "Deux réprouvés," 132; Gabrieli, "Zandaqa," 32; van Ess, *TG*, IV, 300, suggests that he may have brought his Shiʿite sympathies with him from his birthplace in Khorāsān.

[38] Nyberg, "Deux réprouvés," 132–34; Gabrieli, "Zandaqa," 34.

[39] Nyberg, "Deux réprouvés," 133. It is related that Ibn al-Rāwandī's *K. al-Tāj* was devoted to a presentation of the proofs for the eternity of the universe. On this book, see van Ess, *TG*, IV, 335–36; and see below, Ch. 4, apud nn. 27–28.

[40] Niewöhner, "Are the Founders of Religions Impostors?" 239; idem, *Veritas*, 238.

[41] Kraemer, "Heresy," 168, 174; Aʿsam, *History*, 7.

[42] Aʿsam, *History*, 7.

[43] Nyberg, "Deux réprouvés," 134; Plessner, "Heresy," 8–10.

[44] *TG*, IV, 297.

[45] Van Ess, "Image," 9, 26; idem, "Fārābī," 391–96: *TG*, IV, 304.

Ibn al-Rāwandī, although admittedly eccentric and disputatious, was not a heretic at all.

This view was accepted by Ulrich Rudolph,[46] who believes that Ibn al-Rāwandī's heretical reputation was the result of an inner conflict within the Muʿtazilite circle of Baghdad, which then spread to other circles. Like van Ess, Rudolph believes that the image of Ibn al-Rāwandī in Māturīdī's *K. al-Tawḥīd* allows us a more balanced persepective. According to him, this text shows that in Samarkand Ibn al-Rāwandī was perceived as a defender of Islam. Rudolph suggests that this image could not have existed without a solid basis. In other words, he suggests that the heretical Ibn al-Rāwandī must have been painted on top of the real one, and that Māturīdī preserved the real, unadulterated image of Ibn al-Rāwandī. Rudolph believes that the key to understanding Ibn al-Rāwandī lies in the dissociation of his image from that of al-Warrāq.[47]

2. The *Kitāb al-Zumurrud*

Facts and Probabilities

Although the contradictory information about Ibn al-Rāwandī touches on various aspects of his personality, the greatest difficulties concern his book against prophets, the *Book of the Emerald* (*K. al-Zumurrud*), regarding which the sources appear to offer two diametrically opposed accounts.[48] The remaining part of this chapter will deal with this book, which, in modern scholarship, is perhaps the best-known of Ibn al-Rāwandī's compositions. I shall first offer a detailed analysis of one passage from the *Zumurrud*. I shall then review the Muslim sources that are supposedly favorable to Ibn al-Rāwandī, and discuss the connection between Ibn al-Rāwandī's heresy and the form of the *Zumurrud*; and finally an attempt will be made to summarize what we know about the *Zumurrud* and its contents.

I shall begin by setting out some facts that we know for certain about the *Zumurrud*, and some inferences that can be drawn with a

---

[46] Rudolph, *Māturīdī*, 176.

[47] Rudolph, *Māturīdī*, 177.

[48] The sources relevant to this book were grouped, in German translation, in van Ess, *TG*, VI, 457–83.

high degree of probability from such facts. This will be done on the basis of only those sources that explicitly mention the *Zumurrud*.

There is unanimous agreement that the *Zumurrud* was directed against prophecy. Al-Ḫayyāṭ, for instance, describes it in the following words:

> The book known as *Kitāb al-Zumurrud*, in which he [i.e., Ibn al-Rāwandī] mentioned the miracles of the prophets, peace upon them, such as the miracles of Abraham, Moses, Jesus and Muḥammad, God's blessing on them. He disputed the reality of these miracles and claimed that they were fraudulent tricks (*maḥārīq*) and that the people who performed them were magicians and liars; that the Qur'ān is the speech of an unwise being, and that it contains contradictions, errors and absurdities. He included in it a chapter entitled: "Against the Muḥammadans in particular," meaning the community of Muḥammad, God's blessing on him.[49]

Almost all the sources present Ibn al-Rāwandī as the author of the *Zumurrud*. Only two sources deviate slightly from the consensus. In a passage of the *Tathbīt dalā'il al-nubuwwa* of 'Abd al-Jabbār al-Hamadhānī (d. 415/1025), the *Zumurrud* appears twice: once as a book written by Abū 'Īsā al-Warrāq, and two lines later as written by Ibn al-Rāwandī.[50] And Ibn al-Jawzī (d. 597/1200) says that Ibn al-Rāwandī and al-Warrāq each charged the other with writing this

---

[49] *Intiṣār*, 12. See also al-Ḥillī, *Rawḍat al-manāẓir fī aḫbār al-awā'il wa'l-awāḫir* (in the margins of Ibn al-Athīr, *al-Kāmil*, Bulaq 1873), XI, 187, quoted by A'sam (*History*, 215); Ibn al-Murtaḍā, *Ṭabaqāt al-Mu'tazila*, 92; *Muntaẓam*, 109 (= Ritter, 3); *Ma'āhid*, 156; *Šadharāt al-dhahab*, II, 235, on the authority of Ibn al-Jawzī and Ibn 'Aqīl (= A'sam, *History*, 239).

[50] The passage (*Tathbīt*, 51:11–17) is somewhat garbled, partly because of the punctuation marks added by the editor. It should read: "Among the books of al-Ḥaddād on this topic [i.e., against prophets] is his book *al-Jārūf*, and his book *al-Arkān*; [also on this topic is] al-Ḥuṣrī's book *fī taswiyat aṣḥāb al-kalām bi'l-'awāmm*; Also the books of Abū 'Īsā al-Warrāq, including the *K. al-Zumurruda*, *K. al-Gharīb al-mašriqī*, and the *K. Ḥanīn al-bahā'im*. And Ibn al-Rāwandī's *K. al-Tāj* about the eternity [of the world], *al-Zumurruda*, *al-Farīd*, *al-Taṣaffuḥ*, *K. Na't al-ḥikma* attacking the wisdom of God, and *al-Dāmigh* in which he attacked the Qur'ān; and other books of theirs." Van Ess has suggested that in the *K. al-Gharīb al-mašriqī* al-Warrāq describes a person who comes from the east, encounters the adherents of scriptural religions, and is bewildered by their beliefs, a suggestion adopted also by Thomas (*Anti-Christian Polemic*, 25). Van Ess has further suggested that al-Warrāq's *K. al-Gharīb al-mašriqī* was Ibn al-Rāwandī's source for his description of the opinions of the *Barāhima* in the *Zumurrud* (*TG*, IV, 290 and 344). As can be seen from this passage of the *Tathbīt*, however, al-Warrāq's *Book of the Stranger from the East* is perceived by 'Abd al-Jabbār as distinct from al-Warrāq's contribution to a *K. al-Zumurrud*.

book, although they agreed on the criticism of the Qur'ān that the book contained.[51]

The book contained arguments both for and against the existence of prophets. This is stated explicitly in the *Majālis Mu'ayyadiyya* of the Ismā'īlī al-Mu'ayyad fi'l-Dīn al-Šīrāzī (d. 480/1077).[52] The *Majālis Mu'ayyadiyya* is the only source that contains relatively long citations of arguments against prophecy which are identified explicitly as belonging to the *Zumurrud*.[53] Al-Mu'ayyad does not claim to have read the *Zumurrud* itself: he says that he relies on a refutation of the book written by an Ismā'īlī missionary (*dā'ī*) and he claims to quote from this refutation. In the quotation the *dā'ī* says:

> We have come across an epistle composed by Ibn al-Rāwandī. He called it *The Emerald* and attributed it to the *Barāhima*.[54] The epistle is about the rejection of the existence of prophecy. In it he [i.e., Ibn al-Rāwandī] listed arguments advanced by those who support the existence of prophecy, and others presented by those who deny the existence of prophecy.[55]

Books dealing with *kalām* often take the form of a dialogue. There were two standard ways of presenting such a dialogue: one was for the author to respond to a hypothetical interlocutor, so that the form would be *in qāla . . . qulnā* ("if he were to say . . . we would reply . . .") or *in qāla . . . yuqāl lahu* ("if he were to say . . . he should be told . . ."). The other was to present the dialogue as having occurred between

[51] *Muntaẓam*, 109:4–6 (= Ritter, 3): *Wa-qad kāna Ibn al-Rāwandī wa-Abū 'Īsā Muḥammad b. Hārūn al-Warrāq al-mulḥid ayḍan yatarāmayāni bi-kitāb al-Zumurrud, wa-yadda'ī kullu wāḥid minhumā 'alā al-āḫar annahu taṣnīfuhu, wa-kānā yatawāfaqāni 'alā al-ṭa'n fi'l-qur'ān.*

[52] On him, see H. F. al-Hamdani, "The History of the Ismā'īlī *Da'wat* and its Literature during the last phase of the Fāṭimide Empire," *JRAS* (1932), 126–36; idem, "Unknown Ismā'īlī Authors," 375–77; de Smet, "Polémique ismaélienne," 88; Poonawala, 103–9. As demonstrated by de Smet (ibid., 89ff.), al-Mu'ayyad's polemics against Ibn al-Rāwandī reflected his Ismā'īlī philosophy. On the reliability of Ismā'īlī polemicists, see below, Ch. 3, apud nn. 136–43.

[53] The first three hundred *Majālis* of this mammoth collection were published by Muṣṭafā Ghālib, Beirut 1974–1984. The relevant excerpts of the *Majālis Mu'ayyadiyya* (which are drawn from the 417–22 *majālis*) were first published by Kraus, "Ketzergeschichte," 96–109 (reprinted in A'sam, *History*, 120–37). The references to the *Zumurrud* in the *Intiṣār* and by Ibn al-Jawzī are frustratingly short.

[54] On the identity of the *Barāhima* and their actual importance for Ibn al-Rāwandī's argument, see below, Ch. 5.

[55] Cf. Kraus, "Ketzergeschichte," 96:18–20: *waqa'at ilaynā risālatun 'amilahā Ibn al-Rāwandī wa-sammāhā al-Zumurrudha wa-nasabahā ilā al-Barāhima fī daf'i al-nubuwwāt wa-dhakara fīhā ḥujajan yaḥtajju bihā muthbitūhā fī ithbātihā wa-ḥujajan yaḥtajju bihā nāfūhā fī nafyihā.*

two real people.[56] The *Majālis Mu'ayyadiyya* shows that the dialogue in the *Zumurrud* was of the second kind. Discussing a certain passage from the *Zumurrud*, Mu'ayyad cites "the heretic" Ibn al-Rāwandī and "his opponent" (*ḫaṣmuhu*) and reports their exchange of arguments. This passage, whose evidential value on this point has been overlooked in the past, shows that the *Zumurrud* was constructed as a dialogue between two participants, the one arguing for the existence of prophecy, the other against it.[57]

In such dialogues one of the participants was often the author himself.[58] It is therefore likely that one of the participants in the *Zumurrud* was Ibn al-Rāwandī.[59] Some of the sources indeed use the phrase: "Ibn al-Rāwandī said in the *Zumurrud*," or words to that effect.[60] But such phrases do not, in themselves, prove that Ibn al-Rāwandī was represented in the *Zumurrud* as one of the participants in the dialogue. They could also mean that, as the author of the book, Ibn al-Rāwandī was held responsible for its content, regardless of whether or not he presented himself as one of the participants.

As to the other participant, there is reason to suspect that he was Abū ʿĪsā al-Warrāq. As noted above, al-Warrāq is often mentioned as Ibn al-Rāwandī's colleague and mentor,[61] and both ʿAbd al-Jabbār and Ibn al-Jawzī associate al-Warrāq specifically with the *Zumurrud*.[62]

Which role was assigned to which speaker? Given that the *Zumurrud* was a book whose purpose was to disprove the existence of prophecy, and that Ibn al-Rāwandī was its author, one would expect Ibn al-Rāwandī to take the role of the one who denies the existence of prophecy. But this could hardly be the case if his interlocutor was

[56] "*Qāla lī . . . Qultu . . .*" or "*Qāla . . . Fa-qāla lahu. . . .*" On these formulas, see M. Cook, "The Origins of Kalam," *BSOAS* 43 (1980), 32–43.

[57] Kraus, "Ketzergeschichte," 103:17–23. Kraus, who perceived the importance of the passage and noted the sudden introduction of the "opponent," did not realize that the debate was an integral part of the quotations from the *Zumurrud*. (On Kraus's interpretation of this passage, see below, apud n. 77.) Van Ess, who was the first to suggest that the *Zumurrud* was in the form of a dialogue (see below), did not notice the evidence of a dialogue in this passage.

[58] See, for instance, Rāzī, *Rasā'il*, 295ff.; Muqammaṣ, *Twenty Chapters*, 248–51; Rāzī, *Mafātīḥ al-ghayb*, III, 96, 88.

[59] Ibn al-Rāwandī's involvement in "discussions and debates" is recorded by some of the later sources. See, for instance, Ibn Ḫallikān (d. 1283), *Wafayāt al-aʿyān*, I, 94: *wa-lahu majālis wa-munāẓarāt maʿa jamāʿa min ʿulamāʾ al-kalām.*

[60] For instance *Intiṣār*, 12; *Muntaẓam*, 110:9, 15; and, persistently, the *Majālis Mu'ayyadiyya.*

[61] See above, apud n. 36.

[62] Cf. nn. 50–51 above.

indeed al-Warrāq. Al-Warrāq is consistently portrayed in the sources as denying the existence of prophecy, whereas Ibn al-Rāwandī is portrayed as sometimes arguing against the existence of prophecy and sometimes arguing for it. The balance of probability is therefore that in the *Zumurrud* it was al-Warrāq who had the role of the one who attacks prophecy. This would leave the role of the defender of prophecy to Ibn al-Rāwandī.

While the first two points are fairly certain, the third and fourth points rely mainly on the *Majālis Mu'ayyadiyya*, and the fifth point is based wholly on inference. I now want to show that these probabilities can be strengthened by looking at some sources which do not refer to the *Zumurrud* by name.

Māturīdī's Kitāb al-Tawḥīd

The most problematic of these sources is the *K. al-Tawḥīd* of the Ḥanafī scholar Abū Manṣūr al-Māturīdī (d. ca. 333/944).[63] Whereas the *Majālis Mu'ayyadiyya* presents Ibn al-Rāwandī as denying the existence of prophecy, Ibn al-Rāwandī appears in Māturīdī's book as arguing in its favor. For this reason, Schacht, Madelung and (in the past) I myself, assumed that the relevant passages in Māturīdī reflected Ibn al-Rāwandī's views before he wrote the *Zumurrud*, i.e., before his break with Islam.[64]

In sharp contrast, van Ess suggests that the passages cited by Māturīdī are derived from the *Zumurrud*.[65] He further implies that the *Zumurrud* was written in the form of a dialogue in which al-Warrāq presented arguments against the existence of prophecy, while Ibn al-Rāwandī presented arguments in its favor. Van Ess advances his ideas with some caution, saying that "the final decision [concern-

[63] Māturīdī, *Tawḥīd*, 186–202. Kholeif's edition is riddled with inaccuracies, and a slightly better reading of the relevant pages is provided in A'sam, *History*, 55–73. For an outline of the passages relevant to our disscussion, see Rudolph, *Māturīdī*, 229–30. The relevance of the *K. al-Tawḥīd* to the *Zumurrud* was first indicated by Schacht ("New Sources," 41–42).

[64] Schacht, "New Sources," 41: "The polemics in which Ibn al-Rēwandī indulged against Abū 'Īsā at a period when he himself had not yet broken away from Islam." Madelung, in his review of Kholeif's edition of Māturīdī (*ZDMG*, 124 [1974], 150) suggests that Māturīdī's source may have been Ibn al-Rāwandī's book *Ithbāt al-rusul*, presumably one of the books written in Ibn al-Rāwandī's "righteous period" (cf. Fück, 73); and see Stroumsa, "*Barāhima*," 231.

[65] Van Ess, "Image," 15, apud n. 5, and 18. This, if I understand him correctly, is also suggested by Chokr, *Zandaqa et zindīqs*, 165, n. 43.

ing the form and nature of the *Zumurrud*] must be left to a detailed philological analysis."[66] The main reason for van Ess's hesitation is his assumption that the evidence of Māturīdī is inconsistent with the evidence of the *Majālis Mu'ayyadiyya*. Van Ess thinks that the dialogue form of the *Zumurrud* can be seen only in the passages cited by Māturīdī,[67] whereas the *Majālis Mu'ayyadiyya* offers a different picture of the *Zumurrud*, a picture which van Ess believes to be distorted.[68]

In fact, however, as we have just seen, van Ess's points can be inferred with reasonable probability from the *Majālis Mu'ayyadiyya* and other sources, without any reference to Māturīdī. That these points fit well with the relevant passages cited by Māturīdī constitutes in fact an independent argument for the view that the passages cited by Māturīdī are derived from the *Zumurrud*. In the following section I shall offer another argument to the same effect, establishing it on the kind of detailed analysis that van Ess calls for. Through an analysis of one passage from the *Zumurrud* I will show that Māturīdī, like Mu'ayyad, draws from the *Zumurrud* itself, and that the information that each one of them offers complements rather than contradicts what the other has to say.

3. The *Mubāhala* Verse in the *Kitāb al-Zumurrud*

The background

According to Muslim tradition, Q. 2:94 ("Then wish for death, if you are speaking the truth") was revealed when the Jews of Medina claimed arrogantly that the hereafter belonged to them alone. In Muḥammad's presence, the Jews feigned confidence in the bliss awaiting them in the world to come. But God revealed to Muḥammad

[66] Van Ess, "Image," 15, n. 5.

[67] Van Ess, "Image," 15–16; Idem, "Fārābī," 395.

[68] Van Ess suggests that Mu'ayyad saw "only a Mu'tazilī refutation in which Ibn al-Rāwandī's . . . counter arguments were blatantly neglected" (*Bist Guftar*, 9–10); see below, apud n. 139. On another occasion van Ess seems to suggest that the distorted picture was the result of Kraus's interpretation of the *Majālis Mu'ayyadiyya*, and that "Kraus relied too much on al-Ḫayyāṭ who, following his polemical intention, slyly took the doxographical summaries given by Ibn al-Rēwandī for Ibn al-Rēwandī's own ideas" ("Image," 15). Cf. Also Thomas, *Anti-Christian Polemic*, 25, who suggests as the best explanation for the mix-up "a gradual confusion of the two men [i.e., Ibn al-Rāwandī and al-Warrāq]."

that the Jews were in fact afraid of him, and that if he challenged them to wish for death, they would not dare to do so. Muḥammad then presented the Jews with this challenge (*taḥaddī*) and, indeed, the Jews never expressed a wish for death, neither then nor on a second occasion, when Q. 62:6-7 was revealed. The commentators note the fact that the Jews were eager to prove that Muḥammad was not a real prophet. They fought Muḥammad with all their might, sacrificing their property and their dear ones. Yet they were not willing to add to their fight such a simple, easy thing as saying that they wished for death. The fact that Muḥammad predicted this strange behavior is taken by the commentators as proof that he was a prophet.[69]

Like the Jews, the Christians of Najrān did not accept Muḥammad as a true prophet. When they came to sign an agreement with Muḥammad, they had not planned to accept his terms. Muḥammad challenged them to let God be their judge by calling down His curse on the party which lied, a challenge which is alluded to in Q. 3:61 (the *mubāhala* verse).[70] After a consultation with the Jews, the Christians decided not to risk the ordeal and accepted Muḥammad's terms. In Muslim tradition the two confrontations are frequently connected, and both are regarded as proving that Muḥammad was indeed a prophet.[71] As the term *mubāhala* is used by the commentators in reference to both these verses, they can be called "the *mubāhala* cluster."

The commentators mention other verses which they regard as proofs that Muḥammad was a prophet who posessed secret knowledge. Among such verses are Q. 29:48 and Q. 48:27. Muslim tradition regards Q. 29:48 as referring to Muḥammad's illiteracy, and as

---

[69] For a detailed description of the confrontation of Muḥammad with the Jews, see *Tathbīt*, 411–12; and cf. also *Ta'wīlāt ahl al-Sunna*, 193–95 (= *Tafsīr al-Māturīdī al-musammā Ta'wīlāt ahl al-Sunna*, ed. Ibrāhīm Awḍīn and al-Sayyid Awḍīn, Cairo 1981, 223–26).

[70] Q. 3:61. "And to whoever disputes with you concerning him (i.e., Jesus), after the knowledge that has come to you, say, 'Come now, let us call our sons and your sons, our wives and your wives, ourselves and yourselves, then let us humbly pray and so lay God's curse upon the ones who lie.'" Cf. art. "*mubāhala*," *EI*[2], VII, 276–77 (W. Schmucker); L. Massignon, "Le Mubâhala: Étude sur la proposition d'ordalie faite par le prophète Muḥammad aux Chrétiens Balhârith du Nejrân en l'an 10/631 à Médine," in: *La Mubâhala de Médine et l'hyperdulie de Fatima*, Paris 1955, 7–15; and R. Strothmann, "Die Mubāhala in Tradition und Liturgie," *Der Islam* 33 (1958), 5–29.

[71] See, for instance, Rāzī, *Mafātīḥ al-ghayb*, II, 203–207 and IV, 86–92; Ibn Kathīr, *Tafsīr*, I, 113–15. Some commentators say that Muḥammad challenged the Jews to a *mubāhala* in which the parties called on God to bring death upon the party that lied.

implying that Muḥammad's knowledge of the content of the Jewish and Christian scriptures must have come to him through revelation.[72] Q. 48:27 is seen as referring to Muḥammad's prediction of his victory over the Meccans.[73] It should be noted, however, that unlike the verses of the *mubāhala* and the *taḥaddī*, which the commentators constantly mention together, Q. 29:48 and Q. 48.27 are not, as a matter of course, connected with each other or with the first two.

Muḥammad's confrontations with the Jews and Christians are discussed in the *Zumurrud*, and fragments of the discussion are preserved in several sources.

The Majālis Mu'ayyadiyya

Let us first examine quotations from the *Zumurrud* which are explicitly ascribed to it, those found in the *Majālis Mu'ayyadiyya*. As mentioned above, Mu'ayyad claims that his source was a refutation of the *Zumurrud* by an Ismāʿīlī *dāʿī*. This *dāʿī*'s refutation contained arguments against the existence of prophecy, derived from the *Zumurrud*, and responses to those arguments, written by the *dāʿī* himself. Although the *Zumurrud* had already presented arguments for the existence of prophecy, the *dāʿī* preferred to replace them with pro-prophecy arguments of his own.[74] The voice of the proponent of prophecy in the *Zumurrud* is thus usually silenced by that of the *dāʿī*. But in one rather obscure passage this voice emerges unannounced. The difficulty as well as the importance of this passage were already noticed by Kraus. What has remained unnoticed is the fact that this short passage provides the key to a correct understanding of the original structure of the *Zumurrud*.[75]

The exchange of arguments in the dialogue is not quoted in full. Consequently, both the arguments and the identities of those presenting them remain obscure. In what follows, I divide the passage into numbered paragraphs. The following abbreviations are used in identifying the persons referred to: IR is Ibn al-Rāwandī; D is the

[72] Ibn Kathīr, *Tafsīr*, II, 357–58; and cf. also *Ta'wīlāt ahl al-Sunna*, 171:12–14 (p. 183 in the Cairo edition).

[73] Ibn Kathīr, *Tafsīr*, IV, 177–80.

[74] Kraus, "Ketzergeschichte," 96:18–22, 121.

[75] And not the key to the true identity of al-Mu'ayyad's source for the *Zumurrud*, as suggested by Kraus; see below, apud n. 77.

speaker who denies the existence of prophecy; and S is the speaker who supports the existence of prophecy.

> 1. The heretic [IR] refuted the *mubāhala* verse (Q. 3:61) and what is said concerning the circumstances under which it was revealed; he rejected [what is known to be] the [true] meaning of God's saying "Then wish for death, if you are speaking the truth" (Q. 2:94, 62:6); and he also mentioned various other verses of the same kind [and refuted them].
> 2. The *dāʿī* said in response to this that if he [D] and his opponent [S] agree about the meaning of these verses, then the way is open to him [D] to refute them and to reject what they imply. But if his opponent [S] says, "The meaning of these verses is not what the assumptions on which you [D] base your reckoing demand," then his [D's] refutation is invalid and his exertions have been in vain.
> Exactly the same applies also to his [D's] refutation of God's words, "You have not recited a book before . . ." (Q. 29:48) and his commentary on God's words, "You shall enter the holy mosque, if God so wills " (Q. 48:27) where he [D] says that these were mere guesses, and not authoritative announcements of what he [Muḥammad] desired should occur.
> 3. The opponent [D] in his eagerness to refute, lumped all these verses together in his interpretation, without pausing to reflect; the impropriety of this is obvious.[76]

As mentioned previously, much in this passage remains unclear. For example, the text of the *Majālis Muʾayyadiyya* gives no explanation of "the assumptions on which he bases his reckoning."

The passage preserves some remnants of the dialogue form of the *Zumurrud*, but the exchange of arguments in the dialogue is not quoted in full. We can see the presence of two speakers in the *Zumurrud*. The form of introduction which al-Muʾayyad uses here ("If his opponent says . . .") results from the fact that Muʾayyad summarizes rather than quotes, and does not indicate in this case that the dialogue is hypothetical. We may assume that in the *Zumurrud* itself the discussion

---

[76] Kraus. "Ketzergeschichte," 103:17–24 (= Aʿsam, *History*, 130).
1. *Qāla al-dāʿī fī al-jawāb ʿan radd al-mulḥid ʿalā āyati al-mubāhala wa-asbābihā wa-maʿnā qawlihi subḥānahu "fa-tamannaw al-mawta in kuntum ṣādiqīn" wa-mā yajrī hādha al-majrā min al-āyāt <allatī> dhakarahā.*
2. *Innahu in kānat maʿānīha mustaqirra baynahu wa-bayna ḫaṣmihi kāna lahu al-ṭarīq li'l-radd ʿalayhā wa'l-dafʿi fī wajhihā. Fa-in qāla ḫaṣmuhu: "inna maʿāniyahā ghayru mā taḍammanathu šurūṭ ḥisābika," baṭala al-radd kulluhu wa-ḍāʿa taʿbuhu. Wa-ka-mithli dhālika ḥukmu raddihi ʿalā qawlihi "wa-mā kunta tatlū min qablihi min kitāb" al-āya, wa-mā yuʿalliquhu bi-qawlihi "la-tadḫulunna al-masjid al-ḥarām in šāʾa ʾllāh," bi-anna dhālika rajman bi'l-ghayb, lā qaṭʿan ʿalā mā yurīdu kawnahu.*
3. *Fa-min ḥirṣ al-ḫaṣm ʿalā al-radd sāqa taʾwīl al-maqāmāt fī jumlatihi ghayra muʿtabirin, wa-mawḍīʿ al-ʿayb fī dhālika ẓāhir.*

had the form of an actual debate ("He said . . . Then he answered . . .").

As to the identity of the speakers, the *Majālis Mu'ayyadiyya* has been assumed to show that Ibn al-Rāwandī presented in the *Zumurrud* the arguments against the existence of prophecy. But a close reading of this passage—the only passage in the *Majālis Mu'ayyadiyya* where the two speakers are actually heard—calls for a different interpretation. Let us summarize this passage:

1. The *dāʿī* says that the heretic disputed the *mubāhala* verse (Q. 3:61) and Q. 2:94 and the relevant Muslim traditions concerning the circumstances of their revelation.

2. Rather then present the arguments of D and refuting them, as he usually does, in this passage the *dāʿī* summarizes the arguments of both D and S in their discussion of these verses. D attempts to interpret these verses by introducing an assumption that will enable him to criticize them, and his opponent S (*ḫaṣmuhu*) rejects the assumption. D mentions Q. 29:48 and Q. 48:27.

3. The *dāʿī* says that the opponent (*al-ḫaṣm*) grossly misunderstood those verses in his eagerness to refute the Qur'ān.

"The heretic" in the first paragraph, as everywhere in Mu'ayyad's account, is Ibn al-Rāwandī, the author of the *Zumurrud*. We then have three problematic possibilities:

One's first inclination would be to identify Ibn al-Rāwandī with the speaker who criticizes the Qur'ān in the second paragraph (D) and with "the opponent" in the third paragraph. This would mean that Mu'ayyad applies the word "opponent" (*ḫaṣm*) in the same short passage to two people who hold opposite views. For "the opponent" is harshly criticized by Mu'ayyad, whereas "his opponent" is clearly a proponent of the existence of prophecy.

A second possibility, suggested by Kraus, identifies Ibn al-Rāwandī with the speaker who criticizes the Qur'ān in the second paragraph (D). Kraus did not realize that the passage reflects the dialogue in the *Zumurrud*. He suggested that the words "the opponent" and "his opponent" in the second and third paragraphs refer to the same person, and that this person is not Ibn al-Rāwandī, but rather "a Muslim writer who had attempted to refute the *Zumurrud* before the author" (i.e., before the *dāʿī*, whom Kraus believed to have been Mu'ayyad himself).[77] Kraus's suggestion seems rather arbitrary. Nowhere in the

[77] Kraus, "Ketzergeschichte," 114, n. 1, and 358–60; de Smet, "Polémique ismaélienne," 91, tends to accept Kraus's suggestion, mainly because of the *dāʿī*'s style, which is identical with Mu'ayyad's; cf. n. 97 below.

Majālis Mu'ayyadiyya is there any evidence for the existence of such a "previous Muslim writer." Nor is there any other case where al-Mu'ayyad abruptly abandons his attack on Ibn al-Rāwandī and turns to chide someone else; and the language Mu'ayyad uses to berate "the impropriety" of "the opponent" seems much too strong for criticizing the wrong methods of a Muslim who defends the Qur'ān. And though Kraus's suggestion avoids a proliferation of "opponents," it does not address the other difficulties of this passage.

A third possibility is to accept Kraus's idea that the term "opponent" must refer in both occurrences to the same person, but to assume that throughout the passage Mu'ayyad continues to target the *Zumurrud*. "The heretic" of the first paragraph, "his opponent" of the second paragraph, and "the opponent" of the third would then be one and the same person: Ibn al-Rāwandī. This interpretation implies that the pronoun "he" changes its reference in the middle of the passage. Such a change is quite plausible, given the common tendency in medieval Arabic texts to use pronouns in an unclear manner and the obscure wording of this particular passage. This interpretation also means that Ibn al-Rāwandī, who in the beginning of our passage seeks to refute the Qur'ān, defends it in the second paragraph. This would be possible only if we assumed that Ibn al-Rāwandī, the author of the heretical *Zumurrud*, is not the one presenting the attack on the Qur'ān (D) but rather "his opponent" (S).

None of these three possibilities is entirely free of problems, and the *Majālis Mu'ayyadiyya* does not provide us with enough data to choose between them. I believe, however, that the third possibility is the least problematic, and I shall presently try to show that various other sources support this view.

In another passage Mu'ayyad attacks Ibn al-Rāwandī for things that he had said, "citing one of those who reject the existence of prophecy."[78] This could be taken to mean that Ibn al-Rāwandī presented the attack on prophecy as he had heard it from someone else (the "Brahmans," for example). But if our last suggestion is correct, then this passage may indicate that the author of the *Zumurrud* left the role of presenting the arguments against the existence of prophecy to someone else, and that this someone was not only a vague ethnic

---

[78] Kraus, "Ketzergeschichte," 106:19: "*Wa-ammā ḥikāyatuhu ʿan baʿḍi dāfiʿ al-nubuwwāt...*"

group of "Brahmans,"[79] but a specific person, someone known for his disbelief in prophecy. It would follow that Ibn al-Rāwandī took upon himself the role of the proponent of prophecy (S).

The *mubāhala* passage in the *Majālis Mu'ayyadiyya* is too cryptic to allow us to go any further.[80] In order to identify the speaker who attacks the Qur'ān and denies prophecy in the *Zumurrud* and to understand his relation to Ibn al-Rāwandī, we must now turn to Māturīdī's *K. al-Tawḥīd.*

Māturīdī

As mentioned above, Māturīdī does not refer to the *Zumurrud* by name. But his book contains a dialogue on prophecy in which the arguments against the existence of prophecy bear a striking resemblance to those cited by Mu'ayyad and by Ibn al-Jawzī as part of Ibn al-Rāwandī's *Zumurrud.* Māturīdī clearly identifies the participants in the dialogue: Abū 'Īsā al-Warrāq, who attacks the prophets, and Ibn al-Rāwandī, who responds, defending them. It is the passage on the *mubāhala* verse in Māturīdī's work that proves that Māturīdī was using the *Zumurrud.* The relevant passage in Māturīdī reads:[81]

> [We have already set out some of the arguments that show that Muḥammad was a prophet.] In addition to this, he [i.e., Ibn al-Rāwandī] cites Muḥammad's saying to the Jews, "Then wish for death."[82] He [Ibn al-Rāwandī] argues from it in two ways [to show that Muḥammad was a prophet]. First, there is [God's] pledge that if they wish for death, they will die. Second, that although nothing would be easier for them than to wish for death, they will never do so. He [Ibn al-Rāwandī] also used the *mubāhala* with the Christians, and [Muḥammad's] announcement that a [divine] curse would befall them to prove that Muḥammad had already been described in their Scriptures. Al-Warrāq offered two retorts to this. One was that if the Jews had indeed expressed verbally the wish to die, then it would have been said [by Muḥammad] that what he [Muḥammad] meant was that they should wish in their hearts to die. The other was that they [the

---

[79] As suggested by people from Mu'ayyad to Kraus, see "Ketzergeschichte," 109:12, 341–357.

[80] It may also be that the text, which was published by Kraus on the basis of a single manuscript, is corrupt.

[81] Māturīdī, *Tawḥīd*, 195:13–196.10, and see A'sam's emendations to the text, *History*, 66.

[82] Q. 2:94. This verse is also mentioned by Māturīdī, *Tawḥīd*, on 204:1.

> Christians] believed in Moses and Jesus, and that these two had predicted to them [the coming of the prophet Muḥammad] in the same way that an astrologer predicts things.
> The answer to the first argument is that the *mubāhala* will not bear this interpretation. Furthermore, the Jews are clever people: so that if they had been given such a reply, they would have retorted that they also wished in their hearts to die.[83]
> As to the second argument, the answer is that had it been so [i.e., that upon learning of the prophecies of Jesus and Moses, the Christians surrendered to Muḥammad], then they would not have refused to accept Muḥammad when God said, "You shall indeed enter the holy mosque,"[84] and His saying "That He may uplift it above every [other] religion."[85]
> ... He [i.e., al-Warrāq] presented this criticism [of the Qur'ān]. But if this [i.e., what Moses and Jesus prophesized] had been a prediction of the same kind as those made by astrologers, then it would not have been repeated by them [so frequently] ...[86]
> ... And he [i.e., al-Warrāq] attacked the verse "You have not recited a book before,"[87] saying that memory can take the place of a book.[88]

In this passage Māturīdī refers to all the verses mentioned by Mu'ayyad, and almost in similar order.

> Mu'ayyad: Q. 3:61, 2:94, 29:48, 48:27.
> Māturīdī: Q. 2:94, 3:61, 48:27, 29:48.

The fact that neither source offers exact quotations may account for the slight difference in the order of the verses referred to, as well as the different choice of words. Notwithstanding these different elaborations, the similarities of the two texts are remarkable. It is not only that, as has been noted in the past, they contain similar arguments. Rather, the two texts present the same combination of arguments, attached to the same Qur'ānic verses, treated in practically the same order. Such a similarity is too great to be explained as the

---

[83] The last two sentences testify to the blurring of the distinction between Muḥammad's challenge to the Jews (to wish for death) and the *mubāhala* with the Christians.

[84] Q. 48:27.

[85] Q. 61:9.

[86] Compare my previous translation, "The Blinding Emerald," 171. The rest of the sentence is rather obscure, and the text is probably corrupt.

[87] Q. 29:48.

[88] Al-Warrāq means that although Muḥammad could not read, he could still have been instructed by someone else. Compare Māturīdī's reference in *Tawḥīd*, 206:4.

result of quotations from two books on the same subject. In other words, Māturīdī, like al-Mu'ayyad, is probably relying on the *Zumurrud* itself.

This conclusion is strengthened by the fact that reading the two texts together one gets a better grasp of the original argument in the *Zumurrud* than from either one alone. Māturīdī's account leaves much to be desired, but he is still clearer than Mu'ayyad's. Only with Māturīdī's help can we make sense of al-Mu'ayyad's allusion to "the assumption." From Māturīdī's presentation it appears that what is meant by "assumption" is the notion of the heart's intention which al-Warrāq tries to introduce into the discussion of the "wish for death." The information given by Māturīdī and Mu'ayyad is in fact complementary rather than contradictory.

The analysis of the *mubāhala* passage in the *Majālis Mu'ayyadiyya* indicated that in the *Zumurrud* a person whose name we are not told argued against the existence of prophecy and that Ibn al-Rāwandī acted as "his opponent." Māturīdī's version of this paragraph of the *Zumurrud* corroborates this analysis and allows us to identify the first speaker as al-Warrāq. Both the *Majālis Mu'ayyadiyya* and Māturīdī thus show that Ibn al-Rāwandī, the author of the *Zumurrud*, presented in it the views of the proponents of prophecy, whereas the role of the speaker who denies the existence of prophecy was taken by al-Warrāq.

The Tathbīt dalā'il al-nubuwwa

The *mubāhala* cluster appears in still another source, the *Tathbīt dalā'il al-nubuwwa* of ʿAbd al-Jabbār.[89] ʿAbd al-Jabbār cites arguments against the existence of prophecy which closely resemble those of the *Zumurrud*, but he does not identify the source of the arguments he presents against the existence of prophecy, and his formulation of these arguments differs slightly from the formulation in the *Majālis Mu'ayyadiyya*. Scholars are therefore hesitant in identifying these fragments as belonging to the *Zumurrud*.[90] In fact, the *mubāhala* cluster allows us to see not only that the *Tathbīt* cites the *Zumurrud*, but that the information it gives is entirely consistent with what we learn from the two other sources.

---

[89] See *Tathbīt*, index, art. "Ibn al-Rāwandī"; Aʿsam, *References*, I, 15–30.
[90] See van Ess, "Image," 15, n. 5.

In the *Tathbīt*, ʿAbd al-Jabbār deals at length with the *mubāhala* cluster and its relevance to prophecy. He adduces the arguments for and against the existence of prophecy that we have already seen cited by Muʾayyad and Māturīdī. And, most interestingly, ʿAbd al-Jabbār seems unable—or unwilling—to make up his mind as to who presented the arguments against prophecy. Was it one person (to be addressed in the singular) or more than one (to be addressed in the plural)? Was it Ibn al-Rāwandī, or Ibn al-Rāwandī in the name of al-Warrāq, or perhaps both of them together? The relevant passages read:

> 1. . . . . The heretics were confounded by these verses [i.e., Q. 2:94–95, and Q. 62:6–7]. . . . This is why they curse [Muḥammad's] Arab [opponents, saying], "Why did they not produce an equivalent to the Qurʾān?" And [they curse] the Jews, [saying], "Why did they not wish for death? [Had they done so] they would have proved that he [i.e., Muḥammad] was a liar; they would thus have been rid of him, and so would we!" This is what people like al-Ḥaddād and his friend Abū ʿĪsā say, may God disgrace them.[91]
>
> 2. Ibn al-Rāwandī said that al-Warrāq used to say: "The only reason that they did not wish for death is that the Jews and the Christians believed in Moses and in others who claimed to be prophets. In their books these [prophets] had announced the coming of Muḥammad as a prophet, therefore they [i.e., the Jews] did not dare to wish this [i.e., to wish for death].
>
> He was told: This, then, proves the prophethood of those prophets as well as the prophethood of Muḥammad, so you [pl.] are obliged [*lazimakum*] to believe in all their books, whereas you [pl., *antum*] deny all of it!
>
> He [*sic*] said, "The ability of these people [i.e., Moses and Jesus] to predict the coming of Muḥammad is similar to the ability of astrologers to predict the future." He was told, "And since when can astrologers predict something like this, that is to say, something like the advent of Muḥammad, the date of his appearance, the content of his message, his country of provenance, the generation to which he belongs and his ancestry, in the same detailed manner that this was announced? This kind of prediction is not to be found in the announcements of the most proficient astrologers. They cannot even come close to it or do anything of a similar nature. Only by chance they [i.e., the astrologers] may happen to hit on a general, insignificant thing, after a thousand cases of errors and mistakes."[92]

---

[91] *Tathbīt*, 412:13–16.
[92] *Tathbīt*, 413:7–17.

> ... Then these *zindīqs* say: The only reason why they did not wish for death is that, had they expressed such a wish, he [i.e., Muḥammad] would have answered that he had meant them to wish it in their hearts. Had they said to him, "We did wish it in our hearts," he would have answered "Gabriel told me that you did not."[93]
> 3. Abū ʿĪsā al-Warrāq and Ibn al-Rāwandī criticized the story of the *mubāhala*, [saying] that it is a mutual curse, and that these people [i.e., the Christians] withdrew from it of their own accord. He [*sic*] said: "Your claim that he [i.e., Muḥammad] said to them [i.e., the Christians]: 'If you accept to have a *mubāhala* with me, a God-sent revenge will hit you'—this is not in the Qur'ān, it is only one of your *ḥadīths*."[94]
> 4. Ibn al-Rāwandī claimed also that when he [i.e., Muḥammad] had challenged the Christians to the *mubāhala*, or [when he challenged] the Jews to wish [for death], this was not meant to prove that he was a prophet; had this been his intention, they would have been quick to respond.[95]

Many of the central elements in these passages are also to be found in the citations of Mu'ayyad and Māturīdī: ʿAbd al-Jabbār mentions the claim that, due to the prophecies of Moses and Jesus, the Jews and the Christians had foreknowledge of the coming of Muḥammad. He also mentions the comparison of these two prophets to astrologers, and retorts that astrologers can predict the future only in general terms, and that only rarely do they hit on the truth. He dwells on the question of whether the wish for death meant an oral expression alone or had to include also the intention of the speakers. And he mentions the connection of these arguments to Q. 3:61 and 2:94.[96] It appears therefore that although ʿAbd al-Jabbār does not name the *Zumurrud*, he is in fact drawing from it. And ʿAbd al-Jabbār, who is aware of the fact that Ibn al-Rāwandī was citing al-Warrāq, holds both of these heretics accountable for the arguments against the Qur'ān and against the existence of prophecy.

The mubāhala cluster in the Zumurrud

Māturīdī, ʿAbd al-Jabbār and al-Mu'ayyad belong to three different schools of thought. It is not likely that ʿAbd al-Jabbār or Mu'ayyad would have copied from Māturīdī, and it is only remotely possible

[93] *Tathbīt*, 414:10–12; cf. *TG*, VI, 467–69.
[94] *Tathbīt*, 431:3; cf. *TG*, VI, 469–70.
[95] *Tathbīt*, 433:6–7.
[96] Ibn al-Rāwandī's interest in these two verses is attested also in *Intiṣār*, 28:17, see Kraus, "Ketzergeschichte," 113 and 126.

that Mu'ayyad would have copied from 'Abd al-Jabbār. Moreover, notwithstanding the similarities in their treatment of the *mubāhala* cluster, each of the three has some information which is not found in the other two. It is therefore evident that they do not draw their extracts of the *Ẓumurrud* from each other.[97]

We are now in a position to summarize our analysis of the *mubāhala* cluster. Ibn al-Rāwandī wrote the *Ẓumurrud* in the form of a dialogue with al-Warrāq, in which the latter presented the arguments against the existence of prophecy and Ibn al-Rāwandī responded, defending prophecy and presenting the arguments in favor of its existence. The discussion does not seem to have been very systematic. Al-Warrāq presented an argument, and when Ibn al-Rāwandī responded, al-Warrāq either answered again to the same point, or moved to attack the same verse from another angle, or to criticize another verse.[98] Although much of the discussion remains obscure, the three texts taken together allow us to reconstruct one part of it, and to assign each participant his role, as follows:

> Ibn al-Rāwandī: In Q. 2:94 Muḥammad challenged the Jews to wish for death. Muḥammad was confident that if the Jews were to wish for death, they would immediately die. [The Jews knew that too, and therefore they did not wish for death. This shows that, despite their refusal to accept Muḥammad, the Jews knew that he was a true prophet.]
>
> Muḥammad also knew that, despite the certainty they feigned, the Jews would not dare to express the wish for their own death,[99] [and the

[97] It is of course possible that, like Mu'ayyad, they rely on extensive quotations from a refutation of the *Ẓumurrud*, rather than on the book itself. They may have used different refutations (e.g., by Abū 'Alī al-Jubbā'ī and by an Ismā'īlī *dā'ī*), or a single refutation (such as Abū 'Alī al-Jubbā'ī's) may have served them all. Their ultimate source is nevertheless the *Ẓumurrud*. Kraus's intuition that the *dā'ī* was Mu'ayyad's invention (above, n. 77) may thus still be true, although Kraus's reasons for suggesting it are unacceptable.

[98] In this context, 'Abd al-Jabbār complains about the heretics' opportunistic willingness to try first an argument, then its opposite. They first claim that the Jews did not wish for death because they were stupid, then they say that the Jews were actually clever and accomplished (*Tathbīt*, 414:14–16).

This erratic form of presenting arguments was, in fact, quite common in dialogues, although not quite in such extreme form of self-contradiction. Sa'adia's interpretation of *Job* as progressing in straight line, with none of the speakers ever repeating what had already been said, reflects his ideals and ideology more than the reality of his days; see *Sa'adya's Commentary on the Book of Job*, ed. Y. Qafiḥ, Jerusalem 1973, 21 (Judeo-Arabic text).

[99] Māturīdī, *Tawḥīd*, 195:13–15: *Wa-iḥtajja fī ithbāt risālati Muḥammad ṣl'm, ma'a mā bayannā, bi-qawlihi li'l-yahūd "fa-tamannaw al-mawta," bi-wajhayni: aḥaduhumā al-wa'd bi-*

verse says explicitly that they will not do so. This proves that Muḥammad was a prophet, because without God's help, he would not have been able to guess what the Jews would do].

Al-Warrāq: [If the Jews did not accept Muḥammad's challenge to wish for death, this was not because they believed that he was a prophet, but rather because they did not take his challenge seriously.] When Muḥammad challenged the Christians to take part in the *mubāhala* and the Jews to wish for death, the words he used did not imply that he was doing so in order to prove that he was a prophet. Had the Jews and Christians realized that this was his intention, they would have been glad to accept the challenge.[100]

Ibn al-Rāwandī: Q. 3:61 refers to the *mubāhala* with the Christians. Muḥammad warned the Christians that [if they did not submit to him] they would be cursed. [This proves that he was a prophet, because he knew what the outcome of the *mubāhala* would be, should it take place. The Christians preferred to accept Muḥammad's terms rather than perform the *mubāhala* and be cursed. This proves that] the Christians realized that he was a true prophet. They knew it because their scriptures foretold the coming of a prophet whose description is like Muḥammad's.[101]

Al-Warrāq: People who claimed to be prophets, like Moses and Jesus, had indeed foretold the coming of Muḥammad as a prophet. The Christians and Jews believed in prophets; therefore, they did not dare to respond to Muḥammad's challenge.[102]

Ibn al-Rāwandī: But if you agree that Muḥammad is described in the Scriptures of the Jews and Christians, then you must admit that prophecy exists, and that Moses and Jesus, as well as Muḥammad, were prophets.[103]

Al-Warrāq: Moses and Jesus did indeed predict the coming of Muḥammad [but this does not imply the existence of prophecy or that these people were prophets]; any astrologer can make correct predictions.[104] In the same way, the fact that Muḥammad could predict certain events

---

annahum law tamannaw al-mawta la-mātū, wa'l-thānī annahum lā yatamannawna abadan, wa-lā šay'a aysar ʿalayhim min tamannī dhālika.

[100] *Tathbīt*, 433 (in the name of Ibn al-Rāwandī): *mā daʿā al-naṣārā ilā al-mubāhala wa'l-yahūd ilā al-tamannī ʿalā wajhi al-iḥtijāj bi-dhālika li'l-nubuwwa, wa-law kāna ilā hādha qaṣada la-bādarū ilā ijābatihi.*

[101] Māturīdī, *Tawḥīd*, 195:15–16: *wa-bi-mubāhalat al-naṣārā wa'l-iḫbār bi-wuqūʿ al-laʿn, thabata annahu maʿlūm al-naʿt fī kutubihim.*

[102] Māturīdī, *Tawḥīd*, 195:17–18: *Qad āmanū bi-Mūsā wa-ʿĪsā; Tathbīt*, 413:8–9: *kānū yu'minūna bi-Mūsā wa-ghayrihi mimman kāna yaddaʿī al-nubuwwa wa-qad aḫbara hā'ulā'i fī kutubihim bi-nubuwwat Muḥammad fa-lam yuqaddimū ʿalā al-tamannī li-hādha.*

[103] *Tathbīt*, 413:9–10: *Fa-hādha yadullu ʿalā nubuwwati ulā'ika wa-nubuwwati Muḥammad jamīʿan.*

[104] Māturīdī, *Tawḥīd*, 195:18: *Wa-qad aḫbarāhum bi-dhālika kamā yuḫbiru al-munajjima; Tathbīt*, 413:11–12: *Innamā iḫbār hā'ulā' ʿan majī'i Muḥammad (ṣlʿm) ka-mā yuḫbiru al-munajjim ʿan mā yakūnu.* The comparison of the prophets' prediction of the future to that of the astrologers also appears elsewhere in the *Zumurrud*, see n. 194 below.

does not prove that he was a prophet: he may have been able to guess successfully, but this does not mean that he had real knowledge of the future. And certainly the fact that he was able to recount events from the past does not prove that he was a prophet. [He could have read about those events in the Bible] and, if he was illiterate, he could still have had the Bible read to him.[105]

Ibn al-Rāwandī: The Jews and Christians had access to a very detailed description of the Prophet and the circumstances of his future arrival. No astrologer could predict the future in such a precise manner. Astrologers rarely succeed in predicting the future, and then only by chance.[106]

Al-Warrāq: The Jews knew that if they had accepted Muḥammad's challenge [and declared that they were so confident of prospering in the hereafter that they wished for death], then Muḥammad would have said that they did not really wish for death, but only said so.[107]

Ibn al-Rāwandī: Muḥammad's challenge did not include the condition that they have to mean what they say, they only had to say it.[108] Also, the Jews could have replied that they did wish it in their hearts.[109]

Al-Warrāq: If the Jews were to say that they did wish in their hearts, Muḥammad would have answered that Gabriel had revealed to him that they were lying, and that they did not wish it in their hearts.[110]

---

[105] *Majālis Mu'ayyadiyya*, 103:20–22: *wa-ka-mithli dhālika ḥukmu raddihi ʿalā qawlihi: "wa-mā kunta tatlū min qablihi min kitāb," wa-mā yataʿallaqu bi-qawlihi "la-tadkhulunna al-masjid al-ḥarām in šā'a 'llāh," bi-anna dhālika rajman bi'l-ghayb, lā qaṭʿan ʿan mā yurīdu kawnahu.*

Māturīdī, *Tawḥīd*, 195:21: . . . *mā imtanaʿū ʿan muqābalatihi ʿinda qawlihi: "la-tadḫulunna al-masjid al-ḥarām"*; 196:8–10: *wa-ṭaʿana fī qawlihi: "wa-mā kunta tatlū min qablihi min kitāb," inna al-ḥifẓa yaqūmu maqāma al-kitāb.*

[106] Māturīdī, *Tawḥīd*, 196:5: *Law kāna ʿalā ḥukmi al-munajjima lamā taqarrara ʿindahum; Tathbīt*, 413:12–17: *Wa-matā kāna mithlu hādha fī iḫbār al-munajjimīn . . .; mithlu hādha lā yakūnu fī iḫbār ḥudhdhāq al-munajjimīn wa-lā mā yuqāribuhu wa-lā mā yudāyinuhu, wa-innamā yattafiqu lahum al-iṣāba fī šay'in mujmal qalīl yasīr baʿda an yakdhibū wa-yuḫṭi'ū fī alfi šay'in.*

[107] Māturīdī, *Tawḥīd*, 195:17: *Innahum law tamannaw bi'l-lisān la-qīla innamā urīda bi-hi al-qalb; Tathbīt*, 414: 10–11: *Innamā lam yatamannaw al-mawta li-annahum law tamannawhu bi-alsinatihim la-qāla: innamā ʿanaytu umnīyat al-qulūb.*

[108] Māturīdī, *Tawḥīd*, 195:19: *Inna al-mubāhala lā taḥtamilu dhālika; Majālis Mu'ayyadiyya*, 103:20: *Inna maʿāniyahā ghayru mā taḍammanathu šurūṭ ḥisābika.*

[109] Māturīdī, *Tawḥīd*, 195:19–20: *Innahum ahlu baṣar, idhan law ruddū la-qābalū bi-annahum faʿalū dhālika ayḍan bi-qulūbihim.*

[110] *Tathbīt*, 414:11–12: *fa-in qālū la-hu: qad tamannaynā bi-qulūbinā, qāla lahum: qad aḫbaranī Jibrīl annakum lam tafʿalū dhālika.* Compare Zamaḫšarī's commentary, on Q. 2:94–95: "If you say: wishing belongs to the acts of the heart, it is a secret that nobody can uncover, so how do you know that they did not wish it? I say: wishing is not an act of the heart. It is only a person's saying, with his mouth: 'I wish I had such and such.' . . . If wishing belonged to the acts of the heart, and if they had wished it, they would have said: 'We have wished for death in our hearts.' Yet it is not recorded that they said this." (*Kaššāf*, I, 297–8); cf. also, *Mafātīḥ al-ghayb*, II, 205.

4. The Heretic Ibn al-Rāwandī and his *Zumurrud*

The sources supposedly favorable to Ibn al-Rāwandī

Despite the fragmentary state of our sources, we were able to derive from them a consistent account of one paragraph of the *Zumurrud*. This information confirms van Ess's hypothesis as to the form of the *Zumurrud*, namely that it was a dialogue between al-Warrāq, whose role was to present arguments against the existence of prophecy, and Ibn al-Rāwandī, who presented the arguments in its favor.

Another part of van Ess's hypothesis must now be examined. Van Ess believes that Ibn al-Rāwandī was not a heretic, and that "the extreme and shocking statements which have been filtered out of his books cannot be entirely representative of his own ideas."[111] He suggests that in the *Zumurrud* "Ibn al-Rāwandī defended Islam, or perhaps his special interpretation of Islam."[112] He believes that Ibn al-Rāwandī "wanted to waken the spirit of doubt and hesitation in the hearts of the people [of the Baghdād school], without being a skeptic himself."[113] And he further suggests that even al-Warrāq, who presents the arguments against the existence of prophecy in the *Zumurrud*, was not really a heretic, and that "the dialogue between Ibn ar-Rēwandī and Abū ʿĪsā was . . . apparently a dialogue between Muslims both of whom were still right."[114]

This contention is based on the assumption that Mu'ayyad and Māturīdī give inconsistent accounts, which I have argued is not the case. It is further based on preferring the evidence of Māturīdī to the evidence of the many other sources that unambiguously condemn Ibn al-Rāwandī. Van Ess believes that Mu'ayyad and ʿAbd al-Jabbār, Al-Ḫayyāṭ and Ibn al-Jawzī, are all biased, and that they draw a distorted picture of Ibn al-Rāwandī. Although van Ess's mainstay is Māturīdī, he supports his claim by referring to other sources that are supposedly favorable to Ibn al-Rāwandī.

A closer look at these sources shows that the information in most of them is not inconsistent with his being a heretic. Most of them discuss Ibn al-Rāwandī's scholarship, and are thus irrelevant to the

---

[111] Van Ess, "Image," 11.
[112] Van Ess, "Fārābī," 391.
[113] Van Ess, "Fārābī," 393.
[114] Van Ess, "Image," 21.

subject of his heretical beliefs. Tawḥīdī, for example, describes Ibn al-Rāwandī as a competent theologian and praises his mastery of Arabic style.[115] If this were our only evidence of what Tawḥīdī thought of Ibn al-Rāwandī, we might have inferred that Tawḥīdī regarded Ibn al-Rāwandī as a respectable Muslim. But fortunately there are some passages in which the discussion leads Tawḥīdī to the question of prophecy, and there he unequivocally condemns Ibn al-Rāwandī's heretical views.[116] Tawḥīdī saw no inconsistency between on the one hand giving Ibn al-Rāwandī the praise due to his erudition and on the other censuring his heretical ideas.[117]

In still another book, the *Aḫlāq al-wazīrayn*, Tawḥīdī recounts that Ibn al-Rāwandī's books, as well as those of other heretics, stood in the library of al-Ṣāḥib ibn ʿAbbād.[118] Although neither Tawḥīdī (who does not hide his defamatory intentions towards the *wazīr*) nor his source, the disgruntled *kātib* ʿAlī b. Īsā, are wholly trustworthy witnesses, there is no reason to reject their testimony. It shows that in the tenth century Ibn al-Rāwandī's books still circulated. It also shows that both al-Tawḥīdī and his source did not consider Ibn al-Rāwandī's books as suitable for the library of a respectable Muslim.[119]

The same is true of Abū al-Ḥasan al-Ašʿarī (d. 324/935), who several times mentions Ibn al-Rāwandī's opinions in neutral or even favorable tones.[120] This attitude of Ašʿarī's may account for ʿAbd al-Jabbār's sour remark that the Ašʿarites are impressed with Ibn al-Rāwandī.[121] But on another occasion, Ašʿarī refutes Ibn al-Rāwandī's

[115] Tawḥīdi, *Baṣā'ir*, 183; see below, Ch. 6, n. 58. On Ibn al-Rāwandī's interest in linguistics, see further van Ess, *TG*, IV, 298.

[116] Tawḥīdī, *Imtāʿ*, II, 20:9–14.

[117] Tawḥīdī's attitude was not shared by all Muslims. Ibn al-Anbārī (d. 1119) tries to explain "the mystery" of why al-Mubarrad's *K. al-Muqtaḍab* was not as popular as it should have been. His explanation is that this book was transmitted by Ibn al-Rāwandī, who was "notorious for his *zandaqa* and corrupt beliefs . . . and it is as if his ill reputation stuck to the book." (*Nuzhat al-alibbā' fī ṭabaqāt al-udabā'*, ed. A. Amer, Stockholm 1963, 138).

[118] *Aḫlāq al-wazīrayn*, ed. Muḥammad al-Ṭanjī, Damascus 1965, 181–83.

[119] Compare van Ess, *TG*, IV, 297, who takes this information as corraborating the place of Ibn al-Rāwandī's in the eastern, Ḥanafite-Murji'ite tradition. See further below, apud n. 124.

[120] Ašʿarī's remarks appear in his *Maqālāt al-islāmiyyin*, and the same detached, professional tone was also probably maintained in Ašʿarī's book concerning Ibn al-Rāwandī's *K. al-Jadal*. Cf. Aʿsam, *History*, 44–53; Gimaret, "Bibliographie," 247–48; van Ess, *TG*, IV, 305–7; and see below, Ch. 6, apud nn. 49–55.

[121] *Tathbīt*, 90: "Were it not for the regard that the *Ašʿariyya*, the *Rāfiḍa*, the

report of Muslim theology, adding to his name the unmistakable epithet "the cursed one" (*al-laʿīn*).[122] Ašʿarī also wrote refutations of Ibn al-Rāwandī's heretical books.[123]

The sources from the eastern, Ḥanafite-Māturidite school, belong to a special category. According to van Ess, there can be no doubt that Ibn al-Rāwandī himself belonged to the eastern Ḥanafite-Murji'ite tradition,[124] and that therefore Ḥanafite scholars regarded him favorably. On closer look, we find here the same phenomenon noted above: with the exception of Māturīdī, the references to Ibn al-Rāwandī in these sources have no bearing on his attitude to prophecy or religion. Al-Pazdāwī, for instance, quotes Ibn al-Rāwandī's opinion on the relationship between the human inner speech and its oral expression. Al-Pazdāwī does not share Ibn al-Rāwandī's opinion, which he associates with the Ašʿariyya, and he expresses his disagreement in clear terms. Anything beyond this criticism would have been irrelevant to the topic under discussion. The fact that al-Pazdāwī keeps a decorous discussion does not tell us anything about what he may have otherwise thought of Ibn al-Rāwandī.[125]

According to ʿAbd al-Amīr al-Aʿsam, who painstakingly collected all the medieval references to Ibn al-Rāwandī, only two authors attempted to defend him: Al-Šarīf al-Murtaḍā (d. 436/1044) and Ibn Ḫallikān (d. 681/1282).[126] But a closer look at what these two authors have to say shows that they too are well aware of Ibn al-Rāwandī's heretical views. Al-Murtaḍā had an account to settle with the Muʿtazila,

Christians and the *Zanādiqa* have for this man (*yarawna hādha al-rajul bi-ʿayn al-muḥaṣṣilīn*) we would not have mentioned his questions, since they are so weak."

[122] Ašʿarī, *Maqālāt*, II, 222. It is surprising that Nyberg, who characterizes al-Ašʿarī's reaction to Ibn al-Rāwandī as "rather mild" ("assez douce") in comparison to al-Ḫayyāṭ, uses the epithet "the cursed" in his portrayal; see "Deux réprouvés," 134.

[123] Gimaret, "Bibliographie," 229 (a refutation of the *K. al-Tāj*); 245, 265 (a refutation of Ibn al-Rāwandī's views "on the attributes and the Qur'ān." This could either refer to the *K. al-Dāmigh*, as suggested by Gimaret, or to the *Zumurrud*); 267–68 (a refutation of Ibn al-Rāwandī, "concerning his denial of *tawātur*." Gimaret remarks that no such book of Ibn al-Rāwandī's is known. But in fact, the title of this work of Ašʿarī's does not require us to postulate the existence of a book by this name. It could also refer to the discussion of *tawātur* in the *Zumurrud*, see below, n. 187).

[124] *TG*, IV, 297.

[125] Cf. Abū al-Yusr Muḥammad al-Bazdāwī, *K. Uṣūl al-dīn*, ed. H. P. Lins, Cairo 1963, 61; and see van Ess, *TG*, VI, 448. The same holds true regarding al-Ṣābūnī (cf. van Ess, *TG*, VI, 450–51).

[126] Aʿsam, *History*, 8.

and his book *al-Šāfi fi'l-imāma* was written as a retort to ʿAbd al-Jabbār. It is therefore understandable that in this book he blames the Muʿtazila for Ibn al-Rāwandī's heretical writings and plays down Ibn al-Rāwandī's faults. He mentions the claim that Ibn al-Rāwandī's main object in his heretical works was to spite the Muʿtazila. But even al-Murtaḍā sums up this discussion by saying, "There can be no doubt that he sinned in writing these books, regardless of whether he believed in it [i.e., their contents] or not."[127] And when, as an *ilzām* to the Muʿtazila, al-Murtaḍā compares Ibn al-Rāwandī to al-Jāḥiẓ, his clear intention is to incriminate the latter, not to exonerate the former.[128]

In some cases, the fragmentary remnants from the work of certain authors may include only a single short reference to Ibn al-Rāwandī. When such a reference happens to be neutral or even positive, we would be mistaken to deduce from it that the author condoned Ibn al-Rāwandī's views on prophecy. Muslim authors who were appalled by Ibn a-Rāwandī's religious positions were nevertheless quite capable of mentioning his name without additional comments, or even in praise. They kept their diatribes to discussions of prophecy and religion, but if these discussions are not extant, the diatribes would also be lost.

This principle is particularly to be observed when dealing with relatively late sources, which draw selectively from earlier ones. Such is the case of Ibn Ḫallikān, whose entry on Ibn al-Rāwandī describes him as "the well-known scholar." Ibn Ḫallikān goes on to say that Ibn al-Rāwandī wrote a treatise on *kalām*, that he was among the distinguished men (*fuḍalāʾ*) of his time, and that he wrote some one hundred and fourteen books. Of these books, Ibn Ḫallikān names only four, the *Zumurrud* and three other heretical books. Ibn Ḫallikān is aware of the fact that Ibn al-Rāwandī "had peculiar doctrines, which the theologians transmitted in their books." The theologians from whom Ibn Ḫallikān learned of the five heretical books were surely quite explicit in evaluating these "peculiar doctrines." Ibn Ḫallikān's reluctance to repeat the theologian's words bespeaks his own prejudices and preferences rather than those of his sources. And it would certainly be a mistake to rely on Ibn Ḫallikān's information in determining the tenor of Ibn al-Rāwandī's ideas.[129]

---

[127] Murtaḍā, *Šāfi*, 87.
[128] Ibid.
[129] Ibn Ḫallikān, *Wafayāt al-aʿyān*, I, 94. Ibn Ḫallikān's evasiveness and his reluc-

In the same way, when an author discusses prophecy but fails to mention Ibn al-Rāwandī in any way, we would be mistaken to take this omission as evidence that Ibn al-Rāwandī was in good standing. The silence of the Ismāʿīlī Abū Ḥātim al-Rāzī concerning Ibn al-Rāwandī, for example, should probably be explained in the same way as his silence concerning some typical details of Ismāʿīlī doctrines: both were outside the scope of his planned discussion.[130]

H. S. Nyberg argued that the philosophers and physicians "took Ibn al-Rāwandī seriously and sometimes defended him," and in support of his claim he cited the titles of books written by Kindī (d. ca. 252/866) and Ibn al-Haytham (d. 430/1039).[131] But Nyberg's evidence does not prove his claim. Kindī wrote "a discussion with Ibn al-Rāwandī on God's unity."[132] We have no way of knowing what the contents of this book were, but it is likely to have reflected some disagreements with Ibn al-Rāwandī. As to Ibn al-Haytham, he was not at all supportive of Ibn al-Rāwandī: a treatise in which he criticized Abū ʿAlī's refutation of Ibn al-Rāwandī also included his suggestions for a better refutation.[133] Although it is probable that Ibn al-Rāwandī's intellectual ideas were indeed attractive to some philosophers, the philosophers were careful to conceal such dubious influences and they avoided mentioning his name, as the case of Abū Naṣr al-Fārābī shows.[134]

---

tance to condemn Ibn al-Rāwandī are noted by Ibn Kathīr, who says that this was Ibn Ḫallikān's usual way of writing on heretics (*Bidāya*, 113). An error in the text of Ibn Ḫallikān (reading *maḥāsin* instead of *majālis*) may have contributed to the evaluation of his entry on Ibn al-Rāwandī as favorable; (see Aʿsam, *History*, 191, and compare n. 59 above).

[130] Compare Daiber, "Unity and Diversity," n. 6 and n. 17.

[131] "Deux réprouvés," 133.

[132] IAU, I, 212. Van Ess suggests that Kindī responded to a book which, according to al-Ḫayyāṭ (*Intiṣār*, 19) Ibn al-Rāwandī wrote in order to appease his Muslim opponents; cf. *TG*, IV, 301.

[133] IAU, II, 97. Ibn al-Haytham's treatise is described as "a treatise . . . that explains the shortcomings of Abū ʿAlī al-Jubbā'ī in his refutation of some of Ibn al-Rāwandī's books. [The treatise explains that] the things that, according to Ibn al-Rāwandī, follow logically from al-Jubbā'ī's own principles, do indeed follow from them. [The treatise also indicates] the opinion which, if upheld, makes its possible to avoid the objections of Ibn al-Rāwandī." (*fī īḍāḥ taqṣīr Abī ʿAlī al-Jubbā'ī fī naqḍihi baʿḍa kutub Ibn al-Rāwandī wa-luzūmihi mā alzamahu iyyāhu ibn al-Rāwandī bi-ḥasab uṣūlihi wa-īḍāḥ al-ra'y alladhī lā yalzamu maʿahu iʿtirāḍāt Ibn al-Rāwandī*). Ibn al-Haytham wrote a refutation of Abū Bakr al-Rāzī's book against the existence of prophecy, and another book to prove the existence of prophecy. See also below, Ch. 6, apud n. 71.

[134] Compare van Ess, "Fārābī," 394–98. Ibn al-Rāwandī's "intellectualist" argument appears in Fārābī's *Iḥṣā' al-ʿUlūm*, without Ibn al-Rāwandī's name and without the heretical conclusions. On Fārābī's reaction to Ibn al-Rāwandī's *K. al-Jadal*, see further below, Ch. 6, apud n. 68.

The only early source in which Ibn al-Rāwandī appears as a good Muslim in the context of the discussion of prophecy is Māturīdī's *K. al-Tawḥīd*, and we have seen how this account is to be understood. Māturīdī's book is also the source which most clearly shows the role of Abū ʿĪsā al-Warrāq in the *Zumurrud* as well as his Manichean beliefs.[135] Van Ess is probably right is assuming that the peculiarity of Māturīdī's evidence (namely, the fact that he does not attack Ibn al-Rāwandī) was related to his own eastern provenance. This difference based on geography, however, probably does not mean that in the east people were less dogmatic than in Baghdād, and therefore more understanding towards Ibn al-Rāwandī,[136] or that Ibn al-Rāwandī belonged to the eastern Ḥanafite school. Māturīdī lived in Samarkand, where in the tenth century Manicheism was still considered a real threat to Islam.[137] The interest of the *Zumurrud* for Māturīdī lay in the fact that in this book, the Manichean al-Warrāq denies the existence of prophecy. For Māturīdī, the urgent task was to attack the Manichean al-Warrāq rather than to unmask the real face of Ibn al-Rāwandī.

The most recent scholarly discussions of Ibn al-Rāwandī, and in particular the various studies of van Ess, insist on placing the blame for Ibn al-Rāwandī's heretical image on Muʿtazilite slander. David Thomas goes even further, and accuses al-Jubbāʾī of concocting this image almost single-handedly.[138] Before concluding this part of our inquiry, it may thus not be superfluous to recall an elementary fact, the paramount significance of which seems to have been curiously overlooked in these studies, namely, that our most explicit and detailed source concerning the *Zumurrud* is not a Muʿtazilite but an Ismāʿīlī source. Al-Muʾayyad did not rely on Muʿtazilite hearsay. He either read the book himself (if we accept Kraus's hypothesis) or relied on another Ismāʿīlī *dāʿī* who had read it (if we give credit to his own testimony).[139] In either case, this Ismāʿīlī intellectual, who read the *Zumurrud*, had no doubt that Ibn al-Rāwandī denied prophecy. Nor can we seriously suspect that his reading was thoroughly distorted

---

[135] See above.

[136] See van Ess, "Image," 14, 24.

[137] On dualism in Ḫurāsān in the tenth century, see Madelung "Abū ʿĪsā," 232; van Ess, "Fārābī," 396.

[138] Cf. Thomas, *Anti-Christian Polemic*, 25.

[139] See above, n. 77.

by Muʿtazilī propaganda: al-Mu'ayyad's intellectual portrait is well-documented,[140] and there is no evidence in it of a particularly strong Muʿtazilite influence on Mu'ayyad—certainly not to the extent that it would overrule his own judgement.

Ibn al-Rāwandī as a freethinker

If Ibn al-Rāwandī denied the existence of prophecy, why did he choose to present the case of its proponents? In order to understand the motives of the author of the *Zumurrud*, we would need to have a full view of his person and his books, fuller, at any rate, than the available evidence allows. As new material comes to light or old material is analyzed more closely, we may hope to confirm our views or to correct them. Until then, we can only offer a tentative analysis, on the basis of a very scanty information.

One possible suggestion would follow the approach developed (about other medieval thinkers) by Leo Strauss in his *Persecution and the Art of Writing*. According to this approach, one could argue that Ibn al-Rāwandī, afraid to publish his ideas in his own name, decided to attribute them to al-Warrāq.[141] This explanation, however, is not convincing, since it does not account for the fact that Ibn al-Rāwandī did publish other books, no less offensive to Muslims, in his own name. Moreover, it does not account for the apparently close and unbroken friendship between Ibn al-Rāwandī and al-Warrāq.[142]

It is unlikely that al-Warrāq would have volunteered to present Ibn al-Rāwandī's views, or that he would have appreciated finding himself in the position of having to take the blame for these views.

[140] His immense literary production (much of which is still unpublished) also includes an autobiography, *Sīrat at-Mu'ayyad*; cf. Poonawala, 107. Compare van Ess, n. 68 above; and apud n. 97 above.

[141] The idea that Ibn al-Rāwandī used to attribute his ideas to others is found already in the words of al-Šarīf al-Murtaḍā, *Šāfī*, 87, where he says: "Ibn al-Rāwandī . . . wrote the books for which he was disgraced in defiance of the Muʿtazila and in order to spite them. . . . He used to disclaim these books openly, to deny having written them, and to ascribe their authorship to others"; and again, 88: "Ibn al-Rāwandī did not say, in the books for which he was disgraced, 'I believe in the opinions I describe and hold them to be sound.' Rather, he used to say: 'The *dahriyya* say . . . the monotheists say . . ., the *Barāhima* say . . ., the proponents of prophets say . . .' ." The fear in which the heretics lived and the secrecy in which their books were written are vividly described in *Tathbīt*, 129 and 374.

[142] See n. 36 above.

Yet there is no indication that there was any animosity between the two after the publication of the *Zumurrud*.[143]

It is more probable, I think, that the structure of the book reflects the real relationship between al-Warrāq and Ibn al-Rāwandī. Al-Warrāq is usually described in the sources as the one who started Ibn al-Rāwandī on the path of heresy, introducing the young Muʿtazilite to criticism of the Qur'ān and the prophets.[144] While debating with him, Ibn al-Rāwandī used the method of *ilzām* (i.e., counter-attacking by turning the opponent's own arguments against him). He used al-Warrāq's arguments against revealed religions to attack Manicheism, al-Warrāq's own religion. When Ibn al-Rāwandī was eventually won over to al-Warrāq's heretical views, he went even further than his teacher. Whereas al-Warrāq was a Manichean, Ibn al-Rāwandī did not consider Manichean theology to be any better than Muslim theology. The *Zumurrud* may well record dialogues between the two intellectuals, edited by Ibn al-Rāwandī. The dialogues reflect their collegiality as well as their disagreements. They record the process of discussion by which al-Warrāq persuaded Ibn al-Rāwandī to reject prophetic religions, as well as Ibn al-Rāwandī's refusal to see Manicheism as an exception.[145]

One should also note the connection between Ibn al-Rāwandī's character, his style, and his ideas. I know of no other book where the author, as one of the participants in a dialogue, ascribes to himself an idea in which he no longer believes. Yet such an unusual procedure would not be out of character for Ibn al-Rāwandī. I also know of no other author who systematically wrote refutations (*naqḍ*, not just corrections or retractions) of his own works. Ibn al-Rāwandī did: he is known to have systematically written first a book and then its refutation.[146] Van Ess regards this practice as a *péché mignon*, "a

[143] The statement that according to al-Ḫayyāṭ "Ibn al-Rāwandī became disenchanted with Abū ʿĪsā when Abū ʿĪsā started showing sympathies for Manichean ideas" (van Ess, "Image" 15, n. 5) is incorrect. Al-Ḫayyāṭ tells us only that the Muʿtazila rejected al-Warrāq because of his Manicheism (*Intiṣār*, 108:2: *ka-mā faʿalat* [not *faʿalta*, as van Ess apparently reads] *bi-aḫīka al-Warrāq lammā qāla bi'l-manāniyya*).

[144] See n. 36 above.

[145] This is reflected in Ibn al-Rāwandī's mockery of Manichean myth and his criticism of the validity of Manichean traditions (Māturīdī, *Tawḥīd*, 197:1–2); see n. 23 above.

[146] *Maʿāhid*, I, 156, mentions *al-Tāj*, *al-Zumurrud*, *al-Farīd*, and *al-Lu'lu'a* and then adds: "He himself has refuted most of them, as have others" (*wa-qad naqaḍa huwa aktharahā wa-ghayruhu*). Cf. also *Maʿāhid*. 155; Ibn al-Nadīm says, on the authority of Abū al-Qāsim al-Balḫī, that Ibn al-Rāwandī "wrote his own refutation" of the

certain dose of intellectual coquetry."[147] Muslim authors were much less patient in this respect, and they considered Ibn al-Rāwandī's baffling habit a proof of his frivolous, mercenary tendencies.[148] This tendencies are recorded also in *Ma'āhid al-tanṣīs*, which says:

> Ibn al-Rāwandī could not stick to one school of thought, nor could he remain firm in any state. Thus he composed for the Jews the *K. al-Baṣīra*, refuting Islam, for the sum of 400 dirhams, which, as I have heard, he received from the Jews of Samarrā'. Having collected the money, he contemplated writing a refutation of it, until they gave him an additional hundred dirhams, upon which he abstained from writing the refutation."[149]

In contradistinction to both these views (scil. that of van Ess and of the *Ma'āhid*) I think that this practice of writing self-refutations reflects Ibn al-Rāwandī's deep convictions, in his typical provocative way. As the quotations from Ibn al-Rāwandī's books show, he was a true freethinker, in the sense that he rejected the authority of any scriptural or revealed religion. This conclusion is borne out by citations from his other writings, besides the *Zumurrud*: the *K. al-Dāmigh* against the Qur'ān, and "The Futility of [Divine] Wisdom" (*'Abath al-ḥikma*). Ibn al-Rāwandī's complete break with Islam is reflected not only in the contents of these books, but also in their tone. Some of the questions presented by Ibn al-Rāwandī in the *Dāmigh* could perhaps be asked by a Muslim commentator with an inquisitive mind who was trying to reconcile two apparently contradictory verses. But no Muslim would introduce such a question with a declaration that God does not know how to add two to four and make six.[150] A tormented Muslim might inquire into theodicy, but no Muslim would conclude the inquiry with the statement that God behaves like a wrathful,

---

K. al-Zumurrud and of the *K. al-Marjān*. He also started writing his own refutation of the *K. al-Dāmigh*, but have not finished it (presumably because of his death). Since *K. al-Dāmigh* was probably Ibn al-Rāwandī's last book, this would mean that Ibn al-Rāwandī wrote the book and its refutation in a very short interval, almost as part of the same project. See further below, Ch. 6, apud n. 64.

[147] Van Ess, "Image," 13; see further idem, *TG*, IV, 317.

[148] *Ma'āhid*, 155. See also Ibn Ḥajar al-'Asqalānī, *Lisān al-Mīzān*, Ḥaydrabad 1911, I, 323; Ibn al-Murtaḍā, *Ṭabaqāt al-Mu'tazila*, 92.

[149] On *K. al-Baṣīra*, see van Ess, *TG*, IV, 302; VI, 438 (van Ess expresses skepticism about the very existence of this book).

[150] Ibn al-Jawzī, *Muntaẓam*, 113:7–10 (= Ritter, 6:15): *Lam yaqum bi-ḥisāb sitta*. See also *Mughnī*, XVI, 393, and its translation in R. Ward Gwynne, *The "Tafsīr" of Abū 'Alī al-Jubbā'ī: First Steps Toward a Reconstruction*; Ph.D. diss., University of Washington, 1982 (University Microfilms; 1981), 199–200.

murderous enemy.[151] Even a pious Muslim might question the inherent value of rituals during pilgrimage. But for a Muslim, such a question highlights the notion of submission to the superior knowledge of God and of His Prophet, whereas for Ibn al-Rāwandī it implied the primitive nature of Muslim rites.[152] A person who presented Ibn al-Rāwandī's questions in the impudent tone that he adopted, could not remain a Muslim in any meaningful way.[153] Indeed, it was the tone of Ibn al-Rāwandī's arguments as much as their content that infuriated Muslims. This is reflected in Ibn al-Jawzī's comment that Ibn al-Rāwandī was worse than Iblīs, because although Iblīs disobeyed God, at least he addressed Him with respect.[154]

The impudent, disrespectful tone was an integral part of Ibn al-Rāwandī's attack on established religion. In the same way, the form of his writings was closely connected to their skeptical content. Writing both a book and its refutation is a way of speaking with two voices at once. It may have been used by Ibn al-Rāwandī to convey his mistrust of written authority and his complete commitment to a continuous effort to sharpen one's mind. It is possible that in the *Zumurrud* he tried to incorporate the same dual voice into the book itself, through dialogue with a colleague who shared some of his ideas. The fact that in the dialogue he took the role of the proponent of prophecy may have been a device to further enhance the critical approach. If this was his intention, there was probably more than a shade of mockery in it. But there was also the genuine drive to doubt: to doubt the authority of prophets, and then double-check the conclusion by doubting the doubt.[155]

---

[151] Ibn al-Jawzī, *Muntaẓam*, 112:14 (= Ritter, 6:4): *fiʿla al-ʿadū al-ḥaniq al-ghaḍūb*.

[152] See n. 195 below, and compare the saying attributed to ʿUmar: "I know that you are stones that can do neither good nor harm, but I have seen the Messenger of God kiss you, so I kiss you likewise." (Cf. Wensinck, *Concordance*, I, 435); cf. also Tartar, *Dialogue*, 214. For other instances of people who scoff at the Islamic rituals, see the sources quoted by Kraus, "Ketzergeschichte," 112, n. 7; Massignon, *Passion*, I, 275.

[153] The significance of Ibn al-Rāwandī's tone is noted also by van Ess, *TG*, IV, 289.

[154] *Muntaẓam*, 117:8–10 (= Ritter, 9); and see below, Ch. 6, apud nn. 65–67.

[155] Muslim polemicists often present the skeptics as doubting their own doubts, then doubting their doubts about their doubts, and so on. See, for instance, *Mughnī* XII, 41:1–7; Māturīdī, *Tawḥīd*, 153:12–14; See also Muqammaṣ, *Twenty Chapters*, 65, 77.

Despite the peculiarity of the book's structure, Muslim authors were generally able to infer correctly the real beliefs of its author. We may conjecture that some section of the *Zumurrud*—the introduction or the conclusion—recorded Ibn al-Rāwandī's authorship of the book. It is probable that, like many *kalām* debates, this one too ended with one of the participants admitting defeat (*inqiṭāʿ*). We can therefore further conjecture that the conclusion of the book recorded al-Warrāq's victory and Ibn al-Rāwandī's ultimate rejection of the existence of prophecy.

Some of the Muslim sources which mention the *Zumurrud* reveal their awareness of the fact that the book was in some way the result of a collaboration between al-Warrāq and Ibn al-Rāwandī. Thus Ibn al-Jawzī remarks that Ibn al-Rāwandī and al-Warrāq charged each other with the authorship of this book, although they agreed on its criticism of the Qur'ān,[156] and ʿAbd al-Jabbār names both of them as authors of a *K. al-Zumurrud.* In another passage of the *Tathbīt*, ʿAbd al-Jabbār mentions the *Zumurrud*'s argument concerning the presence of angels in Muḥammad's battles,[157] and then adds, "This question was raised by Ibn al-Rāwandī after he had conspired with Abū ʿĪsā al-Warrāq and Ibn Lāwī the Jew."[158] Ibn Lāwī is reported to be the Jew of Ahwāz in whose home Ibn al-Rāwandī took refuge until his death. A similar idea, in a more confused form, is expressed already by Abū al-Qāsim al-Balḫī, who says that Ibn al-Rāwandī "wrote most of his heretical books for the Jew Abū ʿĪsā [*sic*] of Ahwāz, in whose home he perished."[159] This confused report[160] testifies to the notion retained in the Muslim sources of a certain collaboration between Abū ʿĪsā al-Warrāq and Ibn al-Rāwandī in the production of heretical books. This collaboration did not mean that they

---

[156] *Muntaẓam*, 109:4–6.

[157] See below, n. 190.

[158] *Tathbīt*, 407. Ibn Lāwī is usually connected with the composition of the *Dāmigh* rather than the *Zumurrud* (*Muntaẓam*, 112.9 [= Ritter, 5]; *Maʿāhid*, 158); and see below, Ch. 7, apud nn. 25, 27.

[159] Abū al-Qāsim al-Balḫī, quoted by Ibn al-Nadīm, in Fück, 72.

[160] Theoretically, it is of course possible that Abū ʿĪsā was the *kunya* of this Ibn Lāwī. But it is more likely that we have here a conflation of al-Jubbāʾī's report that Ibn al-Rāwandī died in the home of Ibn Lāwī, the Jew of Ahwāz, and of the tradition that connects of Ibn al-Rāwandī with Abū ʿĪsā al-Warrāq. On the role of Ibn Lāwī and his avatars in Muslim literature, see below, Ch. 7, esp. apud n. 74. On Ibn al-Rāwandī's supposed connections with Judaism, see van Ess, *TG*, IV, 301–2.

wrote books together, but that the books written by Ibn al-Rāwandī were the fruit of his discussions with al-Warrāq.[161]

In the final analysis, however, people identified the author of the *Zumurrud* as Ibn al-Rāwandī. The baffling fact that in the *Zumurrud* Ibn al-Rāwandī allotted to himself the role of the defender of prophecy did not mislead those who knew the work at first hand. For them, the *Zumurrud* was a heretical book written by a heretic.

5. The *Zumurrud* and its Contents

The title

The Ḥanbalite author Abū al-Wafā' ibn 'Aqīl (d. 513/1119) reports that when Ibn al-Rāwandī chose to name his notorious book against prophecy "The Book of the Emerald" (*Kitāb al-Zumurrud*), it was not, as Ibn al-Rāwandī's contemporary Abū 'Alī al-Jubbā'ī (d. 303/915) had thought, in order to compare his science to this precious stone. His intention was rather to hint at a comparison between the power of emerald to blind snakes, and the supposed power of the sophistries which the book contained to silence all arguments.[162] Only to hint: this explanation of the book's name did not appear in the *Zumurrud* itself,[163] for Ibn 'Aqīl claims to have seen it "in some other book of Ibn al-Rāwandī's."

After Ibn 'Aqīl, this explanation became widely and generally accepted, and scholars, both Muslims and others, cite it with no apparent second thoughts.[164] In view of the late appearance of this explanation and of its flimsy basis, such general credulity is remarkable. Ibn 'Aqīl (as cited by Ibn al-Jawzī) remains our only source for this explanation, and he is a source of questionable value. He probably did not read the *Zumurrud* itself, but only Abū 'Alī's refuta-

---

[161] A similar collaboration is noted by 'Abd al-Jabbār concerning Ibn al-Rāwandī's *K. al-Tāj*; see below. Ch. 4, n. 31.

[162] On this special quality of the emerald, see the sources in Kraus, "Ketzergeschichte," 122, n. 1; idem, *Jābir ibn Ḥayyān: contribution à l'histoire des idées scientifiques dans l'Islam, Mémoires de l'Institut d'Egypte* 44, Cairo 1943, II, 74 and n. 1. As noted by Kraus, this quality of the emerald was mentioned by Ibn al-Rāwandī and by Jābir b. Ḥayyān in their respective books entitled *K. al-Ḫawāṣṣ*.

[163] And thus escaped Abū 'Alī's knowledge. See Kraus, "Ketzergeschichte," 122; *Ma'āhid*, 156; Ibn al-Jawzī (*Muntaẓam*, 109:8–13 [= Ritter, 3–4]).

[164] See the sources collected in A'sam, *History*, and idem, *References*.

tion of it.[165] In fact, notwithstanding Ibn ʿAqīl's claim, we have no reason to believe that he had direct access to any of Ibn al-Rāwandī's books. Had there been "some other book" of Ibn al-Rāwandī's that was unknown to Abū ʿAlī and which contained this explanation, one would have expected other writers with heresiographical interests to be aware of it. One would also expect that, in the works of these other writers, the mention of the emerald's special power would trigger a reference to Ibn al-Rāwandī's book. This, however, does not seem to be the case. Nāṣir-i Ḫusraw, the Ismāʿīlī author who wrote half a century after Ibn ʿAqīl, was generally sensitive to anti-prophetic sayings.[166] In his *Jāmiʿ al-ḥikmatayn* he refers to the supposed special quality (*ḫāṣṣa*) of the emerald, and he also mentions some of Ibn al-Rāwandī's heretical views.[167] But he does not associate this special quality of the emerald with any aspect of Ibn al-Rāwandī's work. It therefore seems legitimate to maintain a skeptical reserve towards Ibn ʿAqīl's explanation.

It is apparently true that Ibn al-Rāwandī had a keen interest in the special qualities of minerals. His gemological learning may partly account for the fact that he named several of his books after precious stones (although the practice is quite common with other people, too). It is also true that the existence of remarkable natural phenomena, such as the special powers of some minerals, was used by Ibn al-Rāwandī in his refutation of revealed religions. Ibn al-Rāwandī argued that those who claim to be prophets perform their supposed miracles by exploiting their knowledge of such natural phenomena. But we have no proof that Ibn al-Rāwandī actually explained his choice of the title *K. al-Zumurrud* as connected to the special power of the emerald and to his own attempt to confound his opponents. There is a remote possibility that in his *K. al-Khawāṣṣ*, along with the discussion of the emerald's power, Ibn al-Rāwandī included such

---

[165] As Kraus ("Ketzergeschichte," 351) points out; and cf. art. "Ibn ʿAḳīl," *EI*², III, 699–700 (G. Maqdisi).

[166] As can be seen in his reports on Abū Bakr al-Rāzī (cf. Rāzī, *Rasāʾil*, 220 ff.). Ibn ʿAqīl is also the source of Ibn al-Jawzī, who relates that heretics like Ibn al-Rāwandī and Abū'l-ʿAlāʾ discuss the special properties of stones, in an attempt to discredit the prophets' miracles; cf. *Talbīs*, 63.

[167] For his discussion of the emerald, see Nāṣir-i Ḫusraw, 166 (= Shatā, 282). For his reference to Ibn al-Rāwandī, see Nāṣir-i Ḫusraw, 223. Vajda ("Ibn al-Rāwandī") seems to suggest that the quotations from Ibn al-Rāwandī in Nāṣir-i Ḫusraw's *Jāmiʿ al-ḥikmatayn* may also be connected to the *Zumurrud*. But since the material quoted by Nāṣir-i Ḫusraw supports the eternity of the world, it is more likely that it derived from the *K. al-Tāj*; cf. van Ess, "Image," 13, n. 3.

an explanation. But it is more likely that he did not, and that Ibn ʿAqīl himself read this intention into Ibn al-Rāwandī's title, in order to emphasize the latter's malevolence.

The titles of Ibn al-Rāwandī's other heretical books support this conclusion. Only the title of his book against theodicy describes its content (*K. ʿAbath al-ḥikma* or *K. al-Taʿdīl wa'l-tajwīr*). The other heretical books were entitled either after precious objects and gems (*K. al-Lu'lu'a, K. al-Marjān* and *K. al-Tāj*) or after weapons (the *K. al-Qaḍīb* and the *K. al-Dāmigh*. The title of the *K. al-Taṣaffuḥ*, which is mentioned only in one source[168] may have belonged to this category too). Still another title belongs either to the first or to the second group, depending on how we read it (the *K. al-Farīd* or the *K. al-Firind*).[169] But there is no other book of Ibn al-Rāwandī's whose title is said to have had a double entendre, placing it in both the first and the second group.

The Structure of the Book

Kraus believed that al-Mu'ayyad faithfully followed his source, and that the order of the discussion in al-Mu'ayyad's source followed the original order of the *Zumurrud*.[170] Kraus also assumed that the discussion in the *Zumurrud* moved in a systematic way from one subject to the next. This reconstruction is highly questionable. As al-Mu'ayyad himself tells us, the *dāʿī* discarded from the *Zumurrud* most of the arguments for prophecy that it originally contained. Occasionally, the *dāʿī*'s wording marks the places where he omitted a passage from the *Zumurrud*.[171] As to the material he does quote, it is not at all certain that the *Majālis Mu'ayyadiyya* consistently followed the order of the discussion in the *Zumurrud*. Moreover, to the extent that the order of discussion in the *Majālis Mu'ayyadiyya* (as well as in other sources) does preserve the original structure of the *Zumurrud*, it reflects a rather loose and unsystematic structure.

The book was apparently divided into chapters (*abwāb*), for we find two authors referring to a *bāb* of this book. Al-Mu'ayyad men-

[168] See above, n. 50.

[169] On the fragments of this book, see van Ess, *TG*, IV, 341–42, and VI, 487–88.

[170] "Ketzergeschichte," 120.

[171] For example, 99:22 ("and then he said"), 102:7 ("up to the point where he said"), 104:2, 106:2.

tions a chapter in which Ibn al-Rāwandī relied on the *Barāhima*,[172] and al-Ḥayyāṭ mentions a chapter "against the Muḥammadans in particular."[173] Muslim authors would no doubt react particularly to this last chapter, and it is possible indeed that al-Mu'ayyad and Ibn al-Jawzī drew mostly from it. This chapter apparently took the general arguments listed in the earlier parts of the book and showed how they apply to the particular case of Islam.

Nevertheless, there is no reason to assume that the chapter against Islam repeated the arguments in exactly the order in which they had appeared in earlier chapters of the book. Its discussion of some of topics (theodicy and criticism of the Qur'ān) apparently repeated what Ibn al-Rāwandī had said in other books. Ibn al-Rāwandī reiterated his favorite arguments, and I suspect that even when an argument did not fit the plan of the book, he nevertheless used it. Rather than following a rigorous pattern, the *Zumurrud* was, in all probability, a potpourri of arguments. Our reconstruction can at best list the subjects dealt with in the book, but not the order in which they were presented.

The arguments against prophecy

The subjects discussed in the *Zumurrud* can be divided into several main groups:

Arguments relating to the primacy of the intellect

God has bestowed upon human beings the gift of intellect, by which they can judge right and wrong. If what the prophets announce corresponds to what the intellect decrees, then prophets are superfluous. If it contradicts what the intellect decrees, then one should not listen to them.[174] The discussion with the *Barāhima*, the issue of

---

[172] Kraus, "Ketzergeschichte," 109:12. The expression used by al-Mu'ayyad is *fī hādha al-bāb*, which could also mean "in this matter," but since the division into chapters is attested also by al-Ḥayyāṭ, Mu'ayyad's words probably refer to this division. On the *Barāhima*, see below, Ch. 5.

[173] *Intiṣār*, 12, 123.

[174] *Majālis Mu'ayyadiyya* ("Ketzergeschichte," 97:1–7) and Māturīdī (*Tawḥīd*, 200:13–16). Mu'ayyad says that Ibn al-Rāwandī ascribed this argument to the *Barāhima*, and Māturīdī cites it in al-Warrāq's name.

the abrogation of the law, and the question of the possibility of substituting one law for another are also part of this argument.[175] The argument is then applied to Islam in particular.[176]

Connected to the claim of the sufficiency of human intellect is the discussion of various expressions of this intellect.[177] Humans are taught to speak by their parents, from one generation to another, and this has always been the case.[178] Ibn al-Rāwandī is here probably addressing the question of whether human speech is natural or conventional. He seems to favour the solution of *ilhām* (i.e., natural, innate knowledge), although the term itself does not appear. From the *dāʿī*'s answer we can see that Ibn al-Rāwandī gave various examples of innate knowledge (the ability of birds to communicate with each other, the ability of ducks to swim, the ability of infants to suck milk), and that these were mentioned by him as being analogous to speech and understanding.

The sciences are also mentioned by Ibn al-Rāwandī as proof for the sufficiency of the intellect. According to him, people developed the science of astronomy by watching the skies. No prophet was needed to teach them how to watch. Nor were prophets needed in order to teach them how to build lutes. It is absurd to assume that without prophetic revelation people would not have learnt that the intestines of a sheep, when dried and stretched upon a piece of wood, can produce pleasant tones. All these skills are acquired by the assiduous application of the inborn human intellect, discernment, and power of observation.

Kraus thought that this part of the book opened with a paragraph praising the intellect in rhymed prose, one sentence of which is to be found in the *dāʿī*'s refutation.[179] Kraus noted that neither the *dāʿī*

---

[175] On abrogation and its relevance to this argument see below, Ch. 5, apud n. 33ff.; Niewöhner, *Veritas*, 239–42.

[176] *Majālis Mu'ayyadiyya*, "Ketzergeschichte," 99:22.

[177] These arguments complement the "Brahman" intellectualist argument. Therefore, although these arguments are quoted by Mu'ayyad at the very end of his discussion of the *Zumurrud*, thematically they belong to the first part of the *Zumurrud* (cf. Rāzī, *Rasā'il*, 296). This shows again either that Mu'ayyad did not follow the original order of the book, or that the book moved from one subject to another in no particular order, or both.

[178] The *dāʿī* points out the implication of this theory for the question of the eternity of the world. The background to the question of the origins of speech is discussed by Kraus, "Ketzergeschichte," 127–29. Kraus points out the closeness of Ibn al-Rāwandī's position to that of the Muʿtazila, and the fact that a similar position is taken by Abū Bakr al-Rāzī.

[179] "Ketzergeschichte," 97:17–18. The verse in question runs as follows: "One

nor Ibn al-Rāwandī were given to writing rhymes. He argued that the only place where either one of them could be expected to use such a sentence would be in an opening chapter of a conventional nature, where the praise of the intellect is sung before the real discussion begins. He therefore suggested that the *Zumurrud* had a poetic opening in which Ibn al-Rāwandī glorified the intellect, and that the *dāʿī* opened his response with a poetic paraphraprase of Ibn al-Rāwandī's verse.[180]

It is indeed possible that this sentence is taken from an introduction written in flowery style. I do not think, however, that it could come from Ibn al-Rāwandī's pen. Had this been the case, the *dāʿī* would probably have said so explicitly, as he always does when he wants to attack something said by Ibn al-Rāwandī. It is more likely that this sentence was written by the *dāʿī*.[181] Furthermore, in the debate with the *Zumurrud* the proper estimate of the role of the intellect was not a side-issue, but stood at the core of the discussion. It is therefore likely that, rather than being a conventional opening, the reference to "the person who claims to cover the horizons of science with the wings of the intellect" is the *dāʿī*'s direct assault on Ibn al-Rāwandī's intellectual pretensions.

Arguments relating to the Qur'ān

The *Zumurrud* attacks from various angles the doctrine of the unrivaled beauty of the Qur'ān (*iʿjāz*).[182] It suggests several natural explanations for the fact that the Arabs did not produce anything similar to the Qur'ān. It argues that this may be due to the fact that the people of Qurayš were more eloquent than members of other tribes, and that Muḥammad was a particularly gifted individual; or to the fact the Arabs were too busy fighting with Muḥammad, and had no time to invest in poetic competitions; or to the fact that the Arabs were uneducated people. The *Zumurrud* also argues that the Qur'ān

should say to the person who claims to cover the horizons of science with the wings of his intellect, and to build for himself a glorious light-house by his knowledge of things secret."

[180] Kraus, "Ketzergeschichte," 122–23.

[181] Cf. Kraus, "Ketzergeschichte," 104:18, where Muʿayyad begins his answer by quoting a verse from a poem by al-Mutanabbī. On the poetry attributed to Ibn al-Rāwandī, all of which is probably spurious, see A. al-Aʿsam, *al-Šiʿr al-mansūb ilā Ibn al-Rīwandī*, in Aʿsam, *References*, I, 216–56.

[182] On the development of the doctrine of *iʿjāz* and the reactions of heretics to it, see Chokr, *Zandaqa et zindīqs*, 153–64; and see above, Ch. 1, apud n. 43.

is not really all that beautiful; in the sayings of Aktham b. Sayf there are more beautiful sayings than "We gave you in abundance" (Q. 108:1). The Qur'ān is written in a faulty Arabic (*laḥn*) and its language can be corrected by human beings.[183] The Qur'ān is full of contradictory and absurd sayings, and it cannot possibly be the speech of the Wise One.[184] And, at any rate, while the language of the Qur'ān may seem to the Arabs to be proof that Muḥammad was a prophet, it cannot be expected to impress non-Arabs.[185]

The *Zumurrud* also discussed specific Qur'ānic verses, attempting to point out their logical flaws and to call into question the Muslim tradition about the circumstances of their revelation. As the example of the *mubāhala* cluster shows, such discussions also touched on other issues, such as the validity of Muslim tradition and the Prophet's alleged miracles.

Arguments relating to Muslim traditions

According to the *Zumurrud*, traditions concerning miracles are inevitably problematic. At the time of the performance of a supposed miracle only a small number of people could be close enough to the Prophet to observe his deeds. Reports given by such a small number of people cannot be trusted, for such a small group can easily have conspired to lie.[186] Muslim tradition thus falls into the category of flimsy traditions, which are based on a single authority (*ḫabar al-āḥad*) rather than on multiple authorities (*ḫabar mutawātir*).[187] Such religious traditions are lies endorsed by conspiracies.

The *Zumurrud* points out that Muḥammad's own stipulations (*waḍʿ*) and system (*qānūn*)[188] show that religious traditions are not trust-

---

[183] *Muntaẓam*, 110:8–10.

[184] *Intiṣār*, 12; God's lack of wisdom was the main theme of Ibn al-Rāwandī's *K. ʿAbath al-ḥikma*.

[185] *Maǧālis Mu'ayyadiyya*, Kraus, "Ketzergeschichte," 102:5–8; cf. also Māturīdī, *Tawḥīd*, 191:16–192:1.

[186] *Maǧālis Mu'ayyadiyya*, Kraus, "Ketzergeschichte," 101:11–13 (in the name of Ibn al-Rāwandī); Māturīdī, *Tawḥīd*, 196:14, and *Mughnī*, XVI, 416:6–13 (in the name of al-Warrāq). ʿAbd al-Jabbār associates this argument with the miracle of *ḥanīn al-na'qa*, which may or may not be connected with al-Warrāq's book *Ḥanīn al-bahā'im*, see above, nn. 32, 50.

[187] Kraus, "Ketzergeschichte," 125. Ibn al-Rāwandī's preoccupation with this subject is reflected in the fact that al-Ašʿarī wrote a book *fī naqḍ Ibn al-Rāwandī fī ibṭāl al-tawātur*, see n. 123 above.

[188] Literally: "regulation and law," Kraus, "Ketzergeschichte," 115, n. 10, assumes

worthy. The Jews and Christians say that Jesus really died, but the Qur'ān contradicts them.[189] If statements made by so many people cannot be trusted, all the more so the testimony of a handful of people like Muḥammad's followers.

Ibn al-Rāwandī also points out specific Muslim traditions, and tries to show that they are laughable: the tradition that the angels rallied round to help Muḥammad is not logical, because it implies that the angels of Badr were weaklings, able to kill only seventy of the Prophet's enemies. And if the angels were willing to help Muḥammad at Badr, where were they at Uḥud, when their help was so badly needed?[190]

Arguments relating to miracles

An important part of the *Zumurrud* is devoted to arguing that the miracles of the prophets are products of legerdemain. Like magicians, prophets exploit unusual natural phenomena, similar to the magnet but not as well known; they are only fraudulent tricks (*maḫārīq*).[191] A number of Muḥammad's miracles are specifically mentioned: the ablution basin, Umm Maʿbad's sheep, Surāqa, the wolf who talked, the intoxicated sheep who talked, and the *isrā'*.[192] According

that these two words are taken from the technical terminology of the theologians, and that they are used sarcastically. But on 114:15–16, the same argument is repeated, this time with the words *minhaj* and *asas*. In both places the choice of words may be the *dāʿī*'s rather than Ibn al-Rāwandī's.

[189] *Majālis Mu'ayyadiyya*, Kraus, "Ketzergeschichte," 104:3–5; Māturīdī, *Tawḥīd*, 197:1, 198:6. The docetic argument is used by al-Warrāq (in Māturīdī, *Tawḥīd*, 197:1). A similar argument may also lie behind *Tathbīt*, 132, where there is a cryptic reference to the claims of the Jews and Christians concerning the crucifixion, although the development of the argument there is quite different. Kraus, "Ketzergeschichte," 125 notes that the same argument is used by Rāzī; see below, Ch. 3, apud n. 83. *Tathbīt*, 90, mentions Ibn al-Rāwandī's criticism of Q. 20:133, and his saying "If Muḥammad's knowledge [of the ancient scrolls] is proof that he was a prophet, then the Jews' knowledge of these scrolls is proof that they are prophets." This criticism may have been part of the *Zumurrud*, but it may also come from some other book of Ibn al-Rāwandī's, for example *K. al-Farīd*.

[190] *Majālis Mu'ayyadiyya*, Kraus, "Ketzergeschichte," 105:24, 106:2; Māturīdī, *Tawḥīd*, 199:11–16; *Tathbīt*, 407; *Intiṣār*, 122 may also reflect the use of this argument by Ibn al-Rāwandī. Van Ess, "Image," 20, notes that this argument is already used by Ibn al-Muqaffaʿ (Guidi, *La Lotta*, 20:3ff.).

[191] *Majālis Mu'ayyadiyya*, Kraus, "Ketzergeschichte," 101:11–13; Māturīdī, *Tawḥīd*, 186:10–14; *Muntaẓam*, 110:11, 11:3–4; *Intiṣār*, 12.

[192] *Majālis Mu'ayyadiyya*, Kraus, "Ketzergeschichte," 104:1–7; Ḫayyāṭ, *Intiṣār*, 12 and 123.

to the *Zumurrud*, the distance between Mecca and Jerusalem is not very great, and it is conceivable that a person could go from one of these cities to the other and back in one night, so the Prophet's presentation of his ability to describe Jerusalem as a miracle is a fraudulent trick (*maḫraq*).[193] Even the Prophet's ability to predict the future (as in the case of the slaying of ʿAmmār b. Yāsir) is not regarded as a miracle, since it is claimed that any astrologer can do that.[194]

Arguments relating to the Muslim rituals

The *Zumurrud* criticizes prayer, preoccupation with ritual purity, and the ceremonies of the *ḥajj*: throwing stones, circumambulating a house that cannot respond to prayers, running between stones that can neither help nor harm. It goes on to ask why Ṣafā and Marwa are venerated, and what difference there is between them and any other hill in the vicinity of Mecca, for example the hill of Ibn Qubays, and why the Kaʿba is any better than any other house.[195]

This sketch of the arguments contained in the *Zumurud* confirms several of the conclusions presented in previous sections of the present chapter. It shows the close similarity between the *Majālis Muʾayyadiyya*, Māturīdī, and the *Tathbīt*, and thus corroborates the claim that they derive from the same source, the *Zumurrud*. It also strengthens the impression that all three sources summarize rather than quote the *Zumurrud*. And it shows that none of the three sources relied on either of the other two for its information, since in each one of them some elements are present that are lacking in the other two. A comparison of the sources allows us to see that the same arguments are attributed at times to Ibn al-Rāwandī, at times to al-

---

193 *Majālis Muʾayyadiyya*, Kraus, "Ketzergeschichte," 106:14; Kraus, "Ketzergeschichte," 124, n. 2, finds it hard to believe that Ibn al-Rāwandī had no idea of the distances, and suggested that there must have been some misunderstanding of his argument. To my mind, it is quite possible that he did not appreciate the distances. The information was, of course, available, but since geographical accuracy was not foremost in Ibn al-Rāwandī's mind, he probably did not bother to check it.

194 *Muntaẓam*, 110:15. The comparison of the prophets to astrologers was also mentioned as part of the discussion of Muḥammad's challenge to the Jews, see above, n. 104.

195 *Majālis Muʾayyadiyya*, Kraus, "Ketzergeschichte," 99:18ff.; van Ess ("Image," 18), suggests that the *ḥajj* was particularly popular with critics because of the *jāhilī* flavor of the pilgrimage (as opposed to prayer or fasting), and mentions that it evoked similar criticism from the ṣūfī Abū Yazīd al-Bisṭāmī. Cf. also above, n. 152.

Warrāq, and at times to both of them. A correct understanding of the *Zumurrud* must allow for the active participation of both Ibn al-Rāwandī and al-Warrāq in the dialogue, and take into account the heretical convictions of both of them.

Even if Ibn ʿAqīl cannot be trusted concerning the intended meaning of the title *K. al-Zumurrud*, we must admit that his explanation grasps the effect that this book had on its audience. Its "sophistries" cannot be said to have silenced all argument; on the contrary, they drew furious tirades from many Muslims who read or heard them. But the book did seem to puzzle and confuse its readers. In modern scholarship, in particular, the discrepancy between Ibn al-Rāwandī's position and his role in this book combined with the fragmentary state of our data to produce a generally inaccurate picture of the *Zumurrud*.

Among the Muslim authors who mention Ibn al-Rāwandī without attacking him, Māturīdī is the only one who presents him in positive light in the context of the arguments against prophecy from the *Zumurrud*. Yet the specificity and unambiguous nature of Māturīdī's account have led scholars to give it more weight than a single testimony would normally receive. Since Māturīdī's account of Ibn al-Rāwandī was perceived as inconsistent with the accounts in other sources (e.g., the *Majālis Muʾayyadiyya*, Ibn al-Jawzī, and the *Tathbīt*), the evidence of all these other sources was discarded as biased.

The results of the present chapter allow us to reaffirm the evaluation of Ibn al-Rāwandī as a heresiarch, without disregarding Māturīdī's important evidence. A close analysis of Māturīdī's text in the context of the other sources showed that there is no contradiction between Māturīdī's account and those of Mu'ayyad and ʿAbd al-Jabbār. Rather, when the accounts of these three sources are examined in detail, it becomes evident that the three complement each other. The description of the heretical character of the *Zumurrud* and the identification of its author as Ibn al-Rāwandī are presented most clearly by Mu'ayyad. Mu'ayyad and ʿAbd al-Jabbār preserve some indications of the form of the *Zumurrud*, but it is Māturīdī who retains both the dialogue form, the identity of the two participants, and their respective roles. And the cooperation of Ibn al-Rāwandī and al-Warrāq in presenting the arguments of the *Zumurrud* is most explicit in the account of ʿAbd al-Jabbār. Together, these three sources contribute to produce a clearer, fuller picture of the *Zumurrud*.

Read in the context on the information concerning the whole of Ibn al-Rāwandī's literary activity, the *Zumurrud* shows Ibn al-Rāwandī to have been a freethinker as defined in the introduction to the present book. The relevant medieval sources depict Ibn al-Rāwandī as an apostate, who rejected the notion of prophecy and ridiculed the scriptures and the religious traditions. Contrary to the view put forward and defended by van Ess, this depiction of Ibn al-Rāwandī is not limited to Muʿtazilite sources. Rather, various versions of the same depiction appear in Muʿtazilite, Ašʿarite, Ḥanbalite, Ismāʿīlī, and philosophical sources. In the current state of our knowledge, the unusual intellectual career of Ibn al-Rāwandī is still puzzling in many ways, and will probably remain so in the forseeable future. But, relying on the available medieval sources, we can be confident in describing him as a true freethinker.

CHAPTER THREE

ABŪ BAKR AL-RĀZĪ: A "RESPECTABLE" FREETHINKER

In both medieval and modern lists of heretics, Abū Bakr Muḥammad b. Zakariyā' al-Rāzī and Ibn al-Rāwandī figure together as prominent advocates of radically heretical ideas. Medieval Arabic literature contains explicit citations attributed to Rāzī, in which he, like Ibn al-Rāwandī, mercilessly attacks the idea of prophecy in all its manifestations. As in the case of Ibn al-Rāwandī, modern scholarship has found it difficult to accept the authenticity of these citations. But unlike in the case of Ibn al-Rāwandī, Rāzī's ideas concerning prophecy do not dominate scholarly interest in his figure and thought. The impressive studies which Paul Kraus dedicated to each of these two thinkers were probably motivated by his interest in the history of heterodox ideas in Islam. But since Kraus's days, other aspects of Rāzī's thought have received more attention than his anti-prophetic ideas. Unlike Ibn al-Rāwandī, Rāzī is not known to have suffered persecution. In view of the prevalent image of religion in the medieval Islamic world, this last fact seems difficult to reconcile with the outspoken blasphemies attributed to Rāzī. Moreover, unlike Ibn al-Rāwandī, Rāzī had a stable professional and intellectual career that earned him lasting fame. Rāzī's extant scientific writings provide ample field for research, and his heretical religious ideas can be treated as marginal. At the same time, the very existence of these scientific writings make it difficult indeed to ignore Rāzī's intellectual contributions and treat him as nothing but a heretic.

This chapter aims at showing that, notwithstanding the diversity of Rāzī's intellectual and scientific activity and the respectable position he held, he was nevertheless a full-time freethinker. I will first briefly review the relevant parts of his biography, then present and analyze the data concerning his *K. Maḫārīq al-anbiyā'*.

1. Biography

Most of Rāzī's philosophical works are not extant, and the quotations from his books provide a complex picture of his thought. The same holds true concerning his biography, for which several versions exist, all of them rather sketchy. No attempt will be made here to reconstruct his life. Instead, it seems worthwhile to quote the short biography given by Bīrūnī (d. ca. 442/1050), who, alongside some biographical details, provides a vivid picture of Rāzī's personality:[1]

> He was born in Rayy, on the first of Šaʿbān, in the year 251 (= 865). Nothing is known about him accept that he practiced alchemy. [While working], he kept his eyes unblinking, vulnerable to injuries and diseases. The frequent exposure to fire and the stinging vapors obliged him to seek treatment. This situation attracted him to the practice of medicine,[2] and thence to what lies beyond it, to the things for which he is rebuked.[3]
>
> He achieved a lofty position in the [medical] Art. The greatest sovereigns applied to him and invited him, [bestowing on him] great honor. He was forever occupied in his studies, determinately pursuing them. He would place his lamp in a niche on the wall facing him, and lean his book on the wall. This way, when he dozed off, the book would drop from his hands and wake him up, and he would return to whatever he was reading.
>
> This was among the things that damaged his eyesight, together with his passionate craving for beans and his harmful addiction to them.[4] In the end he lost his eyesight, so as to become "blind in the hereafter."[5] At the end of his life he contracted a cataract in both eyes.
>
> A person referred to him by his students came from Ṭabaristān to treat him. Rāzī inquired about the fashion in which he intended to treat him, and the man told him. Then Abū Bakr [al-Rāzī] said: "I declare that you are the best and the most knowledgeable eye-surgeon.[6] But you know that this treatment is not devoid of pains,

[1] For a German translation, see Ruska, "Al-Bīrūnī," 32–33; and see Meier, "Der 'Urknall'," 2.

[2] That Rāzī turned to medicine only late in life is confirmed by Ibn Abī Uṣaybiʿa (IAU, 414). This author also mentions alchemy as Rāzī's first interest (Ibid., 419). He recounts, however, a story according to which Rāzī's medical interest was spurred after a visit to the hospital in Baghdād, see IAU, 414–15. According to Ibn Ṣāʿid, *Ṭabaqāt*, 52, Rāzī began his career as a musician, and only then turned to philosophy and science; cf. also Ibn al-Qifṭī, 271–72.

[3] Lit. "the things his occupation in which is frowned upon."

[4] Cf. also *Fihrist*, 299.

[5] Cf. Q. 17:72.

[6] Lit.: "a unique piercer and the most knowledgeable oculist" (*awḥad al-qaddāḥīn wa-aʿlam al-kaḥḥālīn*).

> from which [my] soul shrinks; it requires enduring difficulties for a long time, something for which [I] have no patience. Perhaps only a short time is left for me to live, and my appointed time is approaching. For a person like myself, it is reprehensible to opt for pains and hardships rather than tranquility in the remainder of his days. Go and be thanked for your intentions and efforts," and he payed him generously.[7]
>
> After this, he did not live long. He died in Rayy on the sixth day of Šaʿbān, in the year 313 (= 925). He lived sixty two lunar years, which equals sixty years, two months and one day.[8]

In the thirties, Paul Kraus collected the fragments of Rāzī's works and the references to him in the works of later authors, accompanying his remarkable anthology with an analysis of the implications of the fragments.[9] Around the same time, Shlomo Pines studied various aspects of Rāzī's philosophy, in particular those related to his physical theories.[10] Despite more than half a century that has elapsed since the publications of these two great scholars, their masterly work is still as relevant as it was then, and it remains the cornerstone of any study of Rāzī.[11]

Several other scholars have studied various aspects of Rāzī's philosophy. Mehdī Mohaghegh has devoted to him a collection of studies entitled "The Philosopher from Rayy,"[12] and has published his *Šukūk ʿalā Jālīnūs*.[13] A. J. Arberry has translated his *Spiritual Physick*.[14] In the last few years, several fine general studies were devoted to Rāzī.[15] These studies concentrate on Rāzī's empirical approach to medicine and alchemy,[16] on his ethical system,[17] his relations to the

---

[7] A shorter version of the story is given in Ibn al-Qifṭī, 272.

[8] *Risāla*, 4–6.

[9] Rāzī, *Rasā'il*; Kraus, "Raziana."

[10] Pines, *Atomism*, 40–107; idem, "Nouvelles Études." On Rāzī's atomism see also Dhanani, *Physical Theory*, 188; Muḥammad Najmābādī, *Mu'allafāt va-muṣannafāt-e Abū Bakr Muḥammad Zakariyā' Rāzī*, Tehran 1339 H, 193, 209, 212.

[11] Their joint entry in the *Encyclopedia of Islam* (Kraus and Pines, *EI*) is still the most penetrating short article available on Rāzī.

[12] *Filsūf-i Rayy*, Tehran 1970.

[13] Rāzī, *Šukūk*; cf. Pines, "Rāzī."

[14] *The Spiritual Physick of Rhazes*, London 1950; cf. also M. Mohaghegh, "Notes on the 'Spiritual Physic' of al-Rāzī," *SI* 26 (1967), 5–22.

[15] See Goodman, "Rāzī"; idem, "Muḥammad ibn Zakariyyā' al-Rāzī;" Iskandar, "Rāzī"; Bar-Asher, "Abū Bakr al-Rāzī"; Urvoy, *Penseurs libres*, 142–52.

[16] Goodman, "Rāzī," 474; Iskandar, "Rāzī"; cf. also A. Z. Iskandar, "The Medical Bibliography of al-Rāzī," *Essays on Islamic Philosophy and Science*, ed G. Hourani, Albany 1975, 41–46.

[17] Goodman, "Rāzī," 475; idem, "The Epicurean Ethic." As noted by Bar-Asher, Rāzī's ethics bears the marks of his visceral aversion to existing religions. This can

Galenic tradition,[18] his creation myth,[19] even his political theory (or rather, the lack thereof).[20] Rāzī is often described as "the greatest non-conformist in the whole history of Islam";[21] "The most free-thinking of the major philosophers of Islam," or "the least orthodox and the most iconoclast" of them all;[22] and "perhaps the single figure most frequently denounced and disapproved as a heretic in the subsequent history of Islamic thought."[23] It is assumed that he was a rather unorthodox Muslim, but that Islamic thought in his time was somehow able to accommodate his lack of orthodoxy. If, however, the sayings attributed to him are authentic, his doctrine emerges as irreconcilable with any kind of Islam, however open-minded. Moreover, since Kraus's study no serious consideration seems to have been given to the most important component of Rāzī's heretical doctrines, anti-prophetism.[24] It is of course true that Rāzī's heterodoxy transpires in all aspects of his thought: his individualistic ethics can hardly coexist with obedient submission to religious leaders, and his creation myth was considered by Nāṣir-i Ḫusraw to be heretical even in comparison to the teaching of al-Īrānšahrī.[25] But in order to pinpoint the source and the precise nature of his non-conformism, we must concentrate on his freethinking in the narrow sense, i.e., on his rejection of revealed religions.

Although the date of Rāzī's death is not certain, it is less controversial than Ibn al-Rāwandī's. Moreover, in Rāzī's case the implications of the exact date of his death are less significant. The usual dating is 250/854–313/925[26] or around 323/935.[27] His *floruit*, then, corresponds to the period in which Islamic thought reached maturity in various domains. The same years saw the peak of Muʿtazilism,

---

be seen in the absence of any religious text from the many sources he quotes to support his ideas; see Bar-Asher, "L'éthique de Rāzī," 144.

18 Bar-Asher, "L'éthique de Rāzī"; Druart, "Rāzī's Conception of the Soul"; idem, "Ethics".

19 Goodman, "Myth;" Brion, "Le temps;" Meier, "Der 'Urknall'."

20 Walker, "The Political Implications"; see below, Ch. 6, apud n. 79ff.

21 Fakhry, *Islamic Philosophy*, 97.

22 Respectively, Goodman, "Rāzī," 474, and idem, "Muḥammad ibn Zakariyyā' al-Rāzī," 202.

23 Walker, "Political Implications," 65.

24 Corbin, Introduction to Nāṣir-i Ḫusraw, *Jāmiʿ al-ḥikmatayn*, 137, describes anti-prophetism as "the most striking feature of Rāzī's philosophy."

25 See above, Introduction, n. 43.

26 According to Bīrūnī, *Risāla*, 6.

27 According to Ibn Ṣāʿid, *Ṭabaqāt*, 33; Ibn al-Qifṭī, 272; IAU. See the discussion in Walker, "Political Implications," 64 and n. 7.

the completion of the first round of "the translation movement," the crystallization of *falsafa* and of Ismāʿīlī Neoplatonism. It is in the context of this intellectual effervescence that Rāzī's philosophy and heretical ideas must be seen.

Little is known of his education or of his formative years, but some of our sources designate the enigmatic al-Īrānšahrī as his teacher.[28] This could suggest some kind of initiation into freethinking.[29] Rāzī is known first and foremost as a physician, "the uncontested Physician of the Arabs,"[30] and his medical approach cannot be separated from his philosophical occupation.[31] Pines and Kraus describe him as "the least dogmatic of the Arab physicians,"[32] and indeed his free spirit is apparent even in his discussion of specific medical problems. Besides material gleaned from his predecessors and case-histories of his patients, the bulky collection of his clinical notebooks, *al-Ḥāwī*,[33] also contains his own observations, reflections, suggestions, and diagnostic tips. These personal contributions he introduced by the authoritative heading: "Of my own:" (*lī*).[34] This heading, repeated countless times over the twenty-odd volumes of the *Ḥāwī*, manifest the spirit which Šahrastānī could have called *istibdād bi'l-ra'y* at its best: bold application of personal opinion—although based on empirical knowledge, and thus by no means arbitrary.

Nevertheless, while his excellence in medicine was universally recognized, in Arabic sources his appellation "the physician" (*ṭabīb* or *mutaṭabbib*) sometimes carries an ambivalent evaluation.[35] Thus the philosopher ʿĀmirī, a near contemporary of Rāzī, says:

---

[28] Nāṣir-i Ḫusraw, *Zād al-Musāfirīn*, 73, 98; Bīrūnī, *Hind*, 4, 326; *Āthār*, 222, 225 (quoted in Kraus and Pines, *EI*, 1134). Kraus and Pines suggest that al-Balḫī who is mentioned by Ibn al-Nadīm (*Fihrist*, 299) as Rāzī's teacher was identical with al-Īrānšahrī. According to IAU, Rāzī's teacher in medicine was ʿAlī b. Rabbān al-Ṭabarī (ibid., 414), whereas al-Balḫī taught him philosophy (Ibid., 416, quoting Ibn al-Nadīm).

[29] On al-Īrānšahrī's freethinking, see above, Introduction, apud nn. 42–44.

[30] Ibn Ṣāʿid, *Ṭabaqāt*, 52; Ibn al-Qifṭī, 271.

[31] As Walker rightly says, Rāzī approached the world as a physician; see "Political Implications," 74. See also Iskandar, "Bibliography"; Sezgin, *GAS*, III, 274–94; VII, 271–72.

[32] *EI*, 1134.

[33] Edited in Hyderabad, beginning from 1955. On this edition, see Iskandar, "Bibliography," 41–43.

[34] See, for example, the case translated in Iskandar, "Bibliography," 43.

[35] See, for example, al-Kirmānī, cited in Kraus, *Rasā'il*, 312; IAU, II 97.

> The extraordinary thing about the people of our own time is that, when they see that a man has read Euclid's book and mastered the principles of logic, they describe him as a Sage (*waṣafūhu bi-ḥikma*) even if he completely lacks [knowledge of] the divine sciences. Thus they ascribe wisdom to Muḥammad b. Zakariyā' al-Rāzī because of his proficiency in medicine—this in spite of his various ravings.[36]

A similar contempt for Rāzī's philosophical pretensions transpires from Maimonides' words, who evaluates Rāzī as "merely a physician," i.e., unworthy of being considered a philosopher.[37]

Rāzī's interest in alchemy was part and parcel of his scientific quest. Whether he was not acquainted with the corpus of alchemical writings attributed to Jābir b. Ḥayyān,[38] or whether, being acquainted with it, he had rejected Jābir's approach,[39] the basic attitude to alchemy revealed in his alchemical work is substantially different from Jābir's. Rāzī's alchemy was thoroughly experimental, and he clearly regarded it as belonging to the scientific study of the world. Rather than pulling him toward the occult, his involvement with alchemy buttressed his skeptical attitude to anything purporting to be preternatural, including prophetic miracles.[40] In the introduction to his "Book on Properties" (*K. al-Ḫawāṣṣ*), Rāzī argued that judgment should be suspended until experience determines the particular qualities of the minerals and other materials of the world,

> since many wicked men tell lies with regard to such properties, and we do not possess decisive means to distinguish the truth of honest men from the false testimony of liars.

Iskandar, who quotes this passage, sees this book as evidence of [Rāzī's] opposition to scientific dogmas.[41] This is certainly true, but

---

[36] E. K. Rowson, *A Muslim Philosopher on the Soul and Its Fate: Al-ʿĀmirī's* Kitāb al-Amad ʿalā l-abad, New Haven 1988, 74 (tr. 75).

[37] See A. Marx, "Texts by and about Maimonides," *JQR*, ns 25 (1934–35), 374–81; S. Stroumsa, "Al-Fārābī and Maimonides on Medicine as Science," *Arabic Sciences and Philosophy* 3 (1993), 235–49.

[38] As suggested by Kraus and Pines, *EI*, 1135.

[39] As suggested by Corbin, in his introduction to Nāṣir-i Ḫusraw, *Jāmiʿ al-ḥikmatayn*, 129. See also IAU, 426:19, which counts among Rāzī's book a poetic rendering of Jābir's *K. al-Uss.*

[40] See below. On Rāzī's alchemy, see further J. R. Partington, "The Chemistry of Rāzī," *Ambix* 1 (1938), 192–96; Ruska, "Alchemie"; Iskandar, "Rāzī," 371–72; Sezgin, *GAS*, IV, 275–82.

[41] Iskandar, "Rāzī," 377. Rāzī wrote several books on properties; cf. Sezgin, *GAS*, III, 285. According to Sezgin, however, the subject of the particular book entitled *K. al-Ḫawāṣṣ* was medicine rather than alchemy; cf. *GAS*, IV, 281.

the anti-prophetic undertone of these lines is also unmistakable. The claim that prophets are charlatans exploiting unusual natural phenomena to impress ignorant people was made by Ibn al-Rāwandī, who also wrote a "Book of properties."[42] The same idea may have been discussed also in Rāzī's *Book on Magnetic Attraction*,[43] and he may have touched upon it also in his *Book of Stones* (*K. al-aḥjār*).[44]

As we can see, Rāzī's scientific interests are closely connected to his philosophical approach. His critical, independent attitude transpires from his medical and alchemical writings. And yet from Rāzī's scientific works we would not have been able to gauge the extent of Rāzī's rebellion against Islam. Rāzī wrote extensively on philosophy, but only his relatively moderate books, those that were more acceptable to Muslims, survived. Of his more radical philosophical books we have only fragments, quoted by his opponents. Nevertheless, these fragments allow us to appreciate Rāzī's wholly un-Islamic philosophy. The clearest information concerning his rejection of Islam as a prophetic religion can be found in the fragments of his *Book on the Prophets' Fraudulent Tricks* (*K. Maḫārīq al-anbiyā'*).

2. The *K. Maḫārīq al-Anbiyā'*

The Title of the Book

Our sources mention essentially three titles of books in which Rāzī criticized the revealed religions: "The Prophets' Fraudulent Tricks" (*Maḫārīq al-anbiyā'*), "The Stratagems of Those who Claim to be Prophets" (*Ḥiyal al-mutanabbīyīn*), and "On the Refutation of Revealed Religions" (*Fī nakḍ al-adyān*).[45] The first two titles are usually assumed to refer to the same book,[46] and this may be true also with regard to the third; medieval bibliographers often list slightly different versions or manuscripts as separate books. On the other hand, Rāzī may have prepared three separate reworkings of the same theme.

[42] See above, Ch. 2, apud nn. 166–67.
[43] *Fihrist*, 301:15; also Ibn al-Qifṭī, 275; Pines, *Atomism*, 162–64.
[44] Ruska, "Alchemie," 280; cf. *Fihrist*, 358 (*K. al-Ḥajar*); Mas'ūdī, *Murūj*, V, 159–60.
[45] Bīrūnī, *Risāla*, 20, identifies this book with the one entitled "On Prophetic Religions" (*Fī'l-nubuwwāt*).
[46] Bīrūnī, *Risāla*, 20, explicitly identifies them as two different titles of the same book.

Since this book (or books) did not survive, no definitive answer to this problem can be given. It is also impossible to determine this book's (or these books') relation to Rāzī's main metaphysical work, the *K. al-ʿIlm al-ilāhī*, which may have been a compendium (or perhaps a summary) of all of Rāzī's philosophical work.[47] Without deciding on these problems, I proceed in what follows on the assumption that Rāzī's main work on prophecies was contained in a book entitled "The Prophets' Fraudulent Tricks" (*Maḫārīq al-anbiyāʾ*).[48]

This title was probably a response to the *dalāʾil al-nubuwwa* literature. The lists of "signs of prophecy" offer criteria permitting to distinguish between the miracles of authentic prophets and the tricks of impostors. The presentation of the prophets' miracles as *maḫārīq*, i.e., fraudulent tricks performed by charlatans, is thus a tested technique to discredit their claim to prophethood.[49] This was clearly perceived by Rāzī's opponent, the Ismāʿīlī *dāʿī* Abū Ḥātim al-Rāzī (d. 322/933), who responded with his own *Book of the Signs of Prophecy* (*K. Aʿlām al-nubuwwa*).[50] Abū Ḥātim al-Rāzī's book is, in fact, our most detailed source for Rāzī's anti-prophetic views.[51] In this book, Abū Ḥātim al-Rāzī summarized a public debate which he had held with Rāzī in the court of Rayy. According to him, Rāzī repeated

[47] Cf. Walker, "Political Implications," 83, 85 and n. 50; Mohaghegh, "*Kitāb al-ʿilm al-ilāhī*," 16; Kraus, *Rasāʾil*, 165–66; And see below, apud n. 54. Bīrūnī, *Risāla*, 16, lists two books by the name *al-ʿIlm al-ilāhī*, a long one, and a short one "following the opinion of Socrates," which Kraus (*Rasāʾil*, 167) suggested to identify with the "*Kitāb al-ʿIlm al-ilāhī*, following the opinion of Plato" mentioned by Ibn Abī Uṣaybiʿa.

[48] Bīrūnī, *Risāla*, 19, lists three books which deal with the imamate (*Fī'l-imām wa'l-muʾtamm; Fī'l-imāma; Fī'l-imāma, ʿalā al-kayyāl*). Ibn al-Nadīm (*Fihrist*, 301:20) and Ibn al-Qifṭī (p. 276) report another title in this vein, still more puzzling: *K. Fī āthār al-imām al-fāḍil al-maʿṣūm*. In theory, these books could have dealt with Rāzī's understanding of his own "*imāms*," Socrates and Plato (see Walker, "Political Implications," 65; and see, for instance, *al-Sīra al-falsafiyya, Rasāʾil*, 99). This was probably the topic of his book "On Those who Truly Deserve the Titles 'Imām' and 'A Follower of an Imām'" (*fī'l-imām wa'l-maʾmūm al-muḥiqqīn; Fihrist*, 301:21; Ibn al-Qifṭī, 276). But it is more likely that they were aimed at a refutation of the Shīʿī, and in particular Ismāʿīlī, notion of *imām*. If so, their content must have repeated at least some of the arguments contained in Rāzī's refutation of the prophets.

[49] On the term *maḫārīq*, see van Ess, *TG*, IV, 332–33.

[50] Cf. Walker, ibid., 83. The editor's rendering of *aʿlām* as "peaks" misses the semantics of the word in the context of the debate (as noted also by Meier, "Der 'Urknall'," n. 41). On Abū Ḥātim, see Kraus, *Rasāʾil*, 291–92; Hamdānī, "Unknown Ismāʿīlī Authors," 365–67; Poonawala, 36–40.

[51] For a French translation of the relevant parts of the debate, see Brion, "Philosophie et révélation;" for a German paraphrase, see Meier, "Der 'Urknall'," 14–21.

in this debate some views which he had previously published in a book:

> He disputed me concerning prophecy, advancing arguments similar to those which he had written down in his aforementioned book.[52]

Elsewhere in the same work, Abū Ḥātim al-Rāzī claims to quote directly from Rāzī's book.[53] These are the only two places in the *Aʿlām al-nubuwwa* where Abū Ḥātim al-Rāzī refers to this book. In both cases the arguments quoted from Rāzī involve his criticism of religion. The book to which Abū Ḥātim al-Rāzī refers here is probably the lost *Maḫārīq al-anbiyāʾ*. It is worth noting in particular that when Abū Ḥātim al-Rāzī relates other parts of Rāzī's philosophy—for example, the doctrine of the five eternal principles, which were discussed in Rāzī's *K. al-ʿIlm al-ilāhī*—he does not say that this was part of "his aforementioned book." This fact further strengthens our previous assumption (although it by no means proves it) that the *Maḫārīq al-anbiyāʾ* was an independent work, and was not part of the *K. al-ʿIlm al-ilāhī*.[54]

Rāzī's critique of revealed religions

In the debate (and supposedly also in the book) Rāzī laid down his theoretical objections to the notion of prophecy, a notion which for him contradicted both metaphysical and ethical axioms.

According to the spokesmen of revealed religions, the only way to ensure full divine guidance to humanity was by the sending of prophets. Rāzī argued that such a divine economy of salvation did not tally with the idea of God's wisdom:

> He [i.e., Rāzī] said: On what ground do you deem it necessary that God should single out certain individuals[55] by [conferring] prophecy on them alone, that he should set them above [all] other people, that

---

[52] Abū Ḥātim, *Aʿlām al-nubuwwa*, 2. A second-hand account of this debate, based on Abū Ḥātim's book, is given in Kirmānī, *al-Aqwāl al-dhahabiyya* (cited in Kraus, *Rasāʾil*, 313–16). On this book, cf. De Smet, *Quiétude*, 8–16, esp. 9.

[53] *Aʿlām al-nubuwwa*, 31: "Among the things he mentioned as arguments in his book is the following: . . ."

[54] See above, apud n. 47.

[55] *Qawm*, hence Goodman's translation: "that God should prefer one people" etc. ("Muḥammad ibn Zakariyyāʾ al-Rāzī," 203). In this context, however, Rāzī argues against the notion of chosen individuals, and not against the idea of a chosen people (although he would undoubtedly have found this last idea equally untenable).

> he should appoint them to be the people's guides, and make people dependent on them? And on what ground do you deem it permissible in the wisdom of the Wise One to choose this method for [the governance of] humanity, namely, that he should incite [some] people against others, or that he should spread hostility between them, increase fightings and thus bring people to their perdition?[56]

Rāzī's idea of the proper divine *modus operandi* was different. According to him,

> The most fitting [behavior] for the wisdom of the Wise One and the compassion of the Compassionate One is that He should inspire (*an yulhima*) all His servants with the knowledge of whatever is beneficial or detrimental to them, in this world and the next. He should not set some individuals over others, and there should be between them neither rivalry nor disagreement which would bring them to perdition. This would safeguard them better than if He should appoint some of them as Imāms[57] to others. [If He were to choose the latter option], each group would declare the veracity of its own *imām* and the falsehood of [all] others, and they would draw swords against each other.[58] There would be a universal disaster and they would perish in the mutual hostilities and fightings. Indeed, many people have perished in this way, as we can see.[59]

Rāzī accepted the basic religious assumption that God is wise and compassionate. His conclusion was not only that God should, but that He actually did inspire all people, as His wisdom and compassion decree.[60]

In Rāzī's view, the untenable theology of the revealed religions was combined to a narrow-minded attitude inherent to them. One could say that according to Rāzī, the senseless image of God upheld by the adherents of revealed religions reflected their general obscurantist bigotry.

According to Abū Ḥātim al-Rāzī, Rāzī's criticism of the revealed religions started from a basic methodological disagreement with them:

---

[56] Abū Ḥātim, *Aʿlām al-nubuwwa*, 2.

[57] Rāzī speaks in general terms, and therefore the word probably retains its initial meaning of "foremen" or leaders. But in the context of a debate with an Ismāʿīlī, and in the record of this debate written by one, the technical meaning of the word *a'imma* (sing. *imām*) is also undoubtedly intended. The anti-Ismāʿīlī intention of Rāzī's use of the term *imām* was already noted by Kraus, cf. "Raziana," 322.

[58] Lit.: "hit each other's face with the sword."

[59] *Aʿlām al-nubuwwa*, 2–3. See also ibid., 12, where Abū Ḥātim reminds Rāzī: "Did you not say that people perished because of enmity and disagreement?"

[60] Cf. *Aʿlām al-nubuwwa*, 2–3.

He [i.e., Rāzī] said: "The followers of revealed religions have learnt their religion by following the authority of their leaders. They reject rational speculation and inquiry about the fundamental doctrines [of religion]. They restrict [this inquiry] and forbid it. They transmit traditions in the name of their leaders, which oblige them to refrain from speculation on religious matters,[61] and declare that anyone who contradicts the traditions they transmit must be branded an infidel.[62]

At this point, Rāzī turns to give a few examples of these traditions, and his examples are clearly recognizable as Muslim [prophetic] traditions:

Among these [traditions] are those which they recount in the name of their forefathers:[63] "Disputation and disagreement in religious matters is infidelity";[64] "Whoever subjects his religion to logical considerations, will remain forever in confusion";[65] "Do not ponder God; rather, ponder His creation";[66] "Predestination is God's secret; do not plunge into it";[67] "Beware of deep contemplation, for previous generations perished because of it".[68]

Rāzī did not restrict his criticism to general observations on religious narrow-mindedness, but directed a poisoned arrow specifically against Islam. To make his intention even clearer, he added:

If the people of this religion[69] are asked about the proof for the soundness of their religion, they flare up, get angry and spill the blood of

[61] Or: "which imposes on them, as a religious duty, to refrain from speculation."

[62] *Inna ahl al-šarā'iʿ aḫadhū al-dīn ʿan ru'asā'ihim bi-taqlīd, wa-dafaʿū al-naẓar wa'l-baḥth ʿan al-uṣūl, wa-šaddū fīhi, wa-nahaw ʿanhu; wa-rawaw ʿan ru'asā'ihim aḫbāran tūjibu ʿalayhim tarka al-naẓar diyānatan, wa-tūjibu al-kufra ʿalā man ḫālafa al-aḫbāra allatī rawawhā; Aʿlām al-nubuwwa*, 30–31.

[63] *Aslāf.* Most of the traditions quoted below do not appear in the canonical collections. If Abū Ḥātim quotes Rāzī faithfully, the choice of the term *aslāf* may indicate Rāzī's awareness that these precise traditions were not recognized as sound prophetic traditions (*ḥadīths*), and did not present them as such. It is also possible (and perhaps more likely) that Rāzī avoided an explicit mention of the Prophet Muḥammad.

[64] *Inna al-jadal fī al-dīn wa'l-mirā' fīhi kufr*; compare Wensinck, *Concordance*, s.v. "*mirā'*" and "*kufr*"; Ibn Ḥanbal, *Musnad*, Cairo 1955, 14, 240, 249.

[65] *Man ʿaraḍa dīnahu li'l-qiyās, lam yazal al-dahra fī iltibās*; compare *Mawsūʿat aṭrāf al-ḥadīth*, VII, 67.

[66] *Lā tatafakkarū fī Allāhi, wa-tafakkarū fī ḫalqihi*; I was unable to find this *ḥadīth* in the authoritative compilations.

[67] *Al-qadar sirru Allāhi, fa-lā taḫūḍū fīhi*; compare *Mawsūʿat aṭrāf al-ḥadīth*, V, 743.

[68] *Iyyākum wa'l-taʿammuq, fa-inna man kāna qablakum halaka bi'l-taʿammuq*; compare *Mawsūʿat aṭrāf al-ḥadīth*, IV, 140.

[69] *Hādhihi al-daʿwā*; although this could refer to the Ismāʿīlī missionary activity, it can also be an allusion to Islam in general.

> whoever confronts them with this question. They forbid rational speculation, and strive[70] to kill their adversaries. This is why truth became thoroughly silenced and concealed.[71]

In this last sentence, Rāzī may be turning the tables on his Ismāʿīlī interlocutor, in whose doctrine the concealment (*kitmān*) of the deeper truth from the uninitiated is glorified as part of the esoteric tradition. For Rāzī, this *kitmān* has no glory, it is only one of the marks of a dark age of persecution. He then continues his tirade:

> They adopted this approach as a result of their being long accustomed to their religious denomination, as days passed and it became a habit. Because they were deluded by the beards of the goats, who sit in ranks in their councils, straining their throats in recounting lies, senseless myths and "so-and-so told us in the name of so-and so..."[72]

This graphically illustrated deprecation of religious dignitaries is probably indeed "the most violent polemic against religion in the course of the middle ages."[73] The view that sees religious habits at the root of prejudice and other ills connected to religion is not unique to Rāzī. It appears later in the writings of both Maimonides[74] and Averroes,[75] and may have had its origin in a commentary on Aristotle or on Alexander of Aphrodisias.[76] In the writings of Maimonides or Averroes, however, this view is expressed in a detached, matter-of-fact way, whereas Rāzī's savage bluntness goes far beyond the candor of objective analysis. In this "attack on the theologians' obscurantism,"[77] he did not aim to censor only a certain class of traditionists, or a specific manipulation of the traditions. Rāzī believed that the very notion of

---

[70] I read: *ḥaraṣū* (the printed text is unclear).

[71] *Aʿlām al-nubuwwa*, 31.

[72] *Wa-innamā ataw fī hādha al-bāb min ṭūl al-ilfi li-madhhabihim, wa-murr al-ayyām, wa'l-ʿāda, wa-ightirārihim bi-liḥya al-tuyūs al-mutaṣaddirīna fī'l-majālis, yumazziqūna ḥulūqahum bi'l-akādhīb wa'l-ḫurāfāt wa-ḥaddathanā fulān ʿan fulān; Aʿlām al-nubuwwa*, 32.

[73] Kraus and Pines, *EI*, 1136.

[74] *Guide*, I, 31, 67 (*Dalāla*, 45).

[75] *K. Faṣl al-maqāl wa-taqrīr mā bayna al-šarīʿa wa'l-ḥikma min al-ittiṣāl*, ed. A. N. Nādir, Beirut 1973, 45–46; G. F. Hourani, *Averroes on the Harmony of Religion and Philosophy*, London 1976, 59.

[76] See *Maqālat al-Iskandar al-afrūdīsī fī'l-qawl fī mabādīʾ al-kull*, in A. Badawi, *Arisṭū ʿind al-ʿArab*, Cairo 1947, 253–77, esp. 276:18ff.; S. Pines, Translator's Introduction to Maimonides, Guide, lxvii–lxviii; S. Stroumsa, "Habitudes religieuses et liberté intellectuelle dans la pensée arabe médiévale," in M. Abitbol, ed., *Monothéismes et tolérance: Actes du Colloque du Centre International de Recherche sur les Juifs du Maroc, Paris 1–4 octobre, 1996*, Paris 1998, 57–66.

[77] Cf. van Ess, *TG*, IV, 333.

prophetic tradition was inherently wrong, and that no use of it could lead to the truth. His aim was, in all likelihood, to attack Muslim traditionalists in principle, as the embodiment of the notion of prophetic tradition, and to undermine their authority. He goes on in his speech, leaving no room for mistake about the addressees of his attack:

> recounting lies, senseless myths and "so-and-so told us in the name of so-and so," deceitfully and falsely, and transmitting contradictory traditions. Among these are traditions which prove that the Qur'ān is created, and others that deny it; traditions that give precedence to ʿAlī, and others that give precedence to other individuals; traditions that deny free will,[78] and others that deny predestination; and anthropomorphic traditions—the heretic brought these traditions, but we abhor [the idea of] burdening the book with them.[79]

The enormity of Rāzī's attack on Islam is reflected in his opponent's reaction, who repeats once more what he apparently perceived as most shocking in Rāzī's expressions:

> The heretic said: "What deluded them is only the length of the goats' beards, and the whiteness of the cloaks of the surrounding company: feeble[-minded] men, [as well as] women and children. [They were also affected by] the length of time that passed, until [the religious practice] became a [second] nature and a habit.[80]

Although Rāzī's criticism targeted Islam in particular—or so it was felt by his fellow Muslims—it did not spare other religions. The fact that these religions disagree with one another was taken by him as a decisive proof that none of them could be true.

> He said: Jesus claimed that he is the son of God, while Moses claimed that He had no son, and Muḥammad claimed that he (i.e., Jesus) was created, like the rest of humanity. Mani and Zoroaster contradicted

[78] *Qadar* usually designates God's unchangeable, primordial decree. The context, however, demands a translation that juxtaposes this part of the sentence to the *ijbār* (compulsion, as part of of predestinarian views) in the second part. One could easily correct the word to *qudra* (i.e., the power of the human agent to act freely). This, however, is not necessarily required, if we assume that Rāzī used *qadar* in the sense of *qudra*. This last possibility is enhanced by the double use of the term *qadariyya*, which at an early stage served (also?) as a name for the predestinarians, but later came to be applied specifically to the Muʿtazila and the adherents of the doctrine of free-will.

[79] *Aʿlām al-nubuwwa*, 32.

[80] Ibid.

> Moses, Jesus and Muḥammad regarding the Eternal One,[81] the coming into being of the world, and the reasons for the [existence] of good and evil. Mani contradicted Zoroaster concerning the two entities[82] and the reasons [for their existence]. Muḥammad claimed that Christ had not been killed, but the Jews and Christians disagree with him, saying that he had been killed and crucified.[83]
>
> He said: Moses said that God is omnipotent, that He is not compound and uncreated, that He has no use for beneficiary [things], nor is He hurt by detrimental things.[84] It is said in the *Torah* that the fat [of the sacrificed animal] should be burnt, for the Lord to inhale its scent.[85] It is said in the *Torah* that the Ancient One is in the form of an old man, white-haired and white-bearded.[86] And it says: "Why do you sacrifice to me the lame and blind [animals]? Do you think that, had you presented it as a gift to your friends, they would accept from you anything but healthy [animals]?"[87] And it says: "Make for me a finely woven silken rug, and a table made of *šimšār* wood."[88] Then the heretic said: "This sounds like the words of the needy rather than of the Laudable Self-sufficient One." He mentioned many things from the *Torah* and criticized them.[89]

Both non-Jews and Karaites could direct somewhat similar arguments against what they perceived to be the anthropomorphisms of Rabbinic Judaism.[90] Rāzī's arguments, however, concentrate specifically on biblical verses, which Karaites, of course, would be loath to criticize. At least two of the arguments here were included in the critique of the Bible by the ninth-century Jewish apostate Ḥīwī al-Balḫī.[91] They were also part of the Zoroastrian critique of the Bible,[92] and

---

[81] *Al-qadīm*; Brion, "Philosophie et Révélation," 145, reads "*al-qadam*" (or understands *qadīm* in that sense).

[82] *Al-kawnayn* is the standard term to indicate the dual creating principle (rather than a dual generation, as translated by Brion, "Philosophie et révélation," 145).

[83] On Ibn al-Rāwandī's use of this argument, see above, Ch. 2, n. 189.

[84] The phrasing and terminology of the whole sentence suggest its origin in some tract of Jewish *kalām*.

[85] Cf. Lev. 17:6; cf. also Davidson, *Hiwi*, 25, 89–90; Saʿadya, *Amānāt*, III, 146.

[86] Compare Dan. 7:9.

[87] Cf. Mal. 1:8.

[88] Compare Ex. 25:23, 26:31; and cf. R. Dozy, *Supplément aux dictionnaires arabes*, Leiden 1881 (reprinted Beirut 1991), II, 787, s.v. "šamšād."

[89] *Aʿlām al-nubuwwa*, 69–70.

[90] See, for example, Qirqisānī, *Anwār*, I, 15, 31–35; and cf. Cook, "ʿAnan and Islam," 176–77; C. Adang, "Éléments karaïtes dans la polémique antijudaïque d'Ibn Ḥazm," *Dialogo filosófico-religioso entre cristianismo, judaísmo e islamismo durante la edad media en la peninsula iberica*, Turnhout 1994, 419–41, esp. 432–34.

[91] See below, Ch. 8, n. 32; and see Rosenthal, "Hiwi," 317ff.; Davidson, *Hiwi*, 89–90.

[92] Cf. J. P. de Menasce, *Une apologétique mazdéenne du IXème siècle, Škand Gūmānik*

they seem to follow the tradition of the refutation of the Bible by late antique Hellenic philosophers, from Celsus to Porphyry.[93] Bible criticism, however, constitutes only a part of Rāzī's attack on revealed religions:

> And he said: The Christians say that Jesus is pre-eternal, and there is no Lord above him, and that he said: "I have come to fulfill the Torah,"[94] but then he abrogated its laws and replaced its strictures and precepts [by others]. The Christians claim that He is the Father, the Son and the Holy Ghost.[95]

These were the most common arguments in the anti-Christian arsenal.[96] Whether Rāzī himself did not think it necessary to develop less banal argumentations or whether Abū Ḥātim al-Rāzī was not interested in quoting them, it is at any rate clear that the refutation of Christianity was not Rāzī's main goal. Abū Ḥātim al-Rāzī continues:

> He mentioned what the Zoroastrians say in the name of Zoroaster, concerning Ahriman and Ahura Mazda], and the claims of Mani, that the Logos became detached from the Father; that it tore the demons to pieces and that it killed [them]; that the sky was made of the skins of the archons (*šayaṭīn*), that the thunder is the rumbling of the imps, and that earthquakes are [caused by] the demons' agitation under the earth; that Mani lifted in the air Sābūr, for whom he had written the *Sāburaqān*, and that he hid him there for a while; that, while standing in front of the [people], Mani would be snatched by his spirit and carried up to the sun, sometimes for an hour and sometimes for days.[97]

According to Abū Ḥātim al-Rāzī, the fact that Rāzī mingled his criticism of monotheistic religions with reports on Manichean mythology

---

Vičar (*La solution décisive des doutes*), Fribourg 1945; Davidson, *Hiwi*, 30–82; Lazarus-Yafeh, *Intertwined Worlds*, 132.

[93] Cf. Kraus and Pines, *EI*; Kraemer, "Heresy," 168, 174; Lazarus-Yafeh, *Intertwined Worlds*, 132; and see below, Ch. 5, apud. nn. 1–12. On Rāzī's attack on the *Torah*, see also Wasserstrom, *Between Muslim and Jew*, 148–53.

[94] Cf. Matt. V, 17–18.

[95] *Aʿlām al-nubuwwa*, 70.

[96] See, for instance *Qiṣṣa*, sections 33, 35, 70 (on Matt. V, 17), and section 27 (on the Trinity); A. Bouamama, *La littérature polémique contre le christianisme depuis ses origines jusqu'au XIIIe siècle*, Algiers 1989, 94–94 (on Matt. V, 17) and 133–89 (on the Trinity).

[97] *Aʿlām al-Nubuwwa*, 70; see above, Ch. 2, apud n. 23. On the Manichean myths recorded here, cf. also Monnot, *Penseurs musulmans*, 152–64, 281; note also *Fihrist*, 299, which mentions a book recording Rāzī's disputation with the Manichean Sīs (cf. Pines, *Atomism*, 83).

was not the result of an innocent intellectual curiosity for comparative religion:[98]

> He adduced these absurdities, fabricated by the Zoroastrians and Manicheans, and mixed them with the content of the revealed scriptures and the traditions of the prophets. He [thus] attributed these [absurdities] to the immaculate apostles of God, who are innocent of all this.[99]

Furthermore, says Abū Ḥātim al-Rāzī, Rāzī's preoccupation with details of various religious doctrines was motivated by his polemical purposes:

> He claimed that [all] this is in their scriptures,[100] and that this [shows] disagreement and contradiction in what they say. He used it to refute [the notion of] prophecy. By these worthless stupidities he intended to buttress his claims and to expose the fraudulent tricks and the senseless myths[101] [of the prophets].[102]

Here, as elsewhere, Abū Ḥātim al-Rāzī's information agrees with other sources. A similar interpretation of Rāzī's motives is found in Bīrūnī, who says that Rāzī referred the readers of his own works to Mani's books, "as a ruse against the [revealed religions], including Islam." Bīrūnī adds that at the end of Rāzī's "Book on the Prophetic Religions," where he derided the people respected and honored by these religions (i.e., the prophets and saints), the author adduced Mani's books as proof for his own claims.[103]

As could be expected, Rāzī refuted at length the prophets' claims to perform miracles. According to Abū Ḥātim al-Rāzī,

> The heretic dedicated a lengthy essay[104] to miracles, and he composed it in the form of questions and answers. In this essay he derided the proofs adduced by those who claim that prophets perform miracles, and he produced weak arguments [to prove his point].[105]

---

[98] On Rāzī's use of comparative religion, see Wasserstrom, *Between Muslim and Jew,* 152.

[99] *Aʿlām al-nubuwwa,* 70–71.

[100] *Rusūm*; lit.: "proscriptions" (as translated by Brion, "Philosophie et révélation," 147).

[101] *Al-maḫārīq wa'l-ḫūrāfāt*; comp. Brion, "Philosophie et révélation," 147, who understands these words as additional epithets with which Abū Ḥātim describes the dualist myths in Rāzī's accounts.

[102] *Aʿlām al-nubuwwa,* 71.

[103] *Risāla,* 3; and see below, n. 127; and Ch. 5, n. 84.

[104] "*Qawlan kathīran.*"

[105] Abū Ḥātim, *Aʿlām al-nubuwwa,* 191.

Assuming that Abū Ḥātim al-Rāzī referred here to the same book[106] as in the rest of his refutation, we may infer either that the *Maḫārīq al-anbiyāʾ* was written in the form of questions and answers, or that this was the form only of the part which dealt with miracles. It is at any rate clear that a large part of the book was dedicated to miracles. This would indeed befit a response to the *dalāʾil al-nubuwwa* literature, in which the subject of miracles was of paramount importance.

Abū Ḥātim al-Rāzī continues with an exposition of the arguments of heretics, who accuse the transmitters of prophetic traditions of conspiracy to lie (*tawāṭuʾ*) concerning the prophets' miracles. After a general presentation of these arguments, he turns to attack Rāzī whose book, he says, offered quite similar arguments, namely:

> that similar feats were performed by people who made no claim of prophethood. He then mentioned people who perform legerdemain and juggleries, such as dancing on spikes,[107] walking on sharp blades [and] over spears. [He further adduced] the rhymed speech [of oracles] and of soothsayers, the magic performed by magicians, and others things with which he presumed to refute whoever claims that the prophets perform miracles.[108]

After his attack on miracles in general, Rāzī proceeded to refute the particular miracle of Islam and the doctrine of *iʿjāz al-Qurʾān*, saying:

> You claim that the evidentiary miracle [that proves Muḥammad's prophecy] is present and available, namely, the Qurʾān. You say: "Whoever denies it, let him produce a similar one."[109] Indeed, we shall produce[110] a thousand similar [compositions], from the works of rhetoricians, eloquent speakers and valiant poets,[111] which are more appropriately phrased and state the issues more succinctly.[112] They convey the meaning better and their rhymed prose is in better meter . . . By God, what you say astonishes us! You are talking about a work which recounts ancient myths,[113] and which at the same time is full

---

[106] Shortly afterward Abū Ḥātim returns to refer to "what the heretic said in his book." (ibid., 192).

[107] Translation *ad sensum*. *Rasan* usually means a rope or a cord, but the plural *arsān* can also signify rugged and hard tract of ground; cf. E. W. Lane, *Arabic-English Lexicon*, London and Edinburgh 1863 (reprinted Cambridge 1984), I, 1086.

[108] *Aʿlām al-nubuwwa*, 192.

[109] Ibid., 227–28.

[110] I read as on 192.

[111] Lit.: "from the valiants and the poets."

[112] *Aʿlām al-nubuwwa*, 227–28; cf. Pines, *Atomism*, 83–91.

[113] *Asāṭīr al-awwalīn*, cf., for example, Q. 6:25, 25:5.

> of contradictions and does not contain any useful information or explanation. Then you[114] say: "Produce something like it!?"[115]

Rāzī repeated the scorn of the idolaters of Mecca, who claimed that the Qur'ān is like the ancient myths. But like the pagan Platonist philosophers of late antiquity, Rāzī also insisted that the Qur'ān compared to myths unfavorably, since it did not fulfill the role of a myth.

Rāzī presented himself, and was presented by others, as a Platonist,[116] although in other instances he was also described as a Pythagorean.[117] His interest in Platonism was restricted to the *Timaeus*, and like his model, he favored myths.[118] Rāzī's own creation myth started from five eternal principles: time, space, matter, soul, and the creator.[119] The soul became infatuated with matter, and seeing its plight, the compassionate creator united it with matter, so that, through learning, the soul may recognize its own folly, attain wisdom, and become truly free from the bonds of matter.[120] Rāzī's rejection of the Qur'ān thus did not stem from an aversion to mythopoesis, but from the fact that, in his view, the Qur'ān did not qualify as a proper mythology. According to the Platonist philosophers, myths and mythologies were allegorical explanations of phenomena which are too difficult to explain otherwise. A proper myth should read like a riddle and be worthy of higher philosophical interpretation.[121] Rāzī believed that the Qur'ān offered no such explanations, and could not be interpreted in this way.

---

[114] Lit.: "they." Rāzī evidently continues to address the people who incite his astonishment; speaking now in the third person highlights the scornful tone of his words.

[115] Ibid., 227–28.

[116] Cf., for example, *Rasā'il*, 27:14, 31:3, 139–40; Fakhry, *Islamic Philosophy*, 99–102; Walzer, *Greek into Arabic*, 16. On the development in Rāzī's Platonism, see Bar-Asher, "L'éthique de Rāzī," 119–25, 143.

[117] Cf. Ibn Ṣā'id, *Ṭabaqāt*, 33; al-Mas'ūdī, *Tanbīh*, 106; cf. Kraus, *Rāsā'il*, 187–88; Pines, *Atomism*, 93, n. 136; and see P. Kingsley, *Ancient Philosophy, Mystery and Magic: Empedocles and Pythagorean Tradition*, Oxford 1995, 388.

[118] Cf. Pines, *Atomism*, 85–86; idem, "Nouvelles Études," 147–48; Goodman, "Myth," 29, and 39, n. 55.

[119] On Rāzī's five eternal principles, see Kraus, *Rasā'il*, 191–279; Pines, *Atomism*, 47–64; Mohaghegh, "*Kitāb al-'ilm al-ilāhī*," 19–23; Meier, "Der 'Urknall'," 5ff. See also Brion, "Le temps."

[120] Cf. Kraus, *Raziana*, 366:15–17; Goodman, "Myth," 25–26; idem, "Razi," 476; Walker, "Political Implications," 71–72.

[121] This argument, voiced already by Celsus, was repeated in Platonic tradition throughout late antiquity; cf. G. G. Stroumsa, *Hidden Wisdom: Esoteric Traditions and the Roots of Christian Mysticism*, Leiden 1996, ch. 1 and ch. 6, esp. 101–102.

For Rāzī, the Qur'ān was of little intellectual value. Rāzī's criticism of the revealed religions, however, was much more virulent than mere intellectual disdain. In his view, religions based on revelation were an actual disaster for humanity. In his *K. Maḫārīq al-anbiyā'* he depicted religious authorities as intolerant and tyrannical,[122] and he insinuated that both prophets and revealed religions were the source of all bloodshed.[123] This opinion was expressed even more forcefully in another part of his work, which has been preserved by both Bīrūnī and Nāṣir-i Ḫusraw. Bīrūnī, who, at the request of a friend,[124] wrote a catalogue of Rāzī's books, felt the need to apologize for this undertaking:

> Were it not for the esteem in which I hold you, I would not have undertaken this project. This is so because I might draw the animosity of Rāzī's opponents, who might mistake me to be one of his partisans. [They might consider me to be] a person who equates sound conclusions, reached through the application of independent reasoning (*ijtihād*), with the objects of his own inclinations, towards which he leans because of his passions and fanaticism, until their pursuit brings him to shame.[125]

In this last sentence Bīrūnī already moves, in a somewhat convoluted way, from referring to a hypothetical partisan of Rāzī's (with whom he did not want to be identified) to the description of Rāzī himself. He admits that Rāzī's independent reasoning bore some sound conclusions. But, he adds, Rāzī has had also some other ideas, to which he inclined because of what we may call the flaws of his character: his passions and fanaticism.[126] Bīrūnī also insinuates that Rāzī was unable to distinguish ideas belonging to the first category (i.e., reached by independent reasoning) from those which belong to the second (i.e., the results of his passions). In the following sentence Bīrūnī abandons the roundabout description of the feared comparison, and proceeds to describe Rāzī directly:

---

122 *Aʿlām al-nubuwwa*, 30–31.

123 See above, apud n. 71.

124 If this introduction is not a mere literary convention.

125 Bīrūnī, *Risāla*, 2; comp. Ruska, "Al-Bīrūnī," 30. On *ijtihād*, see below, n. 150.

126 *Taʿaṣṣub*; This term (or the word *ʿaṣabiyya*) often designates the group-loyalty which marks false religions or false prophets, see above, Ch. 1, n. 61 and apud n. 81; and below, Ch. 6, n. 40. It may, at the same time, be a counter argument to Rāzī's accusation that the prophets' bigotry sets one people against another. If so, Bīrūnī's intention would be to expose Rāzī as suffering from the faults he criticizes.

> His harshness towards religion does not stop at neglect, refraining from religion and turning away from it. He goes further to active slander, imputing to religion evil spirits, and the deeds of the demons (*šayāṭīn*). So much so, that it leads him to recommend [reading] the books of Mani and his followers, in order to undermine revealed religions, including Islam.[127]

A similar obscure reference to "evil spirits" in Rāzī's philosophy is used by ʿĀmirī, who mentions Rāzī's "ravings about the five eternal principles and about the corrupt spirits."[128]

The oblique insinuations of ʿĀmirī and Bīrūnī would have remained opaque for us, were it not for the more straightforward presentation of Nāṣir-i-Ḫusraw, who says plainly that in his *K. al-ʿIlm al-ilāhī* Rāzī had described the prophets as messengers of evil spirits who had become demons. According to him,

> Rāzī said in his *K. al-ʿIlm al-ilāhī* that the souls of wicked people, who had been transformed into demons, appear to some individuals in the form of angels, and command them: "Go and tell people: An angel had appeared to me and told me: 'God has made you His apostle,' and [—they say—] I am the angel sent to you." This is the reason for the dissemination of discord, and many people have been killed as a result of the machination[129] of these souls-turned-devils.[130]

Nāṣir-i Ḫusraw's report agrees with the cryptic remarks of ʿĀmirī and Bīrūnī and clarifies them. The fact that he identifies Rāzī's words as derived from the *K. al-ʿIlm al-ilāhī* (rather than from the *Maḫārīq al-anbiyā'*, which would be the natural peg to hang this quotation on, if he had forged it), adds credibility to his report. It is also noteworthy that Bīrūnī seems to regard "Rāzī's book on prophetic

---

[127] Bīrūnī, *Risāla*, 2–3; compare Ruska's translation ("Al-Bīrūnī," 30), which glides over the role of the demons in this passage. Bīrūnī's interpretation of Rāzī's fascination with Manichean myths tallies with Abū Ḥātim's, see above, apud. nn. 98–103.

[128] *Ṣunūf hadhayānātihi fī al-qudamā' al-ḫamsa wa-fī al-arwāḥ al-fāsida*; See Rowson, n. 36 above; and see Rowson's commentary, 203, 214–17.

[129] *Tadbīr*; if these words are an exact quotation of Rāzī, then Rāzī would be using here the usual technical term for God's providence; cf. Bīrūnī, *Risāla*, 17, which lists a book Rāzī wrote "*Fī naqḍ K. al-Tadbīr*."

[130] *Rasā'il*, 177–78; *Jāmiʿ al-ḥikmatayn*, 137; see Corbin's introduction, 137–38. Nāṣir-i-Ḫusraw says that he offered a full refutation of this idea in his *K. Bustān al-ʿuqūl*—which, unfortunately, is not extant; neither is Ibn Ḥazm's refutation of the same book, the *K. al-Taḥqīq*; see *Fiṣal*, 1, 38, 87, and 2, 197; Mohaghegh, "*Kitāb al-ʿilm al-ilāhī*," 17.

religions" (*kitābuhu fi'l-nubuwwāt*) as distinct from the source of his remarks concerning "spirits" on the one hand and the *K. al-ʿIlm al-ilāhī* on the other hand.[131] The information given by both Nāṣir-i Ḫusraw and Bīrūnī further reduces the possibility that the *Maḫārīq al-anbiyāʾ* was actually a part of the larger *K. al-ʿIlm al-ilāhī*. The evidence seems to fit better with the assumption that Rāzī wrote at least one work which was wholly devoted to the refutation of prophecy, although he treated this subject in his *K. al-ʿIlm al-ilāhī* as well.

From Nāṣir-i Ḫusraw's report it appears that Rāzī's animosity for prophetic religions led him to see in the prophets the source of all human-made evil. His depiction of the prophets as possessed by malevolent spirits shows him in a far more radical light than merely a non-conformist or a somewhat skeptical rationalist. His savage antagonism shows that he regarded contemporaneous prevalent religiosity as a threat to the very life of human beings in general, and perhaps to his own life as a free spirit in particular.

3. The Authenticity of Rāzī's Freethinking

As in the case of Ibn al-Rāwandī, there are voices in modern scholarship which cast doubts upon Rāzī's heterodoxy. In fact, such doubts are not all that new: Ibn Abī Uṣaybiʿa (d. 668/1270) had already rejected Rāzī's heretical image as mere slander.[132] The entry on Rāzī in his biographical dictionary is a laudatory résumé of Rāzī's life and work. In his list of Rāzī's writings we read:

> A book in which he attempted to demonstrate his claims concerning the Saints' faults.[133] I say: If this book was ever written—and only God knows—then perhaps one of Rāzī's malevolent enemies composed it and attributed it to him, so that whoever sees the book or hears of it will think ill of Rāzī. How else, for Rāzī is above endeavoring such a thing or writing in this vein. [I say so] although some people, who blame Rāzī, or even brand him an infidel, . . . call this book "Rāzī's Book on the Prophets' Fraudulent Tricks."[134]

[131] *Risāla*, 3.
[132] IAU, 426.
[133] "*Fī-mā yarūmuhu min iẓhār mā yaddaʿī min ʿuyūb al-awliyāʾ*," IAU, 426:13; cf. also Ibn al-Qifṭī, 276.
[134] "*Kitāb al-Rāzī fī maḫārīq al-anbiyāʾ*"; IAU, 426:11–17.

At the bottom of Ibn Abī Uṣaybiʿa's skepticism regarding this book we find the same mistrust of the sources that brought some scholars to cast doubts upon the supposedly heretical nature of Ibn al-Rāwandī's views.

Most of our information concerning Rāzī's opinions is derived from the writings of his adversaries. Naturally, one must then question their reliability. Even when one assumes that they did not actually impute to Rāzī thoughts that he had never expressed, the possibility must seriously be weighed that in citing Rāzī's words his adversaries transformed them, distorted their meaning, and perhaps gave them a more radical twist than these words had in their original context.[135]

Skepticism regarding the sources relates mostly to the reliability of Rāzī's Ismāʿīlī opponents. It is indeed impossible to ignore the fact that our principal reports concerning Rāzī's heretical views come from Ismāʿīlī sources.[136] Paul Walker has noted that the presentation of Rāzī's arguments as heretical suited the intentions of his opponent Abū Ḥātim al-Rāzī."[137] Walker wonders whether this is the only way to understand Rāzī's arguments, implying that the Ismāʿīlī propagandists somehow distorted Rāzī's views.[138]

It seems to me that the examination of Rāzī's philosophical work—both the extant books and the fragments quoted by his opponents—excludes the possibility of a serious falsification of his ideas by later Ismāʿīlī sources. It also excludes a major misunderstanding of him by his opponents. The whole of his *œuvre* presents a consistent picture of a critical and consciously independent thinker. Not even once in his work does he appear as a conformist who bends his own ideas to suit religious consensus. Not once does he say anything that would betray him as accepting the strictures of any religious authority.[139]

---

[135] On the dubious reliability of late heresiographical sources when citing their adversaries, see M. Schwarz, "Can we Rely on Later Authorities for the Views of Earlier Thinkers?" *IOS* 1 (1971), 241–48.

[136] Cf., for instance, Pines, *Atomism*, 44; Walker, "Political Implications," 67.

[137] Walker, Ibid., 89.

[138] Of course, Walker's question becomes even more pressing when we consider the importance of an Ismāʿīlī source – al-Mu'ayyad fī'l-Dīn – to our knowledge of Ibn al-Rāwandī's heresy.

[139] See also above, n. 17. The one puzzling exception is found in two titles of Rāzī's books, listed by Bīrūnī (*Risāla*, 18): *fī wujūb daʿwat al-nabī, ʿalā man naqara* (sic! but perphaps it should be corrected to *nakira*) *bi'l-nubuwwāt; fī wujūb al-duʿāʾ ʿalā ṭarīq al-ḥazm*. A simiar title appear in IAU, 426:3 (*Fī wujūb al-duʿāʾ wa'l-daʿāwā*). As they are, these titles seem more fitting for an Ismāʿīlī *dāʿī* than for a heretic. Bīrūnī does

From quotations of his works by non-Ismāʿīlī authors, such as Ibn Ṣāʿid and Bīrūnī, Rāzī emerges not only as a free mind, but more specifically as a freethinker in the strict sense as defined in this book. In view of this uniform picture, there seems to be no justification in suspecting the three Ismāʿīlī authors (namely, Abū Ḥātim al-Rāzī, Kirmānī and Nāṣir-i Ḫusraw) of being the sole inventors of the image of Rāzī as an apostate.

There are a few more plausible explanations for the fact that the anti-prophetic aspect of his philosophy comes out most dramatically in Ismāʿīlī writings. It is likely that the particularly strong authoritarian streak of the Ismāʿīliyya triggered an equally strong, even aggressive, defiance by Rāzī. One senses that rather than attempting to find a common ground with his interlocutors, Rāzī was trying to shock and hurt them. It is also likely that his opinions were recorded with particular scrupulousness by Ismāʿīlī writers, as his attack on the notion of prophecy scandalized them even more than it scandalized their fellow Sunnis. As noted by Daniel de Smet, the doctrine of the Imamate was dependent upon the doctrine of prophecy, and the denial of the latter was perceived as direct assault on the former.[140] This feature of the Ismāʿīliyya became even more pronounced after the establishment of the Fāṭimid state, when Ismāʿīlī authors tended to show a more conservative attitude towards the then existing institutions of Islam.[141] Finally, one must not overlook the simple historical fact that Ismāʿīlism was not a side show in Rāzī's time and place, but a forceful intellectual presence on the rise. It should come as no surprise, then, if Ismāʿīlī authors provide a major source of information about a contemporary intellectual event of such dimensions as Rāzī's apostasy must have been.

The three Ismāʿīlī authors who polemicize against Rāzī's views corroborate one another. Moreover, one of them had the foresight to choose as object of his refutation a treatise which would survive the vicissitudes of time. Ḥamīd al-Dīn al-Kirmānī's *al-Aqwāl al-Dhahabiyya* is a systematic refutation of Rāzī's *K. al-Ṭibb al-rūḥānī*. Rāzī's book was published by Kraus on the basis of surviving manuscripts. Even

---

not include them among Rāzī's *kufriyyāt*, nor does he say about them (as, for example, Ibn al-Nadīm says about Ibn al-Rāwandī) that they are "*min kutub ṣalāḥihi*." Nothing further is known of these books. On the (remote) possibility that they indicate interest on Rāzī's part in political philosophy, see below, Ch. 6, n. 79.

[140] Cf. De Smet, "Polémique ismaélienne," 92 and n. 33.

[141] Cf. Hamdānī, "Unknown Ismāʿīlī Authors," 365.

a cursory comparison of the two works shows that Kirmānī quoted systematically and precisely, omitting little and distorting nothing. His quotations are not limited to short sentences, whose meaning can be suspected of being distorted because the sentences are read out of their original context. Rather, the lengthy fragments quoted by Kirmānī provide the full context of Rāzī's words.[142] The existence of both Rāzī's original work and its Ismāʿīlī refutation permits a reconsideration of the almost automatic suspicion of modern scholars of Ismāʿīlī polemicists. It shows that Ismāʿīlī writers did not necessarily alter their opponents' text—Rāzī's, in this case, but this holds true also for Ibn al-Rāwandī's *K. al-Zumurrud* and its refutation by al-Mu'ayyad. The authentic works of their opponents provided the Ismāʿīlī authors with useful material for refutation, which, they believed, would highlight the superiority of their own philosophy. They were happily satisfied to quote these works with care, so that they could relish their refutation.[143]

Besides the mistrust of the Ismāʿīlī sources (which, as we have seen, is unwarranted), modern scholars have found another reason for emitting doubts about Rāzī's apostasy. This reason is related to our current knowledge of the development of ideas in medieval Islam. With the growing amount of texts published, it becomes increasingly obvious that doctrines and ideas which in the tenth and eleventh century were presented as basic tenets of Islam, or belonging to a given theological school, had in fact been originally absent from Islam, or from that same school. It may have taken many years for doctrines to crystallize and become canonized. It is thus possible that some views, which would be considered shockingly heretical in the fifth Islamic century, could still have offered a legitimate option a century earlier.

By the fourth and fifth Islamic centuries, the miraculous beauty of the Qur'ān had become an integral part of the Islamic creed. But

---

[142] For example, *al-Aqwāl al-dhahabiyya*, 23–24 corresponds to *al-Ṭibb al-rūḥānī*, *Rasā'il*, 17–19; *al-Aqwāl al-dhahabiyya*, 33–36, corresponds to *Rasā'il*, 20–22; *al-Aqwāl al-dhahabiyya*, 39–47, corresponds to *Rasā'il*, 24:4–32, and so on. Kirmānī's faithful quotations from Abū Ḥātim's *Aʿlām al-nubuwwa* provide further illustration of his reliability.

[143] Notwithstanding n. 135 above, heresiographers often tended to be conscientious reporters. In similar fashion, the Nag Hammadi discovery has allowed us to appreciate the relative accuracy of the Church Fathers' reports on the Gnostic heretics of the early centuries, and the publication of early *kalām* works highlights Ašʿarī's usual reliability.

we know that in the third century some theologians, whose adherence to Islam cannot be seriously doubted, did not accept this doctrine as binding. One can thus legitimately argue that Rāzī's attempts to prove the existence of texts more beautiful than the Qur'ān does not necessarily point to his rejection of Qur'ānic revelation. Is it possible that his declarations reveal the open, flexible nature of Islam in his time, before the doctrines became frozen into dogmas?

This question was also raised by Walker, who argued that in Rāzī's times "the doctrine of prophetic miracles—and even the specifics of the miracle of the Qur'ān—was not yet fixed."[144] Walker argued that "in fact, the ways in which al-Rāzī goes about questioning the foundation of apostolic miracles in general, looks remarkably similar, and probably belongs, to the attitude of a number of Muʿtazilite theologians of the third/ninth and early fourth/tenth centuries."[145]

Although it is certainly true that some of the notions used by Rāzī have an important place in Muʿtazilite vocabulary, the striking differences in their respective use of these notions reflect, I believe, the abyss separating the believer from the apostate.

Foremost among these notions is the intellect (*ʿaql*). Rāzī argued that since the intellect is part of the definition of a human being, all human beings share its possession. Since they are equal in this respect, no human being can claim to be naturally endowed with a superior intellectual quality which will grant him superior knowledge. According to Rāzī, therefore, the notion of prophecy is not only superfluous, but also biologically and logically false.

The Muʿtazilites likewise accepted the claim that human beings are endowed with intellect, which in principle should suffice to instruct them with whatever they need for their salvation. But they realized that the intellect is not the sole governing principle in human life. Rather, its governance is modified by various other traits of human nature: psychology, which distracts the individual from the better advice of the intellect; the individual characteristics of each person, and the conditions of life which obstruct the achievement of human goals. According to the Muʿtazila, these additional traits render the prophets indispensable. Divine goodness, which ordained the gift of intellect to humanity, also ordained the accomplishment of God's

---

144 Walker, "Political Implications," 87.
145 Ibid., 89.

providence by sending apostles.[146] Both the Muʿtazilites and Rāzī thus start from the basic recognition of the sufficiency and universality of the intellect. But whereas the former found a way to justify the additional necessity of prophets, the latter clung to the initial argument. For him, the sufficiency of the human intellect was final and unconditional. Rāzī's observations on human nature, and his appreciation of the effects of human psychology and of historical circumstances, strengthened rather than modified his opposition to the so-called prophets. According to Rāzī, the prophets, driven by their own passion for power, exploited the weaknesses of their fellow-humans, thus increasing humanity's misery.[147]

In addition to proclaiming the divine origin of the intellect and its sufficiency, Rāzī also argued for the equality of human intellects: all human beings were born with equal intellectual capacity. The universality of the intellect and the equality of human beings are often presented (by Rāzī, and therefore by students of his thought) as two aspects of the same doctrine. But in fact, in Rāzī's thought these two ideas are quite distinguishable. He is totally committed to former, but his position on the latter varies.

In the debate, Rāzī's opponent pointed out that, notwithstanding his declared egalitarianism, Rāzī claimed for himself superiority to other men in philosophical wisdom. To this, Rāzī answered by attributing his superiority to practice and motivation rather than to inborn advantage:

> —I [i.e., Abū Ḥātim al-Rāzī] said: ... You yourself claim that you are singled out in possessing the philosophical sciences which you claim to have.[148] You say that other people are deprived of it, and they depend on you. You thus oblige them to learn from you, and to follow your lead!

---

[146] See, for instance, Zamaḫšarī's commentary on Q. 4:165, *Kaššāf*, I, 583. Zamḫšarī's non-Muʿtazilite commentator, ʿAlī b. Muḥammad al-Jurjānī, presents the Muʿtazilite position even more clearly; see his commentary printed in the margin of Zamḫšarī's words; cf. also Jāḥiẓ, *Ḥujaj al-nubuwwa*, pp. 118, 125. It is noteworthy that for Jāḥiẓ, who attempts to prove the necessity of traditions, this verse, which says "after the prophets" indicate indeed the post-prophetic era, and not the lifetime of the prophet. And see Qāsim b. Ibrāhīm, *Rādd ʿalā al-mulḥid*, 51.

[147] This is the classical anti-clerical argument against "priestcraft" presented by seventeenth-century deists; cf. for instance, P. Harrison, *"Religion" and the Religions in the English Enlightenment*, Cambridge 1990, 77–85.

[148] "*Qad ḫuṣṣiṣta bi-hādhihi al-ʿulūm allatī taddaʿīhā min al-falsafa.*" Brion, "Philosophie et révélation," 138, misunderstands this sentence, and translates: "Tu soutiens ... avoir en propre ces sciences qui, selon toi, relèvent de la philosophie."

—He [i.e., Rāzī] said: I was not singled out from all other [human beings] to possess these sciences, as opposed to others. But I have sought these sciences, whereas they have neglected them. They are deprived of them only because they fail to reflect, not because of some [inborn] deficiency. . . . If [any one of them] had applied himself with resolution to the same things to which I have applied myself, if he had sought what I did, then surely he would have achieved the same [level] as I did.[149]

Abū Ḥātim al-Rāzī tried to corner Rāzī and to extract from him a simple, unequivocal declaration of his egalitarian belief:

—I said: Are [all] people equal in their intellect, their resolution, their sagacity, or are they not?

Rāzī's answer seems, at first sight, to be in the affirmative:

—He said: Had they made an effort[150] and busied themselves with what would serve them,[151] they would have been equal in intellect and in resolution.[152]

On one level, Rāzī indeed believed in the ability of all human beings to strive for wisdom and thus to make intellectual progress. He thus says:

No soul can be purged from the turbidity of this world and escape to the next, except by contemplating philosophy. If a person contemplates philosophy and comprehends anything, be it ever so small, his soul is purged from this turbidity and is saved.[153]

---

[149] *Aʿlām al-nubuwwa*, 3–4; On the egalitarian position of the *Barāhima*, see below, Ch. 5, apud n. 45.

[150] *Law ijtahadū*, which also carries the technical meaning of *ijtihād*, and thus can be translated: "Had they used their own judgment. . . ." On the centrality of *ijtihād* in Rāzī's thought, see Brion, "Philosophie et révélation," 135; Pines, "Shiʿite Terms," 256–58; Cf. also "Taḳlīd," *EI*, 630 (J. Schacht), which points out the juxtaposition of *taqlīd* and *ijtihād*.

[151] *Mā yuʿīnuhum*; Brion, "Philosophie et révélation," 138, translates: "ce qui les libérerait."

[152] *Aʿlām al-nubuwwa*, 5.

[153] Ibid., 12. Rāzī expresses his appreciation of the intellect also in the opening chapter of the *al-Ṭibb al-rūḥānī*, where he describes the intellect as God's gift to humanity, the tool by which a human being can achieve utmost perfection and happiness; cf. *Rāsā'il*, 17–19. This laudatory chapter closely resembles Ismāʿīlī texts (as noted already by Kraus, *Rāsā'il*, 19, n. 19) and is best understood if read as Rāzī's response to the Ismāʿīlīs. I am indebted to M. Bar-Asher for drawing my attention to this chapter.

In his philosophical *Apology* (*al-Sīra al-falsafiyya*), however, Rāzī presented exactly the opposite argument. As everyone knows, he argued, human beings are unequal by nature. Some are stronger or more gifted than others. Consequently, even the consideration given to people's life should not be the same for everyone:

> If two men chance to be in a waterless plain, and one of them has water just sufficient to save his life but not his companion's, in such a case it is right that the water should go to *the one who is of greater benefit* to his fellows.[154]

The implication of the fundamental inequality of humans was that the philosopher must not behave like the common people, nor should he share their pleasures and aspirations.

Which one of these two contradictory views did Rāzī believe in, the one he expressed in the debate or the one formulated so crudely in his *Sīra*? It is of course possible that Rāzī had changed his views over the course of the years.[155] If so, and since we do not know the relative dates of these respective works (scil. the *Maḫārīq al-anbiyā'* and the debate that quotes it, and the *Sīra*), we have no way of knowing Rāzī's final position. It seems to me, however, that rather than reflecting a change of position, the difference between these statements should be explained by the different contexts in which they were made and the different audiences to which they were addressed. Whereas the egalitarian position suited Rāzī's polemical purpose, elitism was his innermost conviction. He was capable of moderating this position or camouflaging it to suit the needs of the hour. But his equation of intellectual universality with intellectual equality was only a polemical stance.

As is clear not only from the *Sīra* but also from his *Šukūk 'alā Jalīnūs*, Rāzī was quite conscious of his own intellectual value.[156] This is apparent also in some of the titles of his medical books: "On the Reasons for People's Preference of Inferior Physicians";[157] "On the

---

[154] *Rasā'il.* 104; Arberry, 122 (italics mine). On the Stoic and Jewish sources of this case, see Pines, "Two who Were Walking in the Desert," *Tarbiz* 16 (1945), 238–40 (in Hebrew; reprintd in *Studies in the History of Jewish Philosophy: The Transmission of Texts and Ideas*, Jerusalem 1977, 9–11).

[155] On the evolution of Rāzī's thought, see Pines, "Nouvelles Études," 143–45; Bar-Asher, "L'éthique de Rāzī," esp. 145; and see, on the other hand, Druart, "Ethics," esp. 56.

[156] Cf. also Walzer, *Greek into Arabic*, 15.

[157] Cf. Sezgin, *GAS*, III, 261.

Reason Why Simple People Despise the Expert Physician."[158] These titles offer matter-of-fact observations on human irrational behavior. The composed self-assurance that transpires from them bespeaks Rāzī's confidence that he himself was a superior, expert physician, whether his patients agreed to his self-perception (as many of them did) or rejected it.[159] Rāzī's perception of himself was that of a gentleman, who avoided disagreements even if it meant renouncing his rights.[160] His opponents never accused him of arrogance, as Ibn al-Rāwandī's opponents so often accused him.[161] And yet his elitist attitude is betrayed by his answer to his adversary in the debate, quoted above. What appeared, at first sight, to be an affirmative answer to Abū Ḥātim al-Rāzī's question, is in fact an elusive, ambivalent, and in the final count, negative answer:

> Had they made an effort[162] and busied themselves with what concerns them, they would have been equal in intellect and in resolution.

Or again:

> If the multitudes, who bring perdition upon themselves and fail to inquire, had contemplated philosophy [even] in the most cursory way, it would have meant their salvation from this turbidity, even if they had comprehended only a small fraction of it.[163]

In theory, people *could have been* equal; in theory, we could say that *if* they had possessed both resolution and sagacity, they *would have been* resolute and wise. But they did not possess these, and therefore they are neither resolute nor wise. The hypothetical, conditional formulations disclose Rāzī's awareness that not all people apply themselves assiduously, not all of them make the effort, and therefore, people are not equal.

We can now return to Walker's abovementioned suggestion that Rāzī's rejection of the authority of traditions and his insistence on the role of the intellect were no more heretical than the ideas of the

---

158 IAU, 426:8–9; cf. also *Fihrist*, 300.

159 For it is probably self-assurance which is reflected in these and other titles of Rāzī's books, rather than "irreverent spirit" which "peeps out . . . puckishly," as suggested by Goodman, "Muḥammad ibn Zakariyyā' al-Rāzī," 199.

160 See *al-Sīra al-falsafiyya*, *Rasā'il*, 110.

161 Ibn al-Nadīm describes Rāzī as "generous and dignified"; See *Fihrist*, 299:7–8 (Dodge, 701–702).

162 *Law ijtahadū*; see above, n. 150.

163 *Aʿlām al-nubuwwa*, 12.

Muʿtazila. On closer look, the difference between them becomes apparent. The Muʿtazilites indeed assumed that all men are created equal, endowed with the divine gift of the intellect. The Muʿtazilites also appreciated, however, the extent of human shortcomings. These shortcomings, inherent to human nature, are common to all human beings. There is thus no way, in Muʿtazilite thinking, for humanity to overcome its difficulties without exterior help. Therefore, the Muʿtazilites argue, God, always acting in the most beneficiary way (*al-aṣlaḥ*), sends additional guidance in the form of prophets.

Rāzī, on the other hand, develops the apparently similar presuppositions in a very different direction. According to him, humanity can expect no divine guidance after the initial gift of the intellect. Human beings must therefore rely on their own wits and resourcefulness. It is precisely the fact that, in reality, human beings are *not* equal, and that some humans are more driven and more resolute than others, which ensures the advance of humanity. Rāzī's similarity with the Muʿtazila is thus only superficial: whereas the Muʿtazila remains well within the tradition of revealed religions, Rāzī presents a system of radical (and, we can say, energetic) free thinking.

The fact that upholding the universality of the intellect is not identical to an egalitarian position can be more clearly understood if we compare Rāzī's opinions to those of the Ismāʿīlīs. Like Rāzī and the Muʿtazilites, the Ismāʿīlīs cherished the human intellect. In a sense, the intellect had an even higher place in their system than it had either for Rāzī or for the Muʿtazila. In Ismāʿīlī doctrine the hypostasized intellect was endowed with a role in creation, and rational reflection had a soteriological function. Unlike for Rāzī, however, and in a more pronounced and systematic way than in Muʿtazilī theology, the human intellect was not considered in the Ismāʿīlī system to be self-sufficient. If the human soul was to be liberated from the body and return to its celestial origin, it needed the guidance and presence of cosmic intellect, which was manifested in the prophets and *imāms*.[164] Prophets and *imāms* are individuals who by virtue of divine assistance (*ta'yīd*) are superior to other human beings. On the ladder of nature, they constitute a species in themselves, above human beings, just as humans are a species above mute animals.[165] Against

[164] Cf. De Smet, "Polémique ismaélienne," 94–95, 97.

[165] See, for instance, *Rasā'il iḫwān al-ṣafā'*, IV, 175ff.; and cf. Pines, "Shīʿite Terms," 232–43. On the concept of *ta'yīd* in Ismāʿīlī thought, see Walker, *Early*

this view Rāzī affirmed the universality of human intellect, in the sense that no human is super-human. But his admittance of the unity of the human species did not oblige him to ignore the existence of individual differences within the species. Relevant to Rāzī's attitude is perhaps also the fact that he did not count the intellect among the five primordial beings, a fact which Walker has called "a glaring omission."[166] In Ismāʿīlī Neoplatonism, the hypostatic intellect was an important part of the chain of emanation which lead to the creation of the world.[167] Rāzī, on the contrary, brought the intellect down to earth, as it were. He gave the intellect human dimensions, and insisted that it was the common, though noble, lot of every man.

In addition to the questions discussed above, a major factor contributing to the hesitations as to the heretical nature of Rāzī's view relates to his biography. Whereas Ibn al-Rāwandī is said to have suffered persecution—at least according to some sources, and at least in the Baghdādī part of his life[168]—our sources are unanimous in saying that Rāzī lived and died in peace.[169] This is particularly noteworthy, since among his adversaries were people like Abū al-Qāsim al-Balḫī, Fārābī and Abū Ḥātim al-Rāzī, who, although not leaders of orthodox Islam, were prominent figures in the intellectual and philosophical scene of their days.[170]

It has thus been argued that "Rāzī's evident, open, and public career and the free dispersal of his ideas and writings, at least while he lived, would suggest that the charge against him of Islamic heresy should be partially discounted or at least reexamined."[171] In view of

Philosophical Shiism, 117–18; and see, for instance, Kirmānī, *al-Aqwāl al-dhahabiyya*, quoted by Kraus, *Rasāʾil*, 19.

[166] Walker, "Political Implications," 75.

[167] See, for example, Walker, *Early Philosophical Shiism*, 87–94.

[168] See above, Ch. 2. Ibn ʿAqīl, for his part, considered the fact that Ibn al-Rāwandī was not promptly executed after writing his heretical books as evidence of the weakness of faith of most people; cf. Ibn al-Jawzī, *Muntaẓam*, 110:1–4 (= Ritter, 3–4).

[169] Note Rāzī's own words quoted above (apud n. 7), where he rejects surgery, with the possibility of a cure and the certainty of pain, and opts for the tranquility (*rāḥa*) of the quiet death which he knew awaits him. The fact that Rāzī was not persecuted is noted also, with surprise, by Kraemer, see above, Introduction, n. 13.

[170] None of them can be seen as clandestine, as suggested by Urvoy, *Penseurs libres*, 144. In particular, Ismāʿīlīs in the tenth century can hardly be seen as clandestine; as new material is being published, the extent of the Ismāʿīlī intellectual impact begins to be appreciated.

[171] Walker, "Political Implications," 66.

the conformity and unanimity of the sources scanned here, it is hardly possible to discard the opinions attributed to him as slanderous falsifications. Rather, we must start from the assumption that Rāzī was indeed a freethinker, and that he really did not hide his views (or at least, did not hide them at all times), and face squarely the difficult question of the fact that these views were tolerated.

Without presuming to offer a definitive answer to this intriguing question, we can point to several factors which may have affected the attitude to Rāzī (and, perhaps, to other people in similar positions). Two such factors were mentioned by Salo Baron, who pointed out the lack of militancy on the part of rationalistically-inclined *zindīq*s.[172] Baron also suggested that the fact that they were only a "relatively small number of intellectuals, living in a rarefied atmosphere of detached philosophic speculation, largely accounted for the relative mildness of the official reaction to their preaching."[173] Rāzī's reputation as a physician undoubtedly added to his respectability and safety.[174] To these factors may be added the volatile political (and therefore, religious) conditions in Rāzī's time. Some insight into the effect that living in such a period could have on an intellectual can be gained from the fate of the thirteenth-century Jewish scholar Ibn Kammūna. Ibn Kammūna wrote a remarkably balanced exposé of the three monotheistic religions.[175] Enjoying the tolerance of the Mongol ruler of Baghdād, Ibn Kammūna apparently felt, at the time, safe to write and publish this book. Shortly thereafter, however, an angry mob forced him to flee the city; the ruler was unable to protect him, but was apparently still tolerant enough to help him escape.

Although Rāzī's situation was different, Ibn Kammūna's case may offer some insight as to the possibilities offered to intellectuals by religiously unstable regimes. Rāzī's Ismāʿīlī interlocutor is said to have succeeded in converting to the Ismāʿīliyya some of the local

---

[172] In theory, even non-militant heretics could still be perceived as posing a real threat to the state and to the social order, as suggested by Kraemer, "Heresy," 167. In practice, however, as I argued above (introduction, n. 13), this was not the case with the freethinkers.

[173] S. W. Baron, *A Social and Religious History of the Jews*, Philadelphia 1957², V, 104.

[174] Cf. Walker, "Political Implications," 70; Iskandar, "Rāzī," 371.

[175] *Tanqīḥ al-abḥāth fi'l-milal al-thalāth*, ed. M. Perlmann, Berkeley and Los Angeles 1967, ix–x; English translation in M. Perlmann, *Sa'd ibn Manṣūr ibn Kammūna's The Examination of the Inquiries into the Three Faiths—A Thirteenth Century Essay in Comparative Religion*, Berkeley 1967; cf. also Urvoy, *Penseurs libres*, 202–14.

dignitaries, among them Aḥmad b. ʿAlī and Mardāwīj. One of these two rulers may have been the prince in whose court the debate took place.[176] But we do not know whether the debate with Abū Ḥātim al-Rāzī took place before or after the ruler's conversion. The same holds true concerning the chronology of Rāzī's books. These conversions, however, reveal a shifting political and religious situation, very different from, for instance, the staunch and relatively stable orthodoxy of thirteenth-century Damascus. While this situation lasted, Rāzī could apparently enjoy safety.

This short examination of Rāzī's anti-prophetic teaching suffices, I hope, to establish him as a true freethinker. The most characteristic feature of his thought is his insistence on his unalienable right to think independently: free from religious traditions and authority, but also free from the authority of past masters. His forceful defense of his right, even respectful duty, to criticize Galen, is a manifesto of Free Thinking in the broad sense:

> The arts of Medicine and Philosophy[177] will not support subordination to leaders[178] and [automatic] acceptance of what they say, nor treating them leniently and refraining from thorough examination of their claims. [Moreover], no philosopher will approve of such an attitude from his students and disciples.
>
> ... Were this noble man [scil. Galen] alive and present here now, he would not have rebuked me for composing this book and it would not have caused him to take offence; for he gave priority to the truth [over everything else], and out of his love for thorough research which gets to the heart of problems.
>
> ... As for him who may criticize me and call me an ignoramus for publishing these reservations [I have about Galen's claims] and for discussing them, I do not heed him, and I do not count him as a philosopher. For he has cast the philosophers' way[179] behind his back, and has adopted the multitudes' way of conforming with the leaders (*taqlīd*) and refraining from criticism of them. The philosophers' way has always consisted of veneration of the leaders together with relentless critical examination, with no leniency.[180]

---

[176] Kirmānī identifies him as Mardāwīj (cf. *al-Aqwāl al-dhahābiyya*, 3); cf. also Daiber, "Unity and Diversity," 88; on the chronological problems that this identification poses, see Walker, "Political Aspects," 63–64, n. 7; Poonawala, 38.

[177] Pines, "Rāzī," 481, reads: "The Art of Medicine is a philosophy."

[178] *Al-taslīm li'l-Ruaʾsāʾ*, which in this context is also "subordination to the masters."

[179] *Sunna*, also: tradition.

[180] *Šukūk*, 1–2; See Pines, "Rāzī," esp. 481–82.

Whereas Ibn al-Rāwandī played the role of freethinker for the theologians, Rāzī introduced radical freethinking into the more philosophically-minded circles, both Sunni and Ismāʿīlī. Ibn al-Rāwandī was pushed, in a sense, to the margins of society, and his freethinking could be discredited also on account of his personality. In Rāzī's case such personal discredit was not so easy, and his basic respectability deepened the challenge presented by his freethinking. It is tempting to indulge in speculation as to what would have been the Islamic reaction to freethinking had there been no Rāzī. One suspects that, had Ibn al-Rāwandī remained a lone figure, it could have been possible to throw his teaching to the commodious heresiographical dustbin, where it would have had the same unthreatening status as other eccentric heretics. As it is, by the middle of the tenth century, the Islamic intellectual world, in all its shades, regarded freethinking as a folly which was anything but inoffensive, and which was liable to hit the best minds. From then on, freethinking would become a major concern of religious thinkers, who endeavored to strike preemptively against any danger, real or imagined, of the resurgence of this folly.

CHAPTER FOUR

THE RELIGION OF THE FREETHINKERS

In the Introduction, I have tried to describe the phenomenon which I propose to call "Islamic freethinking." I have defined it as an independent reflection on metaphysics and ethics, leading to the rejection of prophecy and of the authority of revealed religions in general and of Islam in particular.[1] In this context "Freethinking" expresses, in negative terms, what the freethinkers rejected. It also implies that, like other thinkers, the freethinkers put great stock in the power of rationality. As I have tried to show above, medieval thinkers present Ibn al-Rāwandī and Abū Bakr al-Rāzī as the prototypes of radical heresy. In the present chapter I propose to examine their image in the medieval Arabic sources, in order to see if we can say anything specific, in positive terms, about the nature and the content of their religious beliefs.

1. In Search of Atheists

In current usage, the terms "atheist" and "freethinker" are often used as synonyms.[2] But if we are to identify the precise beliefs held by certain thinkers, it is advisable to draw the parameters of each term with some precision. I shall therefore use the term "atheism" in its strict sense: it refers to the doctrine that denies the existence of God.[3]

Although various Muslim theologians in the early ʿAbbāsid period wrote treatises "Against the Unbelievers," most of these works are lost. The earliest extant work bearing this title is probably the *Radd ʿalā al-mulḥid* of the ninth-century Zaydī theologian al-Qāsim b. Ibrāhīm. In this work, al-Qāsim polemicized with an anonymous

[1] See Introduction, apud n. 33; cf. Stroumsa, "Ecritures alternatives," 270–93, esp. 270.

[2] Cf. Fabro, "Atheism," 479–80.

[3] Ibid.; James, "Atheism," 479–80; Müller-Lauter, "Atheismus," 379.

mulḥid, who begins the discussion by asking for a proof for the existence (*aniyya*) of the Creator.[4] In his concentrated effort to prove God's existence, al-Qāsim also wrote two other books, entitled, respectively "The Small Book of Proof" and "The Great Book of Proof." In his introduction to the latter, he specified that this work was meant to teach the reader "what to answer the heretics and unbelievers (*al-zanādiqa wa'l-mulḥidīn*) when they ask for a proof of the existence of God." One would naturally be inclined to infer from these facts that the "heretics and unbelievers" intended by al-Qāsim were people who denied, or at least questioned, the existence of God, and that al-Qāsim was concerned about the persuasive power of such contemporaneous atheists.

An examination of the theological literature written in Arabic from the ninth century onwards would seem to corroborate such a conclusion. A significant part of *kalām* works, written by Muslim, Christian, and Jewish theologians, is dedicated to the attempt to prove that God does exist. In theological *summae* this discussion, presented as the cornerstone of religious thought, is usually placed close to the beginning of the work, and it is almost always done in a polemical tone.[5] Again, one would be inclined to infer that these theologians were arguing against known atheists, just as in the polemical discussions of theodicy we can sometimes identify the writer's real dualist or predestinarian opponents.

Nevertheless, in the discussions of God's existence the actual opponents are not identified as individuals. As a group, they are some-

---

[4] Cf. Qāsim b. Ibrāhīm, *Rädd ʿalā al-mulḥid*, 17–34 (I am indebted to Hans Daiber for referring me to this edition, and for kindly making a copy of it available to me). See also art. "*mulḥid*," *EI*[2], VII, 546 (W. Madelung); idem, *Qāsim ibn Ibrāhīm*, 100–10. In fact, all the *mulḥid*'s questions suggest (rather meekly) the possibility of a world that was not created in time; he thus seems to be presented as one of the *dahriyya* rather than as someone who doubts the existence of God.

[5] This feature is so common in *kalām* works that it does not really require an example. An illuminating illustration of its importance, however, can be found in Maimonides's polemic against the *mutakallimūn*, who deduct God's existence from the created nature of the world. According to Maimonides, since the created nature of the world is not unanimously agreed upon, it is a serious methodological mistake to base on it the proofs of God's existence. In this context, Maimonides says: "I wish to establish in our belief the existence of God, may He be exalted, through a demonstrative method as to which there is no disagreement in any respect. Thus we shall not cause the true opinion, which is of immense importance, to be supported by a foundation that everyone can shake and wish to destroy, while other men think that it has never been constructed." See Maimonides, *Guide*, I, 71, p. 182 (*Dalāla*, 125).

times referred to as heretics, unbelievers, materialists, or skeptics. These designations often appear together, and they do not always seem to be clearly distinguished in the authors' mind. Thus, for example, in his discussion of theodicy Ibn Fūrak (d. 406/1015) quotes his master, the tenth-century Aš'arī, as having said:

> The methods of our opponents in this context (i.e., theodicy) are all the same, whether they are Mu'tazilīs, unbelievers who deny God's attributes (*al-mulḥida min al-mu'aṭṭila*), heretics (*zanādiqa*), dualists, *barāhima*, or any other person who denies the Creator and rejects revealed religions.[6]

And yet we can search these texts in vain for a specific contemporaneous individual accused of denying the existence of God.[7] As noted above, the notion of atheism seems to have been recognized and to have occupied an important place in medieval Arabic theological literature, but the atheists themselves always remain faceless and nameless. When a name does appear, it is always that of a person accused of some specific heretical doctrine which, the theologians say, is *as bad as* atheism or *may lead to* atheism—never of somebody the core of whose heresy is actually identified as atheism.[8] Furthermore, whereas the freethinkers are usually presented as rejecting the notion of prophecy in blunt and defiant terms, both orally and in writing, no similar declarations of atheism are made by the freethinkers, or, for that matter, by anyone else. Whenever atheistic opinions are discussed in dialogues, they are always expressed by the proponents of orthodox beliefs, who attribute them to their (heretical) interlocutors, while the latter hasten to deny the accusation.[9]

---

[6] Ibn Fūrak, *Mujarrad maqālāt al-Aš'arī*, 143.

[7] Ibn Manẓūr (d. 711/1311) offers a typology of *kufr*, in which he includes *kufr al-inkār*. This kind of disbelief consists in total denial of God's existence, where the *kāfir* does not recognize God [in his heart] nor does he acknowledge His existence in words. Ibn Manẓūr is admittedly a very late source, but it is noteworthy that, like in earlier texts, his typology is also purely theoretical; he does not attempt to identify any such infidel within Islam; see *Lisān al-'arab*, 5, 144 (art. *kufr*), quoted by W. Björkman, art. "Kāfir," *EI*², IV, 408.

[8] Another example of this attitude is reflected in the words of the fourteenth-century al-'Aynī (*'Umdat Al-Qārī'*, I, 288), who argues that everyone is aware of God's watchful eye "unless he be a heretic who denies the [existence of] God (*illā an yakūna zindīqan jāḥidan, lā yuqirru bi-rabb*). Al-'Aynī recognizes the (remote) theoretical possibility of atheism, but his personal experience does not seem to include the encounter with any atheist. I am indebted to Vardit Tokatly for this reference.

[9] See, for instance, Abū Ja'far al-Kulīnī, *al-Uṣūl fī al-kāfī*, Beirut 1401 H, I, 74–76 (cited by Chokr, *Zandaqa et zindīqs*, 110) where the Imām Ja'far al-Ṣādiq accuses Ibn Abī al-'Awjā' that his skies are empty.

The question then arises, whether there were any real atheists in the Arabic cultural sphere in the middle ages? And if such atheists existed, who were they and how were they designated?

A thorough study of this topic would require a fresh examination of the various sects and heresies, both real and imagined. Assuming that for medieval Muslim orthodoxy atheism—to the extent that it was at all thinkable—was the utmost denial of religious beliefs, it might be convenient to look for atheists among those described as having rejected all established religions, the freethinkers. As we have seen in the words of al-Aš'arī quoted by Ibn Fūrak,[10] when Muslim texts refer to "those who deny the existence of God," they often mention them in the same breath with those who deny religions (*al-adyān*), or, more specifically, revealed religions (*al-nubuwwāt*). Unlike the denial of God's very existence, however, the rejection of revealed religions is associated by Muslim medieval authors with specific contemporaneous names. It thus seems appropriate, within a study of the nature of freethinking in the medieval Islamic world, to ask whether the freethinkers also qualify as atheists.

In modern scholarship, Ibn al-Rāwāndī is sometimes described as a "freethinker"[11] and sometimes as an "atheist."[12] Muslim sources usually call him *zindīq*. As noted by Bernard Lewis, this last term, which originally denoted dualists,

> later . . . was generalized to cover all holders of unorthodox, unpopular and suspect beliefs, particularly those considered dangerous to the social order and the state.[13] At the same time it applied loosely to materialists, atheists, agnostics and the like, and came to have the general meaning of free-thinker and libertine."[14]

Quite often Ibn al-Rāwandī is also called *mulḥid*,[15] a term almost as vague as *zindīq*, especially when the two appear together (i.e., *al-zindīq al-mulḥid*). In this context, Lewis says:

---

[10] Above, n. 6.

[11] See, for instance, van Ess, "Image," 11: "free thinker [*mulḥid*]"; Fakhry, *Islamic Philosophy*, 96.

[12] For example, Niewöhner, *Veritas*, 238.

[13] Massignon also defines *zandaqa* as a heresy threatening the security of the state; cf. *Aḫbār al-Hallāj* (Eng.), I, 378–83.

[14] Lewis, "Heresy," 54–55. See also van Ess, *TG*, I, 416–18; Chokr, *Zandaqa et zindīqs*; 'Āṭif Šukrī Abū 'Awḍ, *al-Zandaqa wa'l-zanādiqa*, Amman n.d., 111–12.

[15] According to al-Ḫayyāṭ, Ibn al-Rāwandī himself claimed this title, proclaiming: "Know that I am a *mulḥid*." See A'sam, *History*, 23.

> More or less synonymous with *zandaqa* in its later, generalized application is the word *ilḥād*, originally meaning deviation from the path ... In the first few centuries of Islam the *mulḥid*—deviator—is the man who rejects all religion, the atheist, materialist or rationalist, of the type of the notorious Ibn al-Rāwāndī.[16]

Lewis cites Ibn al-Rāwandī as the example of "the man who rejects all religion, the atheist, materialist or rationalist." Are these terms synonymous in the Islamic medieval world? Did Ibn al-Rāwandī hold all these heretical doctrines, did he adopt only some of them, or perhaps none at all?

Our scant, fragmentary and often contradictory information about Ibn al-Rāwāndī is relatively consistent regarding his rejection of prophetic religions.[17] But almost nothing positive is known of his own religious beliefs. We know what traits of the Scriptural God were repulsive to him: in his view, this God was stupid and petty, vindictive and cruel.[18] Does this mean that he believed in the existence of a different God, or that he did not believe in any God? Ibn al-Jawzī says about him:

> We have never heard anyone defame the Creator and make jest about him as much as this cursed one did.... Had He denied the Creator altogether, it would have been better for him than to admit His existence, then to polemicize against Him and defame Him ...[19]

One must take into account the possibility that Ibn al-Jawzī's words here (scil., that Ibn al-Rāwandī admitted God's existence) reflect some use of *ilzām* by Ibn al-Rāwandī. This widely practiced polemical technique consists of using the opponent's premises in order to refute his claims. From an *ilzām* it is impossible to conclude whether or not the polemicist also accepted his opponent's premises. Ibn al-Rāwandī was using the monotheists' own definition of God to show that, according to their own criteria, the God of their Scriptures was not fit for His position. If so, Ibn al-Jawzī's words do not tell us

[16] Lewis, "Heresy," 56; On the term *mulḥid*, see also van Ess, *TG*, I, 418; Madelung (art. "*mulḥid*," *EI*², VII, 546) defines it as "deviator, apostate, heretic, atheist;" cf. also above, Introduction, nn. 16–17.

[17] See above, Ch. 2.

[18] The depiction of God as cruel and vindictive was probably the main topic of Ibn al-Rāwandī's *K. ʿAbath al-ḥikma*, but it also appeared in his *K. al-Dāmigh*, e.g., in his commentary on Q. 50:9; cf. below, Ch. 7, apud n. 50.

[19] *Muntaẓam*, 117:5–7 (= Ritter, *Philologica*, 9); see further below, Ch. 6, apud n. 65.

whether Ibn al-Rāwandī had a better candidate, a God who did not suffer from these weaknesses, or whether he himself did not believe in God at all. Ibn al-Rāwandī's close association with the Manichean Abū 'Īsā al-Warrāq strengthens the plausibility of the first solution, namely that Ibn al-Rāwandī was not an atheist, but only adopted the dualists' criticism of the Scriptural God.[20]

Besides the terms *zindīq* and *mulḥid*, and the accusation of association with the Manicheans, Ibn al-Rāwandī is often called in the sources a *dahrī*.[21] This term, usually translated as "materialist,"[22] designates the adepts of the doctrine of the eternity of the world. More precisely, it designates those who believe that the world as it is has existed from eternity and was not created *a parte ante*, and in particular the Aristotelian philosophers. The designation of Ibn al-Rāwandī as *dahrī* has led H. S. Nyberg to suggest that Ibn al-Rāwandī was an Aristotelian philosopher.[23] It is of course possible that Ibn al-Rāwandī was influenced by the existing translations of the Aristotelian corpus.[24] But considering his theological background and vocabulary, he cannot be regarded as a precursor of what would be later called *falsafa*, the school of Aristotelian philosophy.[25] In the *K. al-Tāj* he is said to have rejected the *mutakallimūn*'s claim that we can deduce the existence of a creator from the contemplation of the material world:

[20] The closeness of the Gnostic myth (incorporated into Manicheism) to atheism is noted in James, "Atheism," 479–80. James seems to present Gnosticism (which denied worship to the Creator God, and referred to the ultimate being as the unknown One, the unfathomable) as a form of atheism. This, however, is clearly a misleading presentation of Gnosticism as well as of Manicheism in the first Christian centuries.

[21] See, for example, *Intiṣār*, 123.

[22] See below, n. 35.

[23] Nyberg, "Deux réprouvés," 133.

[24] Al-Ḥayyāṭ wrote his *K. al-Intiṣār* around 269 H, and *K. al-Tāj* (where Ibn al-Rāwandī attempted to refute the temporal creation of the world) must have been written before that date. The earlist dateable translations of Aristotle's *Organon* stem from the ninth century; see, for instance, P. Kraus, "Zu Ibn al-Muqaffa'," *RSO* XIV (1933/4), 1–20; Syed Nomanul Haq, *Names, Natures and Things: The Alchemist Jābir ibn Ḥayyān and his* Kitāb al-aḥjār (*Book of Stones*), Boston and London 1994, 25–26 and 230ff. There is also ample evidence for substantial philosophical influence on the ninth-century al-Muqammaṣ (cf., for instance, *Twenty Chapters*, 19–20, 44–63), and Ibn al-Rāwandī could have had access to the same kind of philosophical material; compare van Ess, *TG*, IV, 336.

[25] The *falāsifa* would have surely found it difficult to accomodate Ibn al-Rāwandī's rejection of prophetic religions. On the importance of prophecy in the metaphysics and in the political philosophy of the *falāsifa*, see, for instance, Rahman, *Prophecy in*

> He denied the created nature of bodies and rejected it. He claimed that there is no indication in the affect for the existence of its cause,[26] nor in the act for the existence of its agent. The world and everything in it . . . is pre-eternal; it has neither a maker nor a governor, neither an initiator nor a creator.[27]

It is noteworthy that al-Ḫayyāṭ, who summarizes the contents of the book and endeavors to prove its shockingly heretical nature, does not seem to conclude from these words that the book refuted the existence of God. For him, Ibn al-Rāwandī's purpose in writing the *K. al-Tāj* was only—and that was bad enough—"to prove the pre-eternity of bodies and to disprove the proofs propounded for their created nature."[28] The main point of Ibn al-Rāwandī's attack was to disqualify the validity of the *kalām*'s arguments for creation. This can be seen also in the words of ʿAbd al-Jabbār, who reports that Ibn al-Rāwandī disagreed with what is nowadays known at "the *kalām*'s classical argument for creation:" that a body which is intrinsically bound to (*lā yanfakk min*) newly created attributes, such as combination and separation, movement and rest, is necessarily also newly created.[29] Although Mānkdīm, who cites ʿAbd al-Jabbār, does not name the source of the report, it may well derive from the same *K. al-Tāj*.

Another quotation from ʿAbd al-Jabbār, preserved by Ibn Abī al-Ḥadīd, corroborates this view of the book. The *Qāḍī* is quoted as saying:

> No sensible person has ever held the view that totally denies the existence of a creator to the world. Some copyists (*warrāqūn*), however, came together and invented a thesis, in which none of them believed, namely that the world is pre-existent; that the world has no creator whatsoever; and that it had always existed, having no creator nor cause.[30] Then Ibn al-Rāwandī took this thesis and defended it in his book known as *K. al-Tāj*."[31]

Islam; on the possibility that the freethinkers may have had some influence on the *falāsifa* in this respect, see below, Ch. 6, apud nn. 81–85.

[26] *Mu'aththir*; lit: "that which leaves an impression."

[27] *Intiṣār*, 11–12; cf. van Ess, *TG*, IV, 336 and VI, 454–55.

[28] *Intiṣār*, 123.

[29] Mānkdīm, *Šarḥ al-uṣūl al-ḫamsa*, 113:11–13.

[30] *Mu'aththir*; see n. 26 above.

[31] *Lākinna qawman min al-warrāqīn ijtamaʿū wa-waḍaʿū baynahum maqāla lam yadhhab aḥadun ilayhā, wa-hiya anna al-ʿālam qadīm . . . wa-aḫadha Ibn al-Rāwandī hādhihi al-maqāla fa-naṣarahā fī kitābihi*; Ibn Abī al-Ḥadīd, *Šarḥ nahj al-balāgha*, ed. M. Abū al-Faḍl Ibrāhīm, Cairo 1959, III, 239; the distinction in this text between *maqāla* and *kitāb*

Ibn al-Rāwandī argued that the explanation of the phenomena of the world does not require a creator, and that the contemplation of the world does not lead to a belief in such a Creator. For ʿAbd al-Jabbār as well as for al-Ḫayyāṭ it was clear that the weight of Ibn al-Rāwandī's argument was to prove a pre-eternal world. *K. al-Tāj* was meant to prove that God neither created the world in time, nor intervened in it, but not that there was no God of any kind. We thus do not know whether Ibn al-Rāwandī believed in Aristotle's prime Mover, or whether he developed another concept of a transcendent being that can be called God.

Regarding Rāzī's heterodox ideas we stand on firmer ground than in the case of Ibn al-Rāwandī, and the Muslim sources also seem to be more confident in their description of the nature of his heresy.[32] When a Muslim writer chooses to be specific in describing Rāzī as a heretic, he calls him a *mulḥid* rather than a *zindīq*.[33] Rāzī is also never described as a *dahrī*; in fact, he is said to have been a virulent opponent (*šadīd al-inḥirāf*) of Aristotle and of Aristotelian philosophy.[34]

In Rāzī's creation myth, matter was one of the five eternal principles. He was thus considered as one of the *aṣḥāb al-hayūlā*, who are perhaps better suited for the name "materialists" than the Aristotelian *dahriyya*.[35] But on other scores, this apellation does not quite fit Rāzī, since the soul was also one of the five eternal principles. Moreover, for Rāzī the soul was not only a cosmic principle; its human dimension is evident when Rāzī exposes his belief in the Hereafter:

> There is a state[36] that [awaits] us after death, which may be either praiseworthy or blameworthy, according to our comportment while our souls were with our bodies.[37]

---

seems to indicate that, although several "*warrāqūn*" participated in inventing the thesis, only Ibn al-Rāwandī composed the book; cf. van Ess's translation ("Image," 13; *TG*, VI, 455 "and fabricated a treatise in whose thesis etc. . . ."). We should also note that the maliciousness of the accusation, pointed out by van Ess, does not necessarily mean that the passage is devoid of any truth.

32 On Rāzī, see Ch. 3 above.

33 E.g., *Raziana*, 358; *Rasāʾil*, 243.

34 Cf. Ibn Ṣāʿid, *Ṭabaqāt*, 33.

35 See *Rasāʾil*, 220; On the term "materialist" see Pines, "Nouvelles Études," 143.

36 *Ḥāla*.

37 *Al-Sīra al-falsafiyya, Rasāʾil*, 101. On Rāzī's conception of the soul's survival after death, see Druart, "Rāzī's Conception of the Soul," esp. 248–50.

Rāzī's formulation is very vague, and probably intentionally so. At this stage of his philosophical activity, he clearly advocated the un-materialistic position, that the significance of human bodily acts lasts beyond the life of the body.[38] But while he may be understood to be saying that "after death we will be transferred to a place, which is either pleasant or nasty," he does not actually say so. His vague formulation stays close enough to the Muslim conception of after-life, but he avoids the traditional Muslim vocabulary for Paradise and Hell.[39]

Rāzī also counted the Creator as one of the five eternal principles. If the myth serves as a model for his understanding of the world, as suggested by Goodman, then indeed his model assumes the existence of God.[40] God, although eternal, does not have absolute power over the other eternal beings.[41] He is, however, benevolent, omniscient, compassionate and caring.[42] Rāzī believed that his God cared for the created world and intervened in it. His God was thus not the indifferent God of some philosophers, nor the unknowable God of the Buddhists as perceived by the Muslims, the *Sumaniyya*.[43] And, needless to say, the term "atheist" cannot be applied to Rāzī in any significant way. Whether or not we can actually call him a "Theist," (or a Deist) as Goodman does, would depend on how much weight we give to his ruminations about the presence of evil in the world.[44]

To sum up: neither Ibn al-Rāwandī nor Rāzī is described in our sources as an atheist. The same holds true for the enigmatic Īrānšahrī. Our information on the thought of all three thinkers derives from their Muslim opponents. Usually, these sources are suspected of drawing their opponents in dark colors, describing them as more radical

---

[38] On the evolution of Rāzī's attitude to this question, from Galenic Platonism to pure Platonism, see Bar-Asher, "L'éthique de Rāzī," 119–25.

[39] In another context, Rāzī is said to have claimed that the souls of evil people stay in this world; see *Rasā'il*, 177–78, and see above, Ch. 3, n. 130. On metempsychosis (*tanāsuḫ*) in Rāzī's thought, see Ibn Ṣā'id, *Ṭabaqāt*, 33 (quoted in Kraus, *Rasā'il*, 181); and cf. Meier, "Der 'Urknall'," 11.

[40] Cf. Goodman, "Myth," 34.

[41] Cf. *Rāsā'il*, 170 (quoting Ibn Ḥazm; cf. *Fiṣal*, 191–216).

[42] Cf. *Rāsā'il*, 103:13–14. I owe this reference to Meir Bar-Asher.

[43] On the *Samaniyya* or *Sumaniyya*, see van Ess, *TG*, II, 20–22; and see S. Pines, "A Study of the Impact of Indian, mainly Buddhist, Thought on Some Aspects of Kalām Doctrines," *JSAI* 17 (1994), 182–203, esp. 183–85.

[44] See Goodman, "Myth," 34; See above, Introduction, apud n. 30; and below, apud nn. 47–49.

heretics than they actually were. The fact that, when their opponents enumerate their various sins, they do not include atheism in the list, seems to strengthen the supposition that indeed they were not atheists.

In his seminal work on the history of atheism in the western world, Fritz Mauthner has rightly observed that when a religious belief is rejected, the form of the rejection reflects the religion which triggered it, and that thus pagan atheism is not the same as Christian atheism.[45] The particular kind of disbelief generated by Islam was freethinking. The centrality of prophecy in Islam could not but effect thinkers in the Muslim world, and philosophers who chose to remain within the fold had to accommodate prophecy. The *falāsifa*, for example, granted it a central role in their political thought, identifying the prophet with the philosopher-king. Likewise, the thinkers who rebelled against Islam focused their rebellion on the notion of prophecy.

The implication of this conclusion, however, needs to be spelled out. For, if we want to pursue Mauthner's train of thought (scil., that pagan atheism is different from Christian atheism), then we must state clearly that the kind of disbelief generated by Islam was *not* atheism. Moreover, the study of the two most notorious heresiarchs of medieval Islam suggests that, although the abstract notion of atheism existed in medieval Islam, no individual was associated with this notion.

2. Theodicy

Iblīs, the very first freethinker according to Šahrastānī, was also the first to question the wisdom of God's governance of the world. Šahrastānī demonstrates Iblīs's tyranny of reason (*istibdād bi'l-ra'y*) with a series of questions that he puts in Iblīs's mouth: Since God knew in advance what would become of me, he asked, what is the wisdom[46] of His having created me? Since He created me according to His wish and will, why did He (then) command me to know and obey Him? And what is the wisdom of this command (*taklīf*), since He neither benefits from obedience nor is He harmed by disobedience? Since He created me and made me . . . compelled [to

[45] F. Mauthner, *Der Atheismus und seine Geschichte im Abendland*, Stuttgart and Berlin 1920, I, 10.

[46] Arabic: *ḥikma*, with all its ethical connotations.

act in the way I did], ... why did He curse me and drove me out of Eden? Why did He let me tempt Adam and his consort? It would have been better for them and more appropriate to His wisdom to create them having inborn knowledge of the right behavior (*'alā al-fiṭra*), without leading them into temptation. And lastly, why was I allowed to pester humanity? Would it not have been better to have a world free of evil?[47]

One easily recognizes in this list the ethical questions presented to monotheists by dualists, both Manicheans and Zoroastrians.[48] For Šahrastānī, who counts the Zoroastrians as "definitely belonging to the people of revealed religions,"[49] Iblīs was no dualist. His revolt against God's governance was the direct outcome of his rationalistic rejection of religious authority. And yet the fact that in Šahrastānī's presentation of Iblīs dualist elements combine with radical independent rationalism is not incidental: regarding theodicy, the Muslim presentation of freethinkers often blends with anti-Manichean polemics.

The Muslim sources relating the intellectual history of Ibn al-Rāwandī tell us that, while still a Mu'tazilite, he associated with heretics and infidels. He was especially close to the Manichean Abū 'Īsā al-Warrāq, to the point of being considered al-Warrāq's disciple in heresy, and it seems that al-Warrāq's influence can be detected also in the realm of his criticism of God's justice. Like almost everything in Ibn al-Rāwandī's biography, the picture here is contradictory, and at least one source suggests the opposite: Ibn al-Rāwandī's Mu'tazilī opponent al-Ḫayyāṭ says explicitly that Ibn al-Rāwandī "gives thanks to God for his goodness, charity and benevolence." But of course this statement, if true, may reflect only one stage in Ibn al-Rāwandī's turbulent career. Most of our sources stress Ibn al-Rāwandī's attack on God's justice. We learn from the same al-Ḫayyāṭ that Ibn al-Rāwandī wrote a book on the oppressive and arbitrary nature of God's acts in this world, entitled *K. al-Ta'dīl wa'l-tajwīr* or *K. 'Abath al-ḥikma.* In this book

> [Ibn al-Rāwandī] claimed that he who inflicts illness and sickness upon his slaves cannot be counted as one who treats them wisely, nor can he be said to be looking after them or to be compassionate towards

[47] *Milal*, 5–7. Šahrastānī adds the intriguing remark that these questions are listed in a Commentary on the *Injīl* and in the Torah.

[48] See Stroumsa, "Manichaeism," 47–55.

[49] *Milal*, 2.

> them. The same is true concerning he who inflicts upon them poverty and misery.
>
> Also unwise is he who demands obedience from a person whom he knows will disobey him. And He who punishes the infidel and disobedient in eternal fire is a fool (*safīh*); He is not wise, nor does He know the [appropriate] measures of punishment for sins.[50]

Muslim writers saw this book as fostering dualism (*fī taqwiyat al-qawl fi'l-ithnayn*). Indeed, it is probably true that this book—now lost—reflected the influence of the Manichean al-Warrāq and his preoccupation with the presence of evil in this world. It is noteworthy that every question in Šahrastānī's list of Iblīs's impertinent questions starts with the words: *wa-mā al-ḥikma. . . .* One is tempted to consider the possibility that the origin of this list was Ibn al-Rāwandī's *K. ʿAbath al-ḥikma.*

To the extent that we can capture the historical Rāzī and the historical Ibn al-Rāwandī, concerning theodicy the difference between them is striking. In the list of Rāzī's books preserved by Ibn al-Nadīm, one entry presents a book arguing "that human beings have a wise Creator."[51] In his *K. al-Sīra al-falsafiyya* Rāzī stated:

> We have established the principle that our Lord and King is concerned about us; He looks after us and is compassionate towards us. It follows also that it would displease Him that we should suffer any pain. [It also follows] that, whatever [harm] befalls us which is not by our own doing and choice, but rather by a natural cause, must be for a necessary, inevitable cause.[52]

According to Rāzī, God inspired his creatures from birth with the essential knowledge they need and can develop when necessary. This inspiration (*ilhām)* is God's *modus operandi* for all created beings, from ducks and geese to humans,[53] and there is no need for additional instruction through prophets. According to Rāzī's Ismāʿīlī opponent, Abū Ḥātim al-Rāzī,

---

[50] *Intiṣār*, 12. Note that the sentences are phrased as general rules; only the last sentence discusses an act which can be solely God's (punishing for disbelief in eternal fire). The intention, however, must have been clear throughout. See above, Ch. 2, n. 30.

[51] *Fihrist*, 299.

[52] *Rasāʾil*, 103; cf. also Ibid., 108.

[53] In his catalogue of Rāzī's books (*Risāla*, pp. 21–23, 28) Bīrūnī discusses the origin of the sciences and mentions two attitudes to this question: there are those who hold that the sciences are revealed, each science by a different prophet, and those who hold that all acquired sciences are, in fact, natural to man just as ani-

> Rāzī claimed that it is inconceivable that God, who is mindful of His creatures, and can impose on His servants an easy, available thing, will disregard it and choose to impose on them the more strenuous thing. What Rāzī meant to say is that God did not impose upon them the obedience to prophets and messengers, this being "the more strenuous thing." Rather, He inspired them with what they need, so that they can obtain it by nature.[54]

In other words, it is precisely Rāzī's perception of God as wise and compassionate that justifies his rejection of prophecy. Revealed religions cause fanaticism and thus encourage strife and wars. The good, merciful God must have had a better idea for the salvation of humankind.[55]

Rāzī's belief in a merciful God should not, however, mislead us into thinking that he entertained a cheerful image of the world. Maimonides reports that Rāzī's metaphysics (*ilāhiyyāt*) included a discussion of the predominance of evil in this world, a discussion which, he said, led Rāzī to the conclusion that man's existence on earth is but a painful affliction.[56] Moreover, according to Maimonides, this discussion was part of Rāzī's attack on monotheistic theodicy. Maimonides reported that Rāzī blamed the ubiquitous presence of evil in our world on the malevolent character of the creator of the world, who is not the Pure Good (*al-ḫayr al-maḥḍ*) but a spiteful tyrant.

---

mals have natural, inborn knowledge. Unlike animals, however, man possesses these sciences only *in potentia*, and he acquires them gradually during his lifetime as well as by profiting from the accumulated knowledge of previous generations. According to Bīrūnī, the rationale behind this opinion is the following: "Since the Creator is compassionate and generous towards man, once a person exists [the Creator] will not withhold the benefits [that a man can acquire], if indeed He does not prepare it for him in advance, as He prepared for him his abode on earth, the plants to feed him and the animals to serve him." In all likelihood, Bīrūnī is well aware that this opinion, which he presents at some length, is also Rāzī's; Compare, for example, *Raziana*, 368:24–26, 372:20–27.

[54] *Raziana*, 362:26–33, and see also 367. In a polemical (and not very original) pun, Rāzī's Ismāʿīlī opponent turns the *ilhām* (inspiration) into *ihmāl* (negligence), and says that according to Rāzī, human beings are left to stray, like unattended cattle. By this he means to imply that the notion of "inspiration" without further guidance is not compatible with God's compassionate nature. Abū Ḥātim's polemical claim, however, does not do justice to Rāzī, who, as we have seen above, presents his God as caring and compassionate.

[55] For the religions as source of strifes and wars, see *Raziana*, 367–68.

[56] On Rāzī's "ascetic hedonism," see Goodman, "The Epicurean Ethic," esp. 21. Goodman regards Rāzī's asceticism as stemming from a certain conformism on his part concerning the religious and philosophical conventions of his time. Consequently, according to Goodman, "If he is a freethinker, the term must be applied in a relative sense, and even then with care."

It seems that in the *K. al-ʿIlm al-ilāhī* Rāzī dealt indeed with the presence of evil in the world.[57] It is also undoubtedly true that Rāzī did not hold the material world in great esteem, as is clear from the various extant fragments of his lost books, including the *K. al-Ladhdha*, as well as from his extant *K. al-Ṭibb al-rūḥānī*.[58] But it does not seem probable that his pessimistic attitude was his major concern in his book on metaphysics. And, considering Rāzī's creation myth, it is improbable that he regarded the creator God as anything less than the Pure Good. We have no reason to think that Maimonides intentionally imputed to Rāzī a view that had no basis in Maimonides's sources. It seems more plausible that, for some reason, his view of this particular aspect of Rāzī's thought was distorted. One could legitimately ask why Maimonides emphasizes this particular theme in Rāzī's book.[59] One possible answer would be that Maimonides had decided to focus on the one component of Rāzī's doctrine which he found particularly revolting. But the description of Rāzī's creator as evil is obviously wrong, and it calls for another explanation.[60] The evil creator attributed by Maimonides to Rāzī is very reminiscent of the Manichean King of Darkness. One possibility that must be considered is that, like the Muslims Ibn Ḥazm and Ibn Taymiyya, Maimonides (intentionally or unintentionally) blurred the difference between one freethinker and another. He treated Rāzī as if he were Ibn al-Rāwandī, reading the blasphemous sayings of Ibn al-Rāwandī's K. *ʿAbath al-ḥikma* into Rāzī's *K. al-ʿIlm al-ilāhī*.[61] Maimonides' conclusions may, in fact, reflect his perception of what an apostate of this sort would say, a perception which could have been influenced by Maimonides's familiarity with other freethinkers, perhaps even with Ibn al-Rāwandī.[62]

---

[57] See above, Ch. 3, apud nn. 130–31.

[58] Rāzī wrote a book: "On the Reasons for the Creation of Predators" (*Fī sabab ḫalq al-sibaʿ*; Al-Bīrūnī, *Risāla*, 17), or "Predators and Reptiles" (*al-sibaʿ wa'l-hawāmm; Fihrist*, 301:7–8). See also Qazwīnī' report, that according to Rāzī, as long as the soul remained in the material world, it was doomed to suffer (*Rasā'il*, 206).

[59] Cf. Walker, "Political Implications," 67–68; see also Kraus, *Rasā'il*, 165–90; Mohaghegh, *Filsūf-i Rayy*, 273–76.

[60] Goodman offers an interpretation of Rāzī's myth of the fall of the soul, which accomodates a concern for theodicy in general and Maimonides' description in particular; cf. "Myth," esp. 27–28, 32–33.

[61] Cf. Maimonides, *Guide*, III, 12, p. 441, and compare, for example, Rāzī's words in his *K. al-Sīra al-falsafiyya, Rasā'il*, 103:9–18, 108:4–5. See Kraus, *Rasā'il*, 165–69, 179–80.

[62] On Maimonides's familiarity with the ideas of the freethinkers of Islam, see below, Ch. 8.

3. Cult

Both Ibn al-Rāwandī and Rāzī mock Muslim rituals, which they compare to those of pagan Indians.[63] But this in itself does not mean that they rejected the very notion of cult. The figure who influenced Rāzī and directed him to freethinking was probably Abū al-ʿAbbās al-Īrānšahrī. We have seen that according to Bīrūnī, Īrānšahrī had invented his own religion."[64] Bīrūnī's statement, if correct, implies that Īrānšahrī did not dispense with the idea of religion altogether, but rather that he proposed to replace existing religions with a new one. Īrānšahrī thus fits the stereotype of the philosopher as described by the twelfth-century Jewish thinker Judah Halevi. This model philosopher admitted the usefulness of religion, both as a way of expressing devotion and as a means of maintaining law and order. But this model philosopher did not believe in a personal God revealing Himself in history through His prophets. He therefore suggested to the king of the Khazars that he should invent his own religion.[65] Īrānšahrī's invented religion could well have included some cultic elements. Of the cult of the freethinkers we know nothing. We do not know if they had any form of worship, either ritual or contemplative. We can, however, make some guesses.

It seems that in their attempt to distance themselves from their own monotheistic, scriptural religion, the freethinkers of medieval Islam were attracted by other religious models. Ibn al-Rāwandī attributes his rationalistic argument against the prophetic religions to the Indian *Barāhima*.[66] Bīrūnī records Īrānšahrī's interest in Indian religions, and Rāzī is said to have drawn on the Sabeans.[67] One of Rāzī's main sources on the Sabeans of Ḥarrān, Saraḫsī, has left us detailed and gruesome descriptions of Sabean star worship, which included human sacrifice.[68] But of course, the paradoxical fascination of the spokesmen of rationality with the crudest cults does not allow us to say that, in their rejection of monotheist religions, the freethinkers actually adopted the religion of the pagan philosophers or what they believed it to be. The freethinkers' claim that the ritual

---

[63] Cf. Rāzī, *Rasāʾil*, 105:15–106:3.

[64] See above, Introduction, n. 43.

[65] *Kuzari*, I, 1, p. 5.

[66] Cf. Kraus, "Ketzergeschichte," 341–58; Niewöhner, *Veritas*, 238–40; Stroumsa, "Barāhima."

[67] Cf. Bīrūnī, *Hind*, 206; and see below, Ch. 5, apud n. 82ff.

[68] Cf. Rosenthal, *Sarakhsi*; and see below, Ch. 5, apud nn. 89–96.

of revealed religions is as primitive as Indian cults reveals their basic scorn towards the latter as well as the former. But the repudiation of existing rituals did not necessarily amount to the complete rejection of any religious ceremony.

Whether invented or real, these non-monotheistic religions seem to have fascinated the freethinkers. The (academic) analysis of pagan or Indian religions may have offered them the necessary distance from the religion into which they had been born. The claim that all religions were, in fact, equal (*takāfu' al-adyān*) and that consequently no religion possessed the ultimate truth, played an important role in freethinking.[69] The "counter-religions" were necessary for the freethinkers in order to enable them to leave their own original religion behind them. But there is no indication that their fascination with the non-monotheistic religions led the freethinkers to adopt them.

The preoccupation of Muslim freethinkers with prophecy seems to reflect their own religious background, and could be regarded, paradoxically, as a specifically Muslim contribution to freethinking.[70] It is possible that, when turning to refute the validity of the scriptures and to ridicule the authoritative structure of revelatory religion, the freethinkers themselves found it difficult to remain without some scriptural past to rely on. In a culture where every decision had to be supported by a quotation from the scriptures or by some saying of the prophet, it must have been hard, even for a freethinker, to have no authority to quote. This may account for the fact that, when Rāzī and Ibn al-Rāwandī came across a cultural tradition in which prophecy played no part, they were glad to refer to it and use it in their arguments. This pagan tradition buttressed the philosopher's claim that a society can exist and function without revelation; and it served to demonstrate that, to the extent that every society requires a religion, i.e., both beliefs and rituals, these rituals could be fashioned by human beings as were those of the Sabeans.

---

[69] For a masterly study of this topic, see Niewöhner, *Veritas*, esp. 238–50.

[70] This, despite the line of heretical tradition "from Celsus to al-Rāzī," which is described by Kraemer, "Heresy," 175. Ibn al-Rāwandī's attack on Muḥammad's miracles presupposes a well-established tradition of such miracles, as noted by Kraus, "Ketzergeschichte," 124. See also the pertinent observation of L. Fèbvre, *Le problème de l'incroyance au XVIe siècle: la religion de Rabelais*, Paris 1962, 1, who points out that dis-belief is the other face of belief.

4. Alternative Scriptures

The idea that even freethinkers cannot do without some kind of proof-text is actually not new. In the course of medieval disputations, monotheist spokesmen often accused the freethinkers of practicing *taqlīd*, i.e., conformism and uncritical acceptance of the philosophical traditions,[71] an accusation which was of course denied by the philosophers. Yet although the philosophers did not treat their sources with the same reverence that was given to holy scriptures, it seems that their references to foreign wisdom were done with all the seriousness of treating authoritative texts.[72]

Beside the accusation of philosophical *taqlīd*, monotheist authors sometimes accuse the heretics of writing parodies on the holy scriptures, either in order to compete with the authentic revelation—a particularly significant act in view of the doctrine of the inimitability of the Qur'ān—or simply to ridicule it.[73] Thus al-Qāsim ibn Ibrāhim insinuated that the dualist Ibn al-Muqaffa' had intended a parody of the Qur'ān, when he began his book against Islam with the words: "In the name of the merciful and compassionate Light." 'Abd al-Ḥamīd al-Lāḥiqī is claimed to have said "praise be to Mani" instead of to God.[74] Similarly, the Jewish historiographer Ibn Dāwūd says that the ninth-century Jewish heretic Ḥīwī al-Balḫī "invented his own Torah"—an accusation which, if not totally unfounded, is probably inaccurate.[75]

The notion of verses which attempt to rival the Qur'ān appears already in the *sīra* literature, and it is a standard accusation against heretics of all sorts. Its ubiquitous presence raises serious doubts

[71] Cf. Šahrastānī, *Milal*, 138. Šahrastānī identifies this argument as an *ilzām*, which means that he does not really take it seriously; it is only a convenient way to expose the flaws in his opponent's claim.

[72] Kraemer, "Heresy," 174, notes the Islamic "pagan reaction," but not the particular role of "scripture-substitute" which this reaction assumed in Islamic milieu. Similarly, the Neoplatonist critics of Christianity, feeling the need to have scriptures of their own, found them in the Chaldean Oracles; cf. also below, Ch. 5, and esp. n. 12.

[73] Thus Ibn Abī al-'Awjā', who fabricated traditions and attributed them to the prophet (Mohaghegh, *Bist Guftar*, 192), is also accussed of having attempted a *mu'āraḍat al-qur'ān*; see Abū Ḥayyān al-Tawḥīdī, *Aḫlāq al-wazīrayn*, ed. M. al-Ṭanjī, Damascus 1965, 183.

[74] *Bist Guftar*, 193.

[75] On Ḥīwī, see below, Ch. 8.

about its historicity in each of the cases. Nevertheless it is interesting to note that even in the case of the "freethinkers" who reject the notion of scriptures, Muslim writers do not deem it sufficient to record their criticism of the Qur'ān, but go further and attribute to them compositions of *muʿāraḍat al-Qur'ān*. Although probably fictitious, the following anecdote is revealing:

> Ibn al-Rāwandī and Abū ʿAlī al-Jubbā'ī met one day on the bridge of Baghdād. The former said: "O Abū ʿAlī, wouldn't you like to hear some of my parody[76] and refutation of the Qur'ān?" Abū ʿAlī answered: "I know only too well the shameful fruits of your erudition, as well as those of your colleagues who propound the eternity of the world.[77] But I will let you be your own judge: Have you found in your parody of the Qur'ān the same good taste and suavity, the same good balance and harmony, the same good order and sweetness [as in the Qur'ān]?" Ibn al-Rāwandī said: "No, by God, I have not." Abū ʿAlī then said: "That satisfies me, go wherever you want."[78]

In this anecdote Ibn al-Rāwandī is accused of making a "classic" *muʿāraḍa*, that is: an attempt to imitate the Qur'ānic style, its "harmony and sweetness."[79] Ibn al-Jawzī puts forward an image of a "family tradition" of heretics, saying that Ibn al-Rāwandī's father, who was a Jew, had attempted to "corrupt the Torah," and that the Jews had warned the Muslims that his son would be likely to do the same to the Qur'ān.[80]

The authenticity of all these stories is highly doubtful. We have

---

[76] *Muʿāraḍa*, like the following word *naqḍ*, can also mean refutation. The context, and Abū ʿAlī's answer, indicate the meaning in this case.

[77] "*Ahl dahrika*," a clear reference to Ibn al-Rāwandī's association with the *dahriyya*.

[78] *Maʿāhid*, 157–58. Cf. Gwynne, *The "Tafsīr" of Abū ʿAlī*, 20. A similar confession is attributed in several sources to Ibn al-Muqaffaʿ. This, if anything, renders the apocryphal character of both stories even more apparent; see Chokr, *Zandaqa et zindīqs*, 167; van Ess, *TG*, II, 33–36. As noted by van Ess, the fragments that possibly belong to this last *muʿāraḍa* do not attack the Qur'ān or Islam, nor do they attempt to deceive the Muslims and smuggle pagan ideas into Islam; cf. J. van Ess, "Some fragments of the *Muʿāraḍat al-Qur'ān* attributed to Ibn al-Muqaffaʿ," in W. al-Qadi, *Studia Arabica et Islamica; Festschrift Iḥsān ʿAbbās*, Beirut 1981, 152–63, esp. 160.

[79] Another kind of *muʿāraḍa* which is sometimes attributed to the freethinkers involves, besides the rivalry of style, heretical content: verses which defy the prophetic message, such as are attributed to Maʿarrī. The title of Maʿarrī's book, *K. al-Fuṣūl wa'l-ghāyāt fī tamjīd Allāh wa'l-mawāʿiẓ* was cited by his adversaries as *al-Fuṣūl wa'l-ghāyāt bi-muḥādhāt al-suwar wa'l-āyāt*; see Abū al-ʿAlā' al-Maʿarrī, *Zajr al-nābih*, ed. Amjad al-Tarābulsī, Damascus 1965, 4–6, 15ff.; Margoliouth, *The Letters of Abū al-ʿAlā' al-Maʿarrī*, Oxford 1898, 129–32; art. "al-Maʿarrī," *EI*², V, 932 (P. Smoor); Stroumsa, "Ecritures alternatives," 286.

[80] *Muntaẓam*, 108:4–6; see below, Ch. 7, n. 22.

no evidence that such "alternative scriptures" were ever actually written, and the stories probably reflect the anxiety of the Muslim authors more than the actual activity of the freethinkers. Nevertheless, it is not impossible that outspoken freethinkers felt at times the urge to ridicule the idea of revelation by imitating canonical books. It seems that at least in one case Rāzī indulged in such sarcastic imitation.

One of Rāzī's short treatises, the *Epistle on the Marks of Political Ascendency* (*Risāla fī amārāt al-iqdām wa'l-dawla*), is usually considered to be one of his less virulent and less original writings. It offers a list of characteristics or "marks" by which to recognize a successful political ruler. Indeed, Kraus has suggested that it was actually written for a ruler or a prince, while Walker has dismissed it as uninstructive.[81] Yet if we bear in mind that Rāzī regarded prophets as charlatans, whose real action remained within the realm of political power, this small treatise may have had a purpose very different from that of serving the establishment.

The treatise, like its title, follows the same pattern as the *dalā'il al-nubuwwa* literature, and some of the "signs" which it lists, such as moral qualities of the ruler and the unconditional, disinterested loyalty of his followers, are to be found also in the *dalā'il al-nubuwwa* writings. Another sign, namely that the ruler is "preserved from committing sins," is reminiscent of the concept of the infallability (*'iṣma*) of the prophets. But in Rāzī's treatise, these signs serve to prove not their bearer's divine mission but rather his ability to wield worldly power.[82] Al-Mu'ayyad fī'l-Dīn, in his refutation of Ibn al-Rāwandī's *K. al-Zumurrud*, says indignantly that the ability of prophets to organize and control worldly affairs (although only a by-product of prophecy) should have brought the heretics to treat them with some respect; clearly, Rāzī does not share this view. He stresses the ruler's greed for power, and he adds a touch of sarcasm when he mentions that some rulers take it to be a felicitous sign if their cook managed to guess their whims and prepare the food they were craving for. The sarcasm may be going deeper than in this particular "sign," for it is possible to read the whole treatise as a parody on the "signs of

---

[81] Kraus, *Rasā'il*, 135; Walker, "Political Implications," 78, n. 35.

[82] For *amārāt* in the context of the signs of the veracity of prophecy, see, for example, Ṭabarī, *al-Dīn wa'l-dawla*, 13.

The political context of this epistle was noted by Daiber, "Political Philosophy," 846. Daiber regards the treatise as a serious attempt to present a political theory which can be harmonized with Rāzī's ethics.

prophecy" literature. Moreover, the Ismāʿīlī overtones of the epistle's vocabulary suggest that the treatise was intended in particular as a response to the Ismaʿīlī authoritarian soteriology.

5. Skepticism

In theological literature written in Arabic the theoretical discussion of atheism is often preceded or followed by an epistemological discussion aiming at the refutation of skepticism in its various forms. The tenth-century Jewish theologian Saʿadya translates Psalm 14:1 ("The fool says in his heart: 'There is no God'") as: "The ignorant said to himself: God has no commanding power."[83] The atheist of the Bible is thus understood by Saʿadya as a practical atheist.[84] His atheism is connected to the fact that he is "ignorant" (*jāhil*). This is not involuntary ignorance, but the willful ignorance of the agnostic (*tajāhul*), who rejects both sensory evidence and axiomatic knowledge, or refuses to draw the obvious conclusions from them.[85] Saʿadya's fool is ignorant because he refuses to see in the world the marks of God's governance. The close connection between this brand of atheism and skepticism is made clear in Saʿadya's theological *Summa*, where he criticizes the attitude of people "who think that if they do not believe in God being a Lord, they can get rid of His command and prohibition, His promise and His threats." This attitude Saʿadya describes as "the ultimate ignorance."[86]

Some modern scholars have described Ibn al-Rāwandī as a skeptic.[87] In Arabic texts he is indeed often branded an ignoramus (*jāhil*).[88] This may be, of course, only a term of opprobrium, and it is indeed often accompanied by similar curses, such as "fool" (*safīh*), "wicked" (*ḫabīth*), "impudent" (*mājin*),[89] or "cursed" (*laʿīn*).[90] But *jāhil* could also be understood as meaning "a skeptic."[91] The Muslim sources strive

[83] Ed. J. Qāfiḥ, New York 1966, 72: *Qāla al-jāhil fī nafsihi: Laysa li-'llāhi amran* (sic).
[84] Cf. Fabro, "Atheism:" "*Practical atheism*: living as if God did not exist."
[85] On the various kinds of skeptics mentioned by Saʿadya, cf. H. A. Wolfson, *Repercussions of the Kalam in Jewish Philosophy*, Cambridge, Mass. 1979, 150–62.
[86] Saʿadya, *Amānāt*, 13.
[87] For example, Fakhry, *Islamic Philosophy*, 96.
[88] See for example Aʿsam, *History*, 79, fragment 1 (quoting Tawḥīdī, *Imtāʿ*, II, 20).
[89] Aʿsam, *History*, 34, fragments 29–30.
[90] Ibid., 52, fragment 13 (quoting Ašʿarī, *Maqālāt*, II, 222).
[91] Cf. for example ibid., 24, fragment 10 (quoting al-Ḫayyāṭ).

to depict Ibn al-Rāwandī's attacks on all religions as an exercise in spite or as a mercenary endeavor, rather than as stemming from some deep conviction. He is said to have written refutations of his own works,[92] a fact which, if true, would again strengthen the supposition that his thought included some skeptic components. ʿAbd al-Jabbār says about him that "he became an apostate and an unbeliever and he denied that he had the knowledge of God's existence."[93] This saying of the *Qāḍī*, however, also refers to the Manichean al-Warrāq. Moreover, as al-Aʿsam rightly points out, the ability of human beings to know God was a disputed question among many Muʿtazilīs, and not specifically among those of weak faith. Therefore ʿAbd al-Jabbār's testimony cannot be taken as a proof of Ibn al-Rāwandī's agnosticism, still less of his atheism. To my mind, the sum of our information concerning him does not allow us to state that his heterodox teachings included the repudiation of "the very possibility . . . of a satisfactory rational answer to the question of God's existence."[94]

Rāzī's case is clearer: his opponents never accuse him of being a skeptic. Moreover, in his *Dubitationes contra Galenos* Rāzī severely criticized Galen for his agnostic attitude regarding the created nature of the world, and it is patently clear that he did not adopt an agnostic or a skeptical attitude on this question.[95] In his view, the world was created, and had a Creator.

To sum up, on several questions of major significance, such as theodicy and the creation of the world, the gap between the positions of Ibn al-Rāwandī and Rāzī cannot be bridged. The only aspect of religious thought on which they agreed was their total and consistent rejection of the concept of prophecy and of clerical authority. In the world of medieval Islam, however, this aspect was of cardinal importance. Consequently, despite the religious differences between the two thinkers, their Muslim opponents often regarded them as belonging to the same religious group. Therefore, although we cannot speak of *the* religion of the freethinkers of Islam, their attitude to revealed religions allows us to underline the existence of a common phenomenon.

---

[92] Ibid., 88 (quoting Ibn al-Nadīm); and see above, Ch. 2, n. 146.
[93] Quoted in Mānkdīm, *Šarḥ al-uṣūl al-ḫamsa*, 54:15–17: "*irtadda wa-kafara wa-nafā ʿan nafsihi al-ʿilma bi-'llāhi.*" (cf. Aʿsam, *History*, 93).
[94] *Pace* Fakhry, *Islamic Philosophy*, 96.
[95] Cf. Rāzī, *Šukūk*, 7:17ff.

CHAPTER FIVE

THE PAGAN LEGACY OF THE FREETHINKERS OF ISLAM

1. The Hellenic Tradition

The attack on the monotheistic prophetic tradition and its scriptures did not begin with the freethinkers of Islam. Already in the 1930s Paul Kraus and Shlomo Pines suggested that Rāzī had taken up, to some extent, the arguments of contemporary Manicheans against positive religions. Above all, they claimed, he had been inspired by late antique criticism of religion.[1] In an interesting article published some years ago, Joel Kraemer discusses the political dimensions of the religious criticism propounded by heretics like Ibn al-Rāwandī and Rāzī. Kraemer regards this kind of criticism as essentially a revolt against Muslim domination.[2] He discusses their criticism of scriptures and of revelation, and demonstrates its similarities to the battle fought against monotheistic religions by both Gnostics and Manicheans, and that fought against Christianity by the Hellenic philosophers of late antiquity. According to Kraemer, Ibn al-Rāwandī and Rāzī represent a direct continuation of what Pierre de Labriolle has called: "the pagan reaction" against Christianity in late antiquity.[3] As demonstrated by Kraemer, one can follow the arguments employed by the critics of monotheism "from Celsus to Rāzī."[4]

The two arguments presented by Kraemer (viz., the political character of this heresy and its relation to the pagan reaction) are connected, and therefore should be discussed together. In fact, neither Rāzī nor Ibn al-Rāwandī seem to have been involved in any way in the countless political fights storming around them.[5] Kraemer

---

[1] *EI*, 1136; and see above, Ch. 3, n. 93. On the transmission of Manichean and Hellenistic arguments to the Islamic world, see also Lazarus-Yafeh, *Intertwined Worlds*, 132–33.

[2] Kraemer, "Heresy," 172; see also van Ess, "Image," 8; Aʿsam, *History*, 7.

[3] P. de Labriolle, *La réaction païenne: étude sur la polémique anti-chrétienne du Ier au IVe siècle*, Paris 1934.

[4] Kraemer, "Heresy," 174–75.

[5] Cf. Rāzī's declaration in his *Apologia*, that his association with the sultan had

acknowledges this fact,[6] but does not attempt to reconcile it with his thesis that the two heretics represented a threat to both the state and the social order. In fact, one must insist on the surprising lack of any political dimension to the stand of the freethinkers. If the Caliph searched for Ibn al-Rāwandī and his colleague al-Warrāq, this was not because they were suspected of plotting to topple his rule, but rather because they were perceived by Muslim orthodoxy (which, in this context, included the Muʿtazila) as a religious threat, together with so many other heretics and dualists at the time.[7] Although it is true that in medieval Islam—as, for that matter, in medieval Europe and in Byzantium—"church and state are inseparatedly linked to one another,"[8] the attack of the freethinkers touched on all aspects of Muslim thought. It would thus be misleading to limit our analysis of their thought to its rather limited political implications.

Moreover, the lack of political interest which we can discern in the thought of Ibn al-Rāwandī and Rāzī highlights the marked difference between them and between a Celsus or a Julian, even when they use similar arguments. The pagan Hellenic philosophers set out to defend their ancient cultural tradition, which they perceived as the very foundations of the political system of their society. In this sense, in challenging the traditional values of their own society, the Islamic heretics resemble the Christian intellectuals of late antiquity more than they resemble their pagan adversaries.[9]

The freethinkers of Islam differ from the philosophers of late antiquity also in their relationship to their sources. Hellenic philosophers like Celsus or Porphyry, who criticized the Bible, drew the gist of their anti-Christian ethos from their own cultural heritage. Their rejection of the Bible and of Christian theology was anchored in their perception of the nature of a true religion, the model for which was the religion of their ancestors.[10] For a Muslim freethinker of the

---

no political or military aspects, but remained strictly professional and personal: he served him only as a physician and as a companion; cf. *Rasāʾil*, 109.

[6] Kraemer, "Heresy," 173.

[7] Cf. Ibn al-Jawzī, *Muntaẓam*, 112:7–9.

[8] Lewis, "Heresy," 62.

[9] Cf. de Labriolle's description of this Christian challenge: "Dévouement sans limites à l'Etat, goût des 'honneurs', respect du culte établi, quelle qu'en fût la vérité intrinsèque . . . il n'est aucune de ces valeurs constitutives de la civilisation antique que ne compromît l'esprit chrétien" (*La réaction païenne*, 13).

[10] Cf. G. G. Stroumsa, "Celsus, Origen and the Nature of Religion," in his

third or fourth Islamic century, on the other hand, borrowing arguments from the Hellenic pagan reaction meant distancing himself from his own culture, and must have required a considerable intellectual effort.[11]

Even when striking similarities do exist, one can distinguish in the criticism of Muslim heretics a profound change in emphasis in relation to pre-Islamic Bible criticism. Although pre-Islamic philosophical criticism included rejection of prophethood and revelation, these topics played there a relatively marginal role. In the criticism of the Islamic freethinkers, on the other hand, the subject of prophecy becomes central, as central, indeed, as it was in Muslim orthodox thought.[12] One can indeed say that this insistence on the rejection of prophetic authority was the Muslim contribution to freethinking. In this sense, the pagan Hellenic heritage can offer only a partial explanation to the phenomenon of freethinking in Islam.

Nevertheless, it is indeed true that pagan religions and the pagan heritage fascinated the freethinkers (although, as the abundant Muslim heresiographical literature testifies, such fascination was by no means foreign to other intellectuals). Besides numerous references to Greek philosophy, one can find in their writings indications of their interest in the religious ideas of other pagan cultures. Īrānšahrī's curiosity for the Indian religions seems, in this sense, to be typical.[13]

Barbarian Philosophy: The Religious Revolution of Early Christianity, Tübingen 1999, 44–56.

[11] A "reaction" similar to that of the Hellenic philosophers can be detected, on the other hand, in the attitude of the heretics to their Iranian sources. Ibn Muqaffaʿ, for instance, can be said to be glorifying his Iranian culture (cf. van Ess, *TG*, II, 23, 32), despite the fact that this culture was vanquished by Islam. Even in this case, however, there is a difference: by the third Islamic century, such an "Iranian reaction" was probably less natural, and demanded more effort, than the reaction against Christianity had demanded in the the third century C.E.

[12] Cf. above, Ch. 4, apud nn. 45, 70. Similarly, the pagan philosophers of late antiquity reflect the central ideas of Christianity, in their "passionate attempts to substitute the figure of Christ with an adequately prestigious figure" (de Labriolle, 9–10); cf. also Wright, "Against the Galileans," 315: "In spite of [Julian's] insistence on the inconsistency of the Christians in setting up a Trinity in place of the monotheism of Moses and the prophets, he feels the need of some figure in his own pantheon to balance that of Christ the Saviour, and uses . . . about Asclepius or Dionysus or Heracles almost the language of the Christians about Christ." See G. W. Bowersock, *Julian the Apostate*, London 1978.

[13] Al-Biruni *India*, 4; Cf. W. Halbfass, *India and Europe: An Essay in Understanding*, New York 1988, 26; Lawrence, *Shahrastānī*, 23–24.

2. *Barāhima*

In addition to quite accurate descriptions of the people of India (*ahl al-hind*) and their culture, early Islamic heresiographical traditions repeatedly refer to a certain group called *Barāhima*. These traditions, recounted also by Jewish authors, attribute to the *Barāhima* the rejection of all prophets, on account of the supremacy and sufficiency of the human intellect.

Serious doubts have been raised in modern scholarship concerning the postulated identity of these *Barāhima* with the Indian Brahmans. Paul Kraus, in particular, has attempted to show that all our sources are dependent on Ibn al-Rāwandī's *K. al-Zumurrud*, and that it was Ibn al-Rāwandī himself who invented the *Barāhima* as a cover for his own views.[14]

Shlomo Pines, who discussed the heresiographical tradition, pointed to a slightly different version of what we may call "the *Barāhima* argument against prophecy," found in Saʿadya's *K. al-Amānāt wa'l-iʿtiqādāt*.[15] According to Saʿadya the *Barāhima* did not object to the idea of an initial prophecy, in the sense of inspiration of a human being by the divine. They did, however, reject all prophets after Adam. Pines argued that Saʿadya's account, presumably independent from Ibn al-Rāwandī, was less at variance with Indian religious tradition as we know it. The discovery of this version in a relatively early text, by al-Qāsim b. Ibrāhīm (d. 330/942) enabled Pines to modify Kraus's argument. Pines suggested that "Ibn al-Rāwandī had knowledge of accounts of the *Barāhima* referred to by al-Qāsim and Saʿadya," and that only this knowledge prompted him "to use the name of this Indian group in order to express his own views concerning the function of reason."[16]

Following the publication of Pines's article I have argued, on the basis of another quite early source, the ninth-century Jewish theologian Dāwūd ibn Marwān al-Muqammaṣ, that not only al-Qāsim's

---

[14] Kraus, "Ketzergeschichte," esp. 341–57 Kraus argued that no source mentions the Brahmans as deniers of prophecy before Ibn al-Rāwandī. On the reference to Indians in pre-Islamic literature, see also above, Ch. 1, apud n. 24. Kraus further argued that there is no connection between Ibn al-Rāwandī's Brahmans and the Indian reality. One should note, however, that Ibn al-Rāwandī was not writing about the Indian reality, but rather about the Indian reaction to the reality of Muslim theology.

[15] Pines, "Shīʿite terms," 220–23.

[16] Ibid., 223.

version of the *Barāhima*, but the whole topic of the anti-prophetic *Barāhima* probably predates Ibn a-Rāwandī. Furthermore, al-Muqammaṣ offers a rather detailed summary of a discussion with the *Barāhima*. The position attributed to the *Barāhima* in this discussion can, as I have tried to demonstrate, be reconciled with our knowledge of Indian religion (although not necessarily with Hinduism). Consequently, I argued that the explanation for the description of the *Barāhima* in the earliest Arabic sources should be sought in the impression left on the Muslims by the encounter with Indian religions, rather than in Ibn al-Rāwandī's ploy.[17]

Since the publication of this article, the subject of the *Barāhima* has been discussed by several scholars, who rejected my analysis of the texts and offered their own interpretations. In the following pages I will review my arguments, in an attempt to buttress them and to respond to the objections that were raised against them.

In the wake of Pines's article, al-Qāsim's account became the subject of much speculation. In the *K. al-Radd ʿalā al-rawāfiḍ* al-Qāsim speaks of *Ibrāhīmiyya* rather than *Barāhima*. He explicitly situates the *Ibrāhīmiyya* in India; Pines assumed, therefore, a scribal error that corrupted the original *Barāhima* to *Ibrāhīmiyya*. Indeed, in another work (which was not available to Pines) the *K. al-Radd ʿalā al-rāfiḍa*, al-Qāsim names this sect *al-Barhamiyya*.[18] And yet, the singular appearance of Ibrāhīm's name has led two distinguished scholars to disregard all other evidence and to offer a drastically novel approach to the *Barāhima*. Binyamin Abrahamov believes that the scribal transformation from *Ibrāhīmiyya* to *Barhamiyya* is "no less probable than the other way round." He therefore offers to identify the *Barāhima* as a Sabean sect, whose original name was *Ibrāhīmiyya*.[19] This suggestion ignores al-Qāsim's explicit geographical identification, the evidence of another ninth-century source (to which we shall return shortly) and the principle of *lectio difficilior*. Furthermore, since the identity of the Sabeans is still a mystery, the identification of the *Barāhima* as a Sabean sect does little to solve the *Barāhima*'s enigma.[20]

Norman Calder seeks a solution to the identity of the *Barāhima* in the same direction as Abrahamov. Pointing out that "the conflation

[17] Cf. S. Stroumsa, "The Barāhima in Early Kalām," *JSAI* 6 (1985), 229–41.
[18] See Abrahamov, "Barāhima," 73–74.
[19] Ibid., 74–74.
[20] On the Sabeans, see further below.

of Brahma and Ibrāhīm was easy for Muslim observers,"[21] he suggests seeing in the *Barāhima* an otherwise unknown Jewish-Christian religion of Abraham. Calder's suggestion is introduced by a bold criticism of what he regards as a typical presumptuousness of historians. The demagogical style of Calder's sarcastic attack—directed, besides the present writer, also against Kraus, Pines, and van Ess—is a matter of personal taste, and should not concern us here. But it is important to stress, even briefly, the basic flaws in Calder's method. Historians of early Islam are usually painfully aware of the limitations of the available information and of the tentative nature of their own conclusions, and they tend to phrase these results with caution. Calder regards such cautious phrasings as a pathetic collection of "ifs," "it seems's" and "perhapses." Naturally, however, since Calder relies on the same kind of fragmentary information, when offering his own reading of the texts he is forced to use the same "due measure of 'ifs' and 'perhapses'" that he criticizes in others. To the historians' attempt to see connections between various texts Calder opposes the demand for strict and absolute precision: whatever is not identical (i.e., phrased in the very same words, stemming from exactly the same period, expressed by the same author) is not similar. Calder attacks the "narratives" which the historians create in their search for the Brahmins. His own suggestion, however, which introduces the Abrahamites, replaces such narratives by a truly creative fiction.

Both the suggestions of Abrahamov and Calder offer to replace the little-known *Barāhima* with a new, totally unknown entity. There is indeed some evidence[22] for a fascination with Abraham in pre-Islamic and early Islamic times. Such evidence, however, is not enough to allow us to speak of a distinct Abrahamite religion, let alone to claim the existence of "a movement based on the rejection of the prophecy of Moses, Jesus and, later, Muḥammad," and its survival into the ninth and tenth centuries. Moreover, all existing evidence concerning the *Barāhima* is consistent with relating them to the east and to India rather than to the Jewish-Christian tradition.

As to the Sabeans, their name is never mentioned in pre-Islamic literature. In the Arabic literature, the information concerning them is hopelessly garbled already in the earliest texts in which they appear.

---

[21] Calder, "Literary Construct," 47–48.

[22] Quoted by Calder, "Literary Construct," n. 38.

They are often associated with Ḥarrān (and, therefore, naturally, with Abraham), but in some texts they are connected to Egypt, to India, or even to China. The descriptions of the Sabeans are often replete with Hermetic traditions. The pagan Hellenic elements in those descriptions may explain the fact that, in some of them, the Sabeans are presented as opposed to the idea of revealed religion. But if the mythical Sabeans ever corresponded to a real, specific people or religion, the identification of this people escapes us: their descriptions in Muslim literature, like those of mythical winged animals, contain too many contradictory elements.[23] For the solution of the *Barāhima* enigma, at any rate, they can offer no help.

Pines's original suggestion, that al-Qāsim referred to the same *Barāhima* as Saʿadya, offers therefore the most plausible interpretation. In the following pages I would like to discuss a few other relatively early Jewish and Islamic sources, the analysis of which may contribute to the clarification of the presence of the *Barāhima* in Arabic literature, as well as to their association with Ibn al-Rāwandī.

The *K. al-Tawḥīd* of the Ḥanafite *mutakallim* Abū Manṣūr al-Māturīdī (d. 333/944) contains, in Māturīdī's rough style, a detailed discussion of the human need for prophets. As might be expected, Ibn al-Rāwandī is not absent from this discussion, but the position he takes is rather surprising: instead of rejecting the idea of prophecy, he fights to prove the need for prophets in general and for Muḥammad's prophecy in particular. The denier of prophecy in this discussion is the Manichean Abū ʿĪsā al-Warrāq. It is with him, not with Ibn al-Rāwandī, that Māturīdī holds the debate, even when he does not mention al-Warrāq by name.[24] Ibn al-Rāwandī appears in this chapter only on six occasions,[25] and in each and every one of these occasions he is explicitly quoted as responding to al-Warrāq and refuting him.

Joseph Schacht has suggested that what is attributed here to Ibn al-Rāwandī must reflect his opinions before his apostasy.[26] Schacht

---

[23] On the Sabeans, see T. M. Greene, *The City of the Moon God: Religious Traditions of Harran*, Leiden 1992, esp. 162; Stroumsa, "Sabéens de Ḥarrān et Sabéens de Maïmonide," and see further below.

[24] The heading which the editor of the *K. al-Tawḥīd* chose for this chapter reflects Ibn al-Rāwandī's notoriety, but not the text.

[25] *Tawḥīd*, 187:9–15; 193:15–18; 196:18–19; 197:1–4; 198:15–22; 199:17–20.

[26] Schacht, "New Sources," 41–42. The same view was also expressed by W. Madelung (*ZDMG*, 124 [1974], 150). Van Ess believes Ibn al-Rāwandī's refutation of al-Warrāq to have been mistakenly taken by al-Mu'ayyad fī'l-Dīn to repre-

has also remarked that among the arguments attributed to al-Warrāq by Māturīdī, we find one of those known to us as part of Ibn al-Rāwandī's *K. al-Zumurrud*.[27] In a previous study of these texts, I followed Schacht's hypothesis.[28] But a further study of Māturīdī's text has made it clear that the claim that al-Warrāq used the same arguments which were used elsewhere by Ibn al-Rāwandī does not offer a sufficient explanation for the striking similarity of Māturīdī's text and the identified fragments of the *K. al-Zumurrud*. In Māturīdī's text we find al-Warrāq using Ibn al-Rāwandī's arguments: the very same arguments, with the very same handful of Qur'ānic prooftexts, in the same order, sometimes even using the same formulations. In other words, we are not dealing with similar arguments, but with the same book. Once we realize this fact, the other sources are found to corroborate the hypothesis that Ibn al-Rāwandī and al-Warrāq produced the *Zumurrud* together.[29]

In a previous chapter we have analyzed the relation of the quotations in Māturīdī's *K. al-Tawḥīd* to the *K. al-Zumurrud*, through an examination of the *mubāhala* cluster of verses.[30] We shall now turn to the examination of another passage from the same book, which relates more directly to Ibn al-Rāwandī's rejection of prophecy. Māturīdī writes:

> Al-Warrāq has said that if prophets are sent to preach adherence to things that can be proven by human intellect—[then] the prophets are like ordinary people. If, on the other hand, they come to preach what contradicts those things—God has made those things [to be perceived as] proofs; they will not suit anything else except through altering [of the intellect itself].[31]

---

sent Ibn al-Rāwandī's own ideas ("Image," 5–26; see also van Ess's introduction to *Bist Guftar*, 9).

[27] The argument concerns the presence of angels at the battle of Badr and their absence at Uḥud. On this argument, see above, Ch. 2, apud n. 190. Contrary to what Schacht, "New Sources," seems to infer, Ibn al-Rāwandī's words (although following immediately upon Māturīdī's answer to al-Warrāq) do not respond directly to al-Warrāq's saying about Badr, but refer to prophecy in general.

[28] Stroumsa, "Barāhima," 232, 237.

[29] See above, Ch. 2.

[30] Ibid.

[31] *Tawḥīd*, 200:13–15: *Wa-fī mithlihi dhakara al-Warrāq anna al-rusul, law jā'ū ilā al-tamassuk bi-ḥujaj al-ʿuqūl fa-hum minnā, wa-in jā'ū ilā khilāfihā fa-qad jaʿalahā 'llāh ḥujajan, lam tajuz li-ghayr* (following van Ess's emendation, *TG*, VI, 460, n. 14) *illā bi'l-taghyīr*. *Ḥujaj al-ʿuqūl* might be more than mere logical proofs, perhaps close to the innate axioms of reason. In this paragraph al-Warrāq is mentioned explicitly, but the same argument appears beforehand anonymously: "he who denies the

If we decode Māturīdī's cryptic sentences, we can summarize the argument as follows: God, whom you say had sent the prophets, is also the one who gave us the intellect. If the prophets' message agrees with what the intellect decrees, then they are like ordinary people and are therefore superfluous. If, on the other hand, their message disagrees with the intellect, then ipso facto it contradicts God's decree and should be ignored.

This is exactly the argument attributed to the *Barāhima* in Ibn al-Rāwandī's *K. al-Zumurrud*,[32] although in Māturīdī's text it is clear that Ibn al-Rāwandī's role in this book was to refute this very argument, articulated here by al-Warrāq. In his refutation Ibn al-Rāwandī argues for the need to adapt the law in order to suit changing conditions:

> Ibn al-Rāwandī retorted by [mentioning] that one might perceive [another person's] head as black, then (i.e., later) perceive it as white; is it that one's eyesight has changed or else that the thing perceived has changed? For indeed it is not black when perceived by the eyesight [in the second instance]! The same is true concerning that which the human intellect deems just in one [particular] case [but unjust in another]: . . . there is no need to [postulate] an altering of the intellect so as to explain why it approves [now] of something the contrary of which it had approved of [beforehand]. The same [logic] applies to prophets . . .; it is sometimes permissible . . . to abandon one benefit for the sake of another, greater than the first, and the same [rule] applies to religious codes of law.[33]

Besides allowing us an insight into the structure of the *Zumurrud*, Māturīdī's text thus also enables us to get a better understanding of Ibn al-Rāwandī's argument against prophecy. For in al-Warrāq's words, and even more so in Ibn al-Rāwandī's response, the "intellectual" argument against prophecy is considered to be closely linked to a rejection of the notion of abrogation of the law (*nasḫ al-šarā'iʿ*).[34]

This linkage is confirmed through another source, written by a

---

prophets, saying that the Wise One (*al-ḥakīm*) would not order what the intellect condemns . . ." (*Tawḥīd*, 200:3–4). It appears once more, incorporated in Māturīdī's counter-argument (*Tawḥīd*, 184:3–7).

[32] Kraus, "Ketzergeschichte," 97.

[33] *Tawḥīd*, 200:17–24; cf. van Ess, *TG*, VI, 460.

[34] In other texts, such as ʿAbd al-Jabbār's *Mughnī*, this argument may be introduced within a broader epistemological framework. This does not, however, alter the fact that in Māturīdī's quotations the context is explicitly that of the abrogation of religious law; compare Abrahamov, "Barāhima," 77.

contemporary of Māturīdī who also belonged to the Ḥanafite school of *kalām*. Abū Muṭīʿ Makḥūl al-Nasafī (d. 318/930) includes in his *K. al-Radd ʿalā ʾl-bidaʿ* a sub-group of the *qadariyya* which he calls *al-Rāwandiyya*.[35] Introducing this group he says:

> The *Rāwandiyya* maintains that it is legitimate to practice the law of every book which was revealed by God, regardless of whether it abrogates [previous books] or is abrogated [by a later revelation]; for it is not fitting for the Wise One to say [something], then to regret it or recant.

From these lines one might get the impression that according to Abū Muṭīʿ the *Rāwandiyya* accepted all religions as equally and simultaneously true. However, in the course of Nasafī's further discussion of this heresy it becomes clear to the reader that this is not quite what the *Rāwandiyya* meant:

> Abū Muṭīʿ said: I had a disputation with the *harābidha* at Balḫ; their spokesman (*mutakallimuhum*) said that he does not recognize any prophet except Ibrāhīm, for it is not fitting for the Wise One to send messengers, then to regret [His choice] and send others [in their stead].[36]

There is no indication in this paragraph that Nasafī identified the *harābidha*, or Zoroastrian priests, with the *Rāwandiyya*. Rather, he seems to mention them here by way of association. The reason for this association must have been that according to Nasafī's understanding of the *Rāwandiyya*'s claim, they, like the *harābidha*, rejected the notion of abrogation of prophecy.[37]

---

[35] Following the editor's amendation, Nasafī, *Radd*, 92, n. 1 (based on Ḥanafī's *al-Firaq al-Muftariqa*, 58). The manuscript reads *zawīdiya*. Abrahamov ("Barāhima," 78, n. 36) notes the appearance of a similar scribal error in Nawbaḫtī's *K. Firaq al-šīʿa* (ed. H. Ritter, Istanbul 1931, 37:8, 30:3, 7); On the Shiʿite sect bearing this name, see "Rāwandiyya," *EI*², VIII, 461–63 (E. Kohlberg).

[36] Nasafī, *Radd*, 92, 15a. On the *Rāwandiyya*, and the role that the names of some prophets played in their thought, see van Ess, *TG*, III, 10–17, esp. 16. On the *Harābidha* or Zoroastrian priests, see Lane, *Arabic English Lexicon*, s.v. "*hirbadh*." The Zoroastrian identification of Zoroaster with Abraham, which Nasafī seems to suggest here, may have its origin in actual Zoroastrian traditions; see J. Bidez and F. Cumont, *Les mages hellénisés*, Paris 1938, I, 41–42; II, 48. See also L. Massignon, *La passion de Husayn ibn Mansūr Hallāj*, Paris 1979, I, 164.

[37] This seems to be the best interpretation of the evidence as it now stands. Admittedly, this evidence is not conclusive, since the reading *Rāwandiyya* is based on the editor's correction, and the identity of the sect is far from certain; see Abrahamov, "Barāhima," 78. Calder's suggestion to correct *harābidha* to *Barāhima* is, as he candidly admits, "a creative structure building." But, as noted by Calder, some link of these *harābidha* with the *Barāhima* can be established without this correction; (see

It appears then that ninth-century deniers of prophecy had two sets of arguments in their stock. The first argument, which can be called "intellectual," is considered to be typical of Ibn al-Rāwandī, although it is presented in Māturīdī's *Tawḥīd*, as well as in the *K. al-Zumurrud* from which it is taken, by al-Warrāq. It propounds the sufficiency of human intellect as a tool in the quest for truth. The other argument, which can be called "anti-abrogationist," rejects the notion of change both in God's creation (i.e., the human intellect) and in His laws. God's unchanging law was revealed only once to a prophet, who may be variously identified as Adam, Abraham, Seth, or even Moses.[38] The two arguments complement each other, and appear to be intertwined already in the earliest sources.[39]

This analysis confirms Pines's suggestion that Ibn al-Rāwandī was familiar with al-Qāsim's "anti-abrogationist" *Barāhima*; in fact, he seems to have used this argument himself (which could account for the fact that Abū Muṭīʿ al-Nasafī associated the *Rāwandiyya* with this argument).

On the other hand, what we have already seen makes it quite improbable that it was Ibn al-Rāwandī who had invented the attribution of the "intellectual" argument to the *Barāhima*. The two arguments are intermingled already in al-Warrāq's words in the *Zumurrud*. It stands to reason to assume that, if the "anti-abrogationist" argument was identified with the *Barāhima* before Ibn al-Rāwandī, so was the "intellectual" argument.

This seems to be confirmed by the detailed evidence of al-Muqammaṣ.[40] In the thirteenth chapter of his *ʿIšrūn maqāla*,[41] which concerns the necessity for prophecy, al-Muqammaṣ says:

"Literary Construct," n. 34, and compare ibid., 42). The description of the *Rāwandiyya* as a sect which denies *nasḫ* appears also in Ḥanafī, *al-Firaq al-muftariqa*, 57 (quoted in van Ess, *TG*, VI, 490; cf. also *TG*, IV, 304).

[38] According to Ibn al-Jawzī (*Muntaẓam*, 108) Ibn al-Rāwandī incited the Jews to say: "Moses had said: there would be no prophet after me." Whereas the tradition which sees in Adam the only prophet fits the core of Ibn al-Rāwandī's argument, the versions about Abraham, and certainly the one about Moses, seem, in the context, to be polemical, *ad hoc* versions. It is interesting to note that the refusal to admit a change in God's revelation is found already in Julian's "Against the Galileans," 357 (but with no reference to Brahmans).

[39] As testified also by ʿAbd al-Jabbār, *Mughnī*, XV, 113:16; 127:2–4; and not, as Kraus ("Ketzergeschichte," 355) supposes, concerning Saʿadya version, that the issue of *nasḫ* is an *ad hoc* addition to the issue of the *Barāhima*.

[40] See above, Ch. 1, apud nn. 77–78, 82. Vajda ("Finalité," 61, n. 1) dates him as not later than 870. Various stylistic and contextual considerations make an earlier date (around the first half of the ninth century) more plausible. This would

We say that there are two groups of monotheists: those who affirm that the sending of prophets by God is necessary, and those who deny it. Those who deny it are the *Barāhima.* They argue that there are only two possibilities concerning the prophet sent by God: either he deserved to become a prophet, or he did not. If he deserved to become a prophet, then it is either by virtue of perfect knowledge (*ʿilm tāmm*) or of perfect conduct (*ʿamal tāmm*), or both. . . . If it is by virtue of perfect knowledge and conduct, then there are only two possibilities concerning this knowledge and this conduct: either they[42] are something which he perceived with his intellect and understood through his innate intelligence (*adrakahu bi-ʿaqlihi wa-fahimahu bi-qarīḥatihi*), or they are something that was announced to him by [another] prophet. Here, however, we are not concerned with a prophet who is preceded by another prophet,[43] but only with the case of a prophet with no predecessor. Consequently, the perfect knowledge and conduct must be something that the prophet had perceived with his intellect and understood through his innate intelligence. This being the case, there are only two possibilities if we compare the prophet to other people:[44] either his constitution and nature are like theirs, or they are different. Now, no one can ever claim that one human being differs from another in humanity,[45] since both belong to the same species; and since all humans belong to the same species, all of them and each one of them possesses the properties that pertain to the nature of this species . . . The properties of this species are intellect, cogitation, discernment, preference for truth over lying, for the morally beautiful over the morally ugly, and for righteous works over their contrary—wicked works. Given this, anyone can be like a prophet and achieve his status, deserve what

make al-Muqammaṣ, in any case, a contemporary of al-Warrāq and also a (senior?) contemporary of Ibn al-Rāwandī; see above, Ch. 2, n. 4, and compare Lawrence, *Shahrastānī*, 77–78, and n. 269.

[41] See Vajda, "Prophétologie," 228–31; Muqammaṣ, *Twenty Chapters*, 255–63.

[42] That is, the knowledge (*ʿilm*) and the conduct (*ʿamal*), although the text has "*immā an yakūn šayʾ*." Vajda, ("Prophétologie," 229) explained the singular by complementing the verb with an additional subject ("Le message dont il se trouve chargé"). This interpretation, as Vajda himself has noted, gives the discussion a new turn. It also replaces the minor linguistic problem with a major theological one. For this would mean a prophetic mission which is not divinely sent but originates in some sort of prophetic nature or prophetic inheritance.

[43] *Nabī qablahu nabī*; Vajda, "Prophétologie," 229 ("un prophète qui serait antérieur") is incorrect.

[44] The manuscript is not quite clear. I now read *lā yaḫlū* [*kawnu*]*hu min al-nās* (which differs from my previous reading, see "Barāhima," n. 23). Vajda ("Prophétologie", 229) followed Yahuda's reconstruction of this word, and, taking the question to be a new turn in the discussion, interpreted it as referring to the person *to whom* this prophecy is addressed. This interpretation does not fit the context, and Vajda himself abandoned it in his translation of the rest of the chapter.

[45] On the egalitarian argument in an anti-prophetic context, see Pines, "Shiʿite Terms," 199–205; and see above, Ch. 3, apud nn. 148–67.

> he deserves, and, in fact, be a prophet himself. When humanity reaches this point, it can dispense with prophets, since it no longer needs them.
>
> On the other hand, if someone were to claim that a certain man became a prophet without deserving it, this would imply that all human beings could become prophets without deserving it. . . .

Like the *Barāhima* of Ibn al-Rāwandī, those described by al-Muqammaṣ argue that the intellect is the only tool for achieving human perfection. But there is a slight difference in the arguments presented by the two sources:[46] Ibn al-Rāwandī concentrates on the actual content of prophecy, whereas al-Muqammaṣ deals with the prerequisites for the appointment of the prophet.

A third version, somewhere between these two, is reflected again in Māturīdī's *K. al-Tawḥīd*:

> There are only two possibilities: either he (i.e., the person who rejects the notion of prophecy) relies on what his intellect[47] requires of him, or else he has [to argue] that he holds [views] stemming from what someone else, more intelligent than himself, has shown him.
>
> If the first possibility is true, we have to argue for the equality of [all] intellects, and to admit that everyone can achieve [the degree he has achieved], since he says what his intellect has shown him. This amounts to the admission of the achievement [of truth] by every person of [any] religion (*kull dhī dīn*) who relies on his intellect. But this is absurd in view of the contradictions between [different] opinions and beliefs.[48] If, on the other hand, [he supports] the second possibility, then his (i.e., the opponent's) teacher[49] can be regarded as a prophet, sent to humanity by God. He (i.e., the secular teacher), like the prophet, needs a proof, which will indicate to us his identity, and there is no difference between a proof which supports the truthfulness of what he announces in the name of God, and [a proof which supports his claim for] achieving the truth in his teachings, when he relies [only] on his intellect.[50]

---

[46] Therefore, contrary to Vajda, "Ibn al-Rāwandī," 905, it is impossible, in the present state of our sources, to point to Ibn al-Rāwandī as the source of al-Muqammaṣ. There is also no ground for Vajda's (circular) suggestion that we should see in the *ʿIšrūn maqāla* a proof for an early dating of Ibn al-Rāwandī. On the other hand, Schacht's suggestion ("New Sources," 51) that Māturīdī's information provides "an implicit proof for the late dating of Ibn al-Rāwandī" is not compelling.

[47] Read: *ʿaqluhu* instead of *ghafla*.

[48] Read: *wa'l-aqwāl* instead of *qawl*.

[49] Translation *ad sensum*. The text has *ʿaqluhu*, which is probably corrupt; perhaps *muʿaqqiluhu*.

[50] *Tawḥīd*, 182:13–19.

Using the method of *ilzām*, Māturīdī tries in these lines to push his adversary to the absurd implications of his own contentions. On the one hand, the argument presented here focuses, like Ibn al-Rāwandī's, on the content of the achieved truth. On the other hand, it shares with al-Muqammaṣ's version the consideration given to the possibility of a previous teacher (or prophet). As everywhere else in this chapter of the *K. al-Tawḥīd*, Māturīdī aims at refuting al-Warrāq. As a typical *ilzām* this paragraph must reflect, then, an argumentation which had been developed by al-Warrāq himself.

Now Māturīdī never mentions the *Barāhima*, but his *summa* contains in a nutshell different arguments against prophecy, which are found dispersed in other extant sources. Of these sources, two, which are contemporaries of al-Warrāq (al-Muqammaṣ and al-Qāsim b. Ibrāhīm), expressly attribute their anti-prophetic teaching to the *Barāhima*. Therefore, in a previous study of this topic I have suggested that al-Warrāq himself must have also connected the *Barāhima* with arguments against prophecy, or at least, that he was familiar with such a connection.[51] The evidence collected since the publication of that article corroborates my suggestion and makes it more specific.[52] For, as I have argued above, a close analysis of the disparate evidence, and in particular of the *mubāhala* cluster, shows that the quotations in al-Warrāq's name in Māturīdī's *K. al-Tawḥīd* are taken from Ibn al-Rāwandī's *K. al-Zumurrud* itself.[53] Māturīdī ignored the *Barāhima*, just as he failed to mention Ibn al-Rāwandī's authorship of the *K. al-Zumurrud*. The reasons for both these facts are hard to guess. But if we assume that Māturīdī extracted his quotations directly from the *K. al-Zumurrud*,[54] then the peculiarity of his evidence may be related, here too, to his oriental provenance. In tenth-century Samarkand the need to combat Manicheism may have been more pressing than the desire to refute the *Barāhima*.[55]

The argument about the supremacy of human intellect thus seems to have been attributed to the *Barāhima* independently of Ibn al-Rāwandī. This does not, however, solve the riddle of the *Barāhima*.

---

[51] "Barāhima," 237.

[52] *Pace* Calder, "Literary Construct," 42.

[53] See pp. 57–58 above.

[54] And not from some later work dependent on it—a possibility which, at the present state of our knowledge, cannot be ruled out.

[55] See p. 70 above.

It can still be argued that the *Barāhima* were only a literary invention, invented by an author of Ibn al-Rāwandī's generation or of the generation of his teacher al-Warrāq. Van Ess, for example, has argued that al-Warrāq's *K. al-Gharīb al-mašriqī* was a fiction in which anti-prophetic claims were voiced by a stranger from India.[56] Could al-Warrāq be the source for all "intellectual" descriptions of the *Barāhima*, including that of his contemporary al-Muqammaṣ?

This question is rendered more pressing by some other passages from the *ʿIšrūn maqāla*,[57] which deal with the moral aspect of God's retaining the prophet in this world of woe. The argument of al-Muqammaṣ's opponent runs as follows: assuming that for every good deed God had promised a reward, the prophet, having attained perfection before his prophecy, is entitled to receive complete and immediate reward. Therefore, to withhold this reward, even for a short while, and to prolong his trials (in the *dār al-imtiḥān*) until he fulfills his mission, is both pointless (*ʿabath*) and unjust.

To this al-Muqammaṣ retorts that indeed the prophet, having achieved human perfection (*ṣalāḥ ʿāmm*), deserves (immediate) reward in the hereafter (*dār al-jazāʾ*). Nevertheless he lingers in this world for the benefit of others, and as a result of this mission (which is the prophet's special perfection, *ṣalāḥ ḫāṣṣ*)[58] an additional, special reward is prepared for him.

---

[56] *TG*, IV, 289–90. The little we know of this book, however, does not warrant such a description; see p. 43 above.

[57] These passages survived in two Geniza fragments, the first, T-S Ar. 43.198, was identified by Paul Fenton as belonging to al-Muqammaṣ's work. The second, T-S Ar. 12.5, belongs to the same manuscript as the first. Each of these two fragments holds two pages, one of which belongs to the fifteenth chapter of *ʿIšrūn maqāla*. The other page in each fragment does not correspond to any part of the main manuscript of the *ʿIšrūn maqāla*, but it seems to belong to the thirteenth *maqāla*. In theory, these fragments could belong to a different *maqāla*, hence my cautious formulation in a previous publication ("Barāhima," 238; but see my introduction to the edition of the text, *Twenty Chapters*, 38). This caution, however, misled Calder, who categorically states that "these passages do not appear in any known manuscript of that work" and treats the fragments as anonymous ("Literary construct," 42–43). It therefore seems necessary to stress here that al-Muqammaṣ's authorship of these fragments and their appartenance to the *ʿIšrūn maqāla* are beyond any doubt.

[58] *Wa'l-ṣalāḥ al-ḫāṣṣ huwa mā yalzamu al-nabī wa'l-kohen wa'l-levi min al-*[*ʿināya bi-*] *taʿlīm ghayrihi wa-tawqīf siwāhu.* The reference to the Temple priests and the Levites gives this answer a definite Jewish character, which should be explained by the Jewish audience to whom the whole exposition is addressed. There is no reason to infer that al-Muqammaṣ's adversary himself was a Jewish heretic.

At the end of the twelfth chapter of *ʿIšrūn maqāla* we find a short exposition of the topics which are to be discussed in the following chapters. This exposition includes, inter alia, the following questions:

> Of how many kinds are [the divine commandments and prohibitions]? Do they change [in the course of time] so that they come to contradict each other? Does God convey them to mankind by direct speech? Or does He do so by revelation to a prophet sent by Him? Or does He do so by means of the intuition and discernment with which the human mind is endowed? In particular [the next chapter will deal with] the logical disputation with the *Barāhima*, and with their denial of prophecy in the context of reward and punishment.[59]

As George Vajda already noticed, "this table of contents . . . does not enumerate the themes in the order in which they are in fact treated."[60] The irregularity would be explained if we assumed that these lines introduce only the chapters dealing with prophecy (i.e., chaps. 13–15), and are not an exposition of the rest of the whole work. In this case, the Geniza fragment discussing the *thawāb wa-ʿiqāb* would indeed fit into the scheme of the debate with the *Barāhima*, as would also the concept of abrogation of the law.

An intriguing trait of this fragment is the fact that al-Muqammaṣ quotes verbatim single sentences from his interlocutor's claims, introducing each quotation with "as to his saying" (*wa-ammā qawluhu*). Could it be that here al-Muqammaṣ holds a debate with one of the two Muslim heresiarchs whose names are connected to the *K. al-Zumurrud*, Ibn al-Rāwandī or al-Warrāq? In that case al-Warrāq or Ibn al-Rāwandī might also be the ultimate source of al-Muqammaṣ.

The fragmentary state of our evidence concerning all three of the above-mentioned thinkers makes it impossible to reach a definite answer. Yet, as I shall try presently to show, this question can be answered in the negative with a fair degree of probability.

To the best of my knowledge the argument incorporated in the Geniza fragment is not known from any source recording Ibn al-Rāwandī's views, including Māturīdī's quotations from al-Warrāq. On the other hand, this very argument constitutes a topic of theological

[59] *Al-amr wa'l-nahy min allāh ʿalā kam naḥw yakūn, wa-hal yajūzu an yaḫtalifa wa-yatanāqaḍa am lā, wa-hal dhālika bi-muḫāṭaba am bi-waḥy wa-nabī mursal aw bi-mā fī al-fikr wa'l-fiṭan min al-maʿrifa wa'l-tamyīz, wa-bi-ḫāṣṣatin fī munāẓarāt al-barāhima wa-jaḥdihim al-nubuwwa fī dhikri al-thawāb wa'l-ʿiqāb; Twenty Chapters*, 255.

[60] "Prophétologie," 228.

discussion between Hinayana and Mahayana Buddhism.[61] For the Mahayana, in opposition to the Hinayana, an essential feature of the Boddhisattvas is their undertaking to remain in this world for the benefit of mankind:

> Why then do they undertake such infinite labor? For the good of others, because they want to become capable of pulling others out of this great flood of suffering.[62]

The prophetic *ṣalāḥ ḫāṣṣ* (versus *ṣalāḥ ʿāmm*), referred to by al-Muqammaṣ in the Geniza fragment, would thus seem to correspond to the Mahayana altruistic view of the Boddhisattva's role. The term *ṣalāḥ* is otherwise foreign to al-Muqammaṣ's vocabulary. As the context shows, this term implies for him both personal virtue and social involvement. It is definitely an accomplishment *of* the prophet, not a benefit bestowed on him by God (for which al-Muqammaṣ reserves the term *thawāb*).[63] In (later) *kalām* literature this term appears as part of the conventional vocabulary for the discussion of prophetology.[64] Its appearance in al-Muqammaṣ's text—the earliest extant *summa* in *kalām* literature—in this context suggests the possibility that it reflects the use of Buddhist terminology in order to counter Buddhist conceptions.[65]

Moreover, it should be mentioned in this context that al-Muqammaṣ is also the author of a work, no longer extant, which he describes as a "refutation of *aṣḥāb al-budūd*, both the unbelievers and the monotheists among them."[66] This title brings to mind Šahrastānī's termi-

---

[61] I am indebted to the late Shlomo Pines for calling my attention to this possible connection.

[62] Quoted from the *Abhidharmakosa* by E. Conze, *Buddhism, its Essence and Development*, New York 1959, 125–26. See also A. Bareau, *Les sectes bouddhiques du petit véhicule*, Saigon 1955, 61–62, 301–302.

[63] Compare Abrahamov, "Barāhima," 82.

[64] See, for instance *Mughnī*, XV, 72–73 (who relies in this discussion on Abū ʿAlī al-Jubbāʾī's refutations of Ibn al-Rāwandī); Abū Ḥātim al-Rāzī, in: Kraus, "Rāzīana," II, 42–43. Tahānawī, *Kaššāf*, I, 821, defines *ṣalāḥ* as "[the characteristic of] following the right path, and some say: the integrity (*istiqāmat al-ḥāl*) prescribed by both the intellect and the scriptures."

[65] Vajda, "Prophétologie," 231, remarks that the ways al-Muqammaṣ uses the pair *ʿilm-ʿamal*, are "sinon essentiellement philosophiques, du moins teintées de philosophie." The choice of these terms might also be explained by the Buddhist view of the way to enlightenment, in which "one needs a vast accumulation of knowledge and merit, innumerable heroic deeds . . ." (Conze, *Buddhism*, 126).

[66] *Al-Radd ʿalā aṣḥāb al-budūd, mulḥidihim wa-muwaḥḥidihim* (*Twenty Chapters*, 225). It might not be superfluous to recall here that for al-Muqammaṣ the *Barāhima* must

nology in his taxonomy of the *Barāhima*, where the first sub-group, the *aṣḥāb al-bidada*, clearly refers to the Buddhists.[67] The connection with the Buddhists is further strengthened by a passage in Pazdāwī (d. 1099), who, as Calder rightly noted, attributes to the *Sumaniyya* the anti-prophetic position usually associated with the *Barāhima*.[68] *Sumaniyya*, as Calder says, "was an established term indicating *Buddhists*."[69] *Barāhima, Sumaniyya, aṣḥāb al-budūd* or *al-bidada*, all apparently served as generic names for Indian religions, rather than as names for the particular religion which we associate today with the Brahmans. The Geniza fragment, together with the rest of Muqammaṣ's thirteenth *maqāla*, thus seems to reflect some authentic knowledge of Indian religious groups, generically referred to as *Barāhima*.

Ibn al-Rāwandī was an elusive figure; even after he had definitely turned his back on the Mu'tazila, his opinions remained hard to classify, and the heresiographers have attributed to him various religious affiliations. It was, therefore, logically sound to assume, as

also be counted among monotheists (*Twenty Chapters*, 255); and see Māturīdī, *Tawḥīd*, 179:9–10. Ibn Ḥazm, *Fiṣal*, I, 137, and Baghdādi, *Farq*, 26, also regard the *Barāhima* as monotheists who deny prophecy (cf. Kraus, "Ketzergeschichte," 344). Their argument against prophecy, however, is not Ibn al-Rāwandī's, but sounds more like what we could expect from Abū Bakr al-Rāzī; They say that it is pointless to send a prophet to people who will not believe him, and that it would have been better to instill the content of the desired teaching directly in the people's intellect. Bāqillānī may also have shared the assumption that *Barāhima* are monotheists, but the fact that he mentions the *mu'tazila* together with the *Barāhima* testifies only to a well-rehearsed polemical technique, and cannot be taken as proof that for him "they are closely allied to the *Mu'tazila*" (as suggested by Calder, "Literary Construct," 46 and n. 25, quoting *Tamhīd*, 104:3, 107:20, 110:1).

[67] *Milal*, 446; cf. Lawrence, *Shahrastānī*, 100–114, esp. 105, 108. Al-Jāḥiẓ also mentions the *'ibād al-bidada min al-hind*, whom Rescher rightly identified with the Buddhists; see Jāḥiẓ, *Al-Radd 'alā al-naṣārā*, *Thalāth rasā'il*, ed. J. Finkel Cairo 1926, 12; Kraus, "Ketzergeschichte," 345–46, n. 6; see also Ibn al-Nadīm, *Fihrist*, 346–47.

Calder scoffs at the identification of the *aṣḥāb al-budūd* with *aṣḥāb al-bidada* and with the Buddhists ("Literary Construct," 43). Until we retrieve at least parts of the *K. al-Radd 'alā aṣḥāb al-budūd*, their identity remains indeed conjectural. In theory, one could entertain the thought that the three centuries separating al-Muqammaṣ from Šahrastānī warrant a transference of the name from one religion to another, as Calder apparently suggests. Since, however, no other religion has undergone such a change, I prefer my own conjecture to Calder's. See also below, n. 69.

[68] *Kitāb Uṣūl al-dīn*, ed. H. P. Linss, Cairo 1963, 90–91. Calder accepts Pazdāwī's evidence, but he rejects the evidence of Šahrastānī and Ibn Ḥazm as too late ("Literary Construct," 43 and 46). It should be noted, however, that Pazdāwī is a younger contemporary of Ibn Ḥazm and only a few decades older than Šahrastānī.

[69] "Literary Construct," 47; cf. also Lawrence, *Shahrastānī*, 106–107. Calder, who cites Pazdāwī, fails to see that his evidence actually strengthens the Buddhist connection which he endeavors to reject; see above, n. 66.

Kraus did, that he had tried to conceal his ideas. Abū ʿĪsā al-Warrāq's religious identity, on the other hand, has always been much clearer. It is hard to believe that al-Warrāq, the notorious Manichean, would have tried to hide behind the thin cloak of the *Barāhima*. It seems more plausible to assume that it was al-Warrāq's well-known heresiographical interest which brought him to collect and hand down the opinions of the *Barāhima*, as he knew them.

One of Kraus's points of departure in his quest for the inventor of the *Barāhima* was the discrepancy between the views attributed to them and all that is known to us about the Indian Brahmans.[70] In his abovementioned article, Pines has shown that there is no such discrepancy in al-Qāsim b. Ibrāhim's version of the *Barāhima*. However, this should apply ipso facto also to Ibn al-Rāwandī's argument (which we have shown to be the same as al-Warrāq's). The "anti-abrogationist," Qāsimite argument (which rejects all but a single prophetic revelation) is, from its first appearance, interwoven with the "intellectual," Rāwandian argument (which rejects the very notion of a prophet). One should also note that the two versions of the *Barāhima* presented in our sources (namely, the one that totally rejects prophecy and the one that limits it to a single revelation), which may appear at first sight to be mutually exclusive,[71] are, in fact, two variations developed from the same basic presupposition.[72] They both assume an unchangeable human nature, in a given and unchangeable world. This static world-view implies the rejection of the notion of divine interference in the course of history.

We know that Indian-Moslem encounters, including theological debates, took place in the formative years of Moslem *kalām*.[73] Bruce

---

[70] "Ketzergeschichte," 341–42.

[71] As suggested by Abrahamov, "Barāhima," 72–73, 76, and by Calder, "Literary Construct," 47.

[72] The explanation for the existence of different variations should probably be sought in the changing polemical situation.

[73] Cf. Lawrence, *Shahrastānī*, 84. Despite the paucity of evidence available to us, it would take an unusual lack of common sense to proceed from the contrary assumption (namely, that there were no Indian-Muslim encounters in the formative years of Muslim theology; compare Calder, "Literary Construct," 43). Moreover, such a contrary assumption would contradict the evidence we do have. See S. Pines, "A Note on an Early Meaning of the Term *Mutakallim*," *IOS* 1 (1971), 224–50 (reprinted in S. Pines, *Collected Works*, III, Jerusalem 1996, 62–78). See also D. Gimaret, "Bouddha et les bouddhistes dans la tradition musulmane," *JA* 257 (1969), 273–316, esp. 293; Nomanul Haq, "The Indian and Persian background," in Seyyed Hossein Nasr and O. Leaman, eds., *History of Islamic Philosophy*, London 1996, I, 52–70, esp. 55; IAU, Beirut 1965, 473ff.

Lawrence convincingly argued that "although the opinions of the Barāhima, as set forth by Ibn al-Rāwandī, do not correspond to the information about Brahmans in Indian sources, the tag of 'rationalists' is not as far-fetched as Kraus would have us believe." According to Lawrence, to Muslims participating in these debates "Indian Brahmans would indeed have seemed like stark rationalists."[74] Furthermore, it is also known that according to the rules in these debates, the participants had to restrict themselves to common logical and intellectual ground, and avoid reliance on the particulars of their respective religions, especially their scriptures.[75] Therefore, one should not expect to find in the records of these debates explicit reference to Indian mythological and theological conceptions made by the *Barāhima*. Such references would have been against the rules of the game.[76]

On the other hand, it is not difficult to imagine that in the course of these early theological debates, the Indian participants (probably philosophers) would have chosen various arguments to refute prophecy, using at times the notion of a first and only prophet whose law would never change, while preferring on other occasions to focus on reason as a "divine gift" to humanity, a view firmly held by the Muʿtazila. It would thus have been by way of *ilzām* that both arguments were put forward; therefore, the polemical and heresiographical texts in which they appear should not be expected to be an accurate reflection of actual Indian conceptions. What they do authentically reflect, however, is the reality of Indian-Muslim debates, and the lasting impression those debates made on Islam.[77]

It is nevertheless important to note that, whatever real contact there may have been between Indian and Muslim thinkers in the formative period of *kalām*, when the term entered Islamic thought, the Muslim memory of this contact soon faded. The *Barāhima* became a *topos*, where the motifs which "are manipulated in these highly stylized defences of faith bear some relationship to reality, but not

[74] Cf. Lawrence, *Shahrastānī*, 84.

[75] See Ḥumaydī's indignant account of *kalām* meetings, *Jadwā al-muqtabas fī dhikr wulāt al-andalus*, ed. al-Tanjī, Cairo 1352, 101–102.

[76] But it is only natural, within this system, to find the Indian views reflected in al-Muqammaṣ's words; see above, p. 158.

[77] Another evidence of such lasting impression is cited by Nomanul Haq, "The Indian and Persian background," 55. According to him, a passage in the *K. Sirr al-ḫalīqa* attributes to the *Barāhima* certain views which are plausibly derived from the doctrines of classical Vedic philosophy.

much."[78] This *topos* served the theologians for a particular theological construction, always in the context of prophecy. The *Barāhima* always appear as deniers of prophecy, or of the notion of abrogation (*nasḫ*), or of the authority of divinely-guided people (like the *imāms*).[79] They thus became the byword for ideas related to freethinking, even in texts where no specific freethinker is mentioned by name.

3. Sabeans

In the same way that the *Barāhima* are associated with Ibn al-Rāwandī, the Sabeans are associated with Rāzī. In the remaining part of this chapter I propose to examine this association, in the attempt to uncover its real or imagined character and its significance for Rāzī's thought.

The first modern scholar to focus attention on Ibn al-Rāwandī and Rāzī was Paul Kraus. Kraus did not offer a systematic comparison of these two thinkers, but his studies of them follow a similar pattern. In both cases, Kraus attempted to reconstruct their thought from citations by their opponents, and endeavored to identify the doctrines attributed to them, their sources, and their place in their respective thought. Kraus analyzed Ibn al-Rāwandī's argument against prophecy and its connection to the *Barāhima*, suggesting that Ibn al-Rāwandī had decided to attribute his views to the safely distant Brahmans when he realized the dangers involved in their propagation.[80]

Apocryphal literature of all kinds was, of course, a very common phenomenon, in early Islam as before, and Kraus has studied this literature extensively. His research encompassed, besides the freethinkers, also the vast corpus of alchemical literature attributed to Jābir b. Ḥayyān as well as the Neoplatonist Arabic pseudepigrapha.[81] It is thus not surprising that Kraus had developed a sensitivity to false attributions. It was generally assumed that in medieval Islamic

[78] Calder, "Literary Construct," 43.

[79] An interesting case in point are Maʿarrī's vegetarian convictions, which his opponents connected to his *tabarhum*, and than to his denial of prophecy.

[80] Kraus, "Ketzergeschichte," esp. 341–57.

[81] For representative samples of Kraus's output, cf. Paul Kraus, *Alchemie, Ketzerei, Apokryphen im frühen Islam: Gesammelte Aufsätze*. Edited, with an Introduction, by Rémi Brague, Hildesheim 1994.

society the dissemination of unorthodox ideas could prove dangerous, and that such ideas were likely to circulate under cover. Kraus's implicit methodological assumption was that whenever a sect was presented as upholding an unorthodox idea which it had not been known previously to uphold, if this sect was associated with a Muslim heretic, one could suspect the heretic of falsely attributing his own ideas to this sect. It was thus natural that the appearance of what seemed to be a previously unknown brand of Brahmans in Arabic sources would suggest to Kraus the possibility that Ibn al-Rāwandī had invented them as a cover for his own heterodox ideas.

A similar approach can be detected in Kraus's treatment of Rāzī. As mentioned above,[82] one of the peculiarities of Rāzī's philosophy was his doctrine of the five eternal principles: the Creator, matter, the soul, time, and space. As to the origin of this theory, our sources offer contradictory information. According to Bīrūnī, in constructing his philosophy Rāzī drew from the teachings of Īrānšahrī. Some sources, such as Ibn Ṣāʿid al-Andalūsī (d. 685/1286),[83] mention Greek philosophy as Rāzī's main source of influence, while Ibn Ḥazm and Ibn Qayyim al-Jawziyya (d. 751/1350) refer to the dualists,[84] and Masʿūdī as well as Qazwīnī (d. 675/1276) and Ibn al-Jawziyya mention the Sabeans of Ḥarrān.[85] Nāṣir-i Ḫusraw also associated Rāzī with the Ḥarranians, but more precisely with Rāzī's anti-prophetic teachings. According to him, the Ḥarranians, like Rāzī, believed that the souls of evil people haunt the deserts, instructing people to harm their fellow men.[86] Kraus rejected both dualist thought and the Sabeans of Ḥarrān as possible influences on Rāzī. According to

---

[82] See Ch. 3, apud n. 119.

[83] *Ṭabaqāt*, 33.

[84] *Fiṣal*, I, 86–87 and II, 197; and see Meier, "Der 'Urknall'," 5. In his *Risāla* Bīrūnī mentions Rāzī in relation with Manicheism, but he does not suggest Manichean influence on Rāzī (compare Pines, *Atomism*, 83; Meier, "Der 'Urknall'," 2); Bīrūnī only blames Rāzī for having encouraged the reading of Manichean books; see Ch. 3, apud nn. 103, 127.

[85] See Masʿūdī, *Murūj*, II, 394–95; Kraus, *Rasāʾil*, 188, 203; Pines, *Atomism*, 69–81 (quoting Qazwīnī, *al-Mufaṣṣal fī šarḥ al-muḥaṣṣal*). Pines thought that the ascription of this doctrine to the Ḥarranians (or Sabeans) is "of little importance, in view of the vague notions that are associated with that name in the Islamic tradition;" cf. *Atomism*, 81. On the five eternal beings, cf. also Šahrastānī, *Milal*, 143 (Pines, *Atomism*, 81–82). On the Sabeans, see Pines, "Translator's Introduction," Maimonides, *Guide*, cxxiii–cxxiv; van Ess, *TG*, II, 442–445; Fakhry, *Islamic Philosophy*, 99; Stroumsa, "Sabéens de Ḥarrān et Sabéens de Maïmonide."

[86] *Jāmiʿ al-ḥikmatayn*, 137; cf. Corbin's introduction, 137, and see above, Ch. 3, apud n. 130.

Kraus, Rāzī attributed his ideas to other people because he feared that their publication in his own name would endanger him.[87]

Even more than in the case of Ibn al-Rāwandī, there is no reason to accept Kraus's judgment that Rāzī was overly cautious; Rāzī was quite outspoken when ridiculing all religious authorities, from prophets to religious dignitaries, whom he called "the bearded goats." And even more than in the case of Ibn al-Rāwandī's *Barāhima*, Kraus's suggestion that dualists and Sabeans were cover-up religions for Rāzī's ideas is puzzling from the outset. Unlike the *Barāhima*, the dualists were not a distant sect from the far East; their religious beliefs were well known and their presence was felt. It is unlikely that Rāzī would have attributed to them beliefs totally alien to them. Moreover, the dissemination of dualist notions was probably not the best strategy for avoiding religious persecution under the Islamic caliphate.

The Sabeans present a more complex problem. We know who the dualists were, namely either Zoroastrians or Manicheans, and we have some notion as to where to look for Brahmans. The Sabeans of Ḥarrān, however, remain a much more elusive community. Some modern scholars have suggested to see in these "Sabeans" a reference to Alexandrian philosophical traditions, which had found their way to Baghdād via Antioch and Ḥarrān.[88] But these explanations are not enough to account for Rāzī's philosophy; the Alexandrian tradition which reached Fārābī, for example, presumably through the same Ḥarranian connection, was very different from the one adopted by Rāzī. The question that concerns us here is, therefore, not who the Sabeans really were, but who they were for Rāzī.

A clue to the identification of Rāzī's Sabeans may perhaps be found in the titles of his books. From the list of titles of his many books, we know that he wrote against Aḥmad ibn al-Ṭayyib al-Saraḫsī[89] and against Kindī. Kindī, the "philosopher of the Arabs,"

---

[87] See Kraus, *Rasā'il*, 191–94.

[88] The philosophers' fascination with the Sabeans has led Michel Tardieu to his ingenious theory on the Sabean philosophical community of Late Antiquity; see his "Ṣābiens coraniques et 'Ṣābiens' de Ḥarran," *JA* 274 (1986), 1–44. For criticisms of this theory, see D. Gutas, "Plato's *Symposion* in the Arabic Tradition," *Oriens* 31 (1988), 36–60; J. Lameer, "From Alexandria to Baghdad: Reflection on the Genesis of a Problematical Tradition," G. Endress and R. Kruk, eds., *The Ancient Tradition in Christian and Islamic Hellenism: Studies on the Transmission of Greek Philosophy and Sciences, dedicated to H. J. Drossaart Lulofs on his Ninetieth Birthday*, Leiden 1997, 181–91; and Stroumsa, "Sabéens de Ḥarrān et Sabéens de Maïmonide."

[89] On him, see Rosenthal, *Saraḫsī*; art. "al-Sarakhsī," *EI*², IX, 35 (F. Rosenthal).

is, in a way, a Muʿtazilite among the philosophers, in the sense that his philosophy reflects some of the religious sensitivities of the theologians.[90] He tried to prove the creation of the world *ex nihilo*, and he wrote at least one, and possibly two books, in order to establish the necessity and legitimacy of prophethood. Kindī's disciple, Saraḫsī, was executed by the Caliph al-Muʿtaṣim. According to a story told by Ṣafadī and others, Saraḫsī had been accused of propagating heretical ideas, and of expressing doubts concerning the prophets' honesty and the validity of prophethood. Since Saraḫsī is otherwise known to have been a faithful follower of Kindī, Franz Rosenthal did not accept the historicity of this story in his monograph on Saraḫsī.

The interesting point from our own perspective is the fact that both Kindī and Saraḫsī wrote extensively about the Sabeans, Saraḫsī following in his teacher's footsteps.[91] The extracts of Saraḫsī's book preserved by Ibn al-Nadīm and by Maqdisī deal mostly with rituals, sacrifices, and calendary computation. On first view, they seem to include some references to prophets as well.[92] In Saraḫsī's account it is stated that the Sabeans believed in messengers (*rusul*), who would offer "the people of discernment . . . guidance" and who would themselves be "free from blemishes of the soul and from defects of the body." On closer examination, however, we find that these messengers do not fit exactly the biblical or Qurʾānic notion of a succession of prophets. Their mission appears to be devoid of precisely the characteristics of prophecy which Ibn al-Rāwandī had found so objectionable. They are required to summon people to believe in the *ḥanīfiyya*,[93] that is to say, the true, uncorrupted religion of old, the

---

[90] Cf. A. L. Ivry, *Al-Kindi's Metaphysics*, Albany 1974, 22–34.

[91] Ibn al-Nadīm (*Fihrist*, 260:6) records a book of Kindī "On the debate between (*mā jarā bayna*) Socrates and the Ḥarranians." (It is conceivable that this book was similar to Šahrastānī's fictitious debate between Sabeans and *Ḥunafāʾ*.) He also mentions Saraḫsī's book on the description of the Sabean religions (*Fihrist*, 292:21). See also *Fihrist*, 318:14ff. (Dodge, II, 746ff.). My translations below are based on those of Rosenthal and Dodge, with slight changes. Tartar, *Dialogue*, 39 notes that Bīrūnī (*Āthār*, 205) quotes Kindī (the Christian!) about the Sabeans and their human sacrifices. The possibility of a confusion between the two Kindīs cannot be ruled out. On the Christian Kindī, see below, Ch. 7.

[92] *Fihrist*, 318:16–18 (Dodge, 746): "He has shown them the way, sending apostles (*rusul*) for their guidance and for confirmation of proof, ordering them both to call [people] to be approved by Him and to warn them of His anger."

[93] *Fihrist*, 318:19 (Dodge, 746). The Sabeans are quite often linked to the *ḥunafāʾ*; cf. van Ess, *TG*, II, 446–47. In Šahrastānī's *Milal*, however, *ḥunafāʾ* and Sabeans are represented as two opposites: the *ḥunafāʾ* represent the followers of revealed religions, while the Sabeans advocate the sufficiency of human intellect and the equality

one that was revealed once, and has not been changed or abrogated since. Saraḫsī also stresses that all the Sabeans agree concerning their religion, and that this religion agrees with human innate knowledge.[94] Their prophets—Hermes, Agathodaimon—are sometimes called "messengers" and sometimes "angels." As we have seen above, in one of the versions of the argument of the *Barāhima* against prophecy, the *Barāhima* do not object to the very notion of a prophet, but to the notion of a succession of prophets sent by God. In this version, which is recorded by Saʿadya as well as by al-Qāsim b. Ibrāhīm, those who reject the revealed religions admit a single revelation, given to humanity either through an angel or through a prophet like Adam. In this pristine revelation the Creator guides humanity to the right path. But they object to the idea of abrogation of the law (*nasḫ*). It seems that a similar view is held by the Sabeans according to Saraḫsī (and hence probably also according to Kindī), and that it is in this sense that the Sabeans upheld the religion of the *ḥunafāʾ*. In other words, Saraḫsī's Sabeans, like Saʿadya's *Barāhima*, believed in some kind of original divine guidance or inspiration, while rejecting the idea of revelation in its historical sense.

In all likelihood, Rāzī was familiar with either Kindī's or Saraḫsī's accounts of the Sabeans. If so, it is also possible that these accounts were instrumental in the shaping of his philosophy not only concerning the five Eternals, but also concerning prophecy. For these Sabeans are not very distant from Rāzī himself, both intellectually or religiously. Like them, Rāzī did not object to the very notion of divine guidance, but only to the possibility of multiple interventions of God in the course of human history through chosen human beings. In Rāzī's view, such interventions could only cause discord and wars. The Sabean religion as Rāzī knew it emphasized the invariability of God's command to humanity, and it must have appealed to him as a religion free of the dangers of revealed religions.

The suggestion offered here, that Rāzī was influenced by Saraḫsī's

---

of all humans in the use of this intellect. Šahrastānī's ingenious polarisation of the two is not typical of other sources; cf. Lawrence, *Shahrastānī*, 63–74; and see Pines, "Shīʿite Terms," 196–210.

[94] *Fihrist*, 318:21: "the preaching of all these people is one and the same, their traditions and laws do not vary . . . they have all adopted one direction of prayer; . . . by this, they sought to search for wisdom, and they rejected what contradicts the innate knowledge" (reading: *mā naqaḍa al-fiṭar* instead of *al-qiṭar*, as suggested by Chwolson [cited by Flügel, *Fihrist*, 158] and Rosenthal, 43; compare Dodge, 746).

account of the Sabeans, might have further significant implications. First of all, it would mean that Rāzī did not attribute his own ideas to the Sabeans, as suggested by Kraus, but rather that he could have found their attribution to the Sabeans in his sources.[95] It would also imply an unexpected connection between the teachings of Kindī, the theologically more "orthodox" among the philosophers, and Rāzī, the most heterodox of them all. Of course, Kindī and Rāzī can in no way be regarded as belonging to the same school; in some respects, however, they seem to have grown on the same cultural ground.[96]

Finally, the suggestion that Rāzī was influenced by Saraḫsī's description of the Sabeans would imply a close similarity between Rāzī's Sabeans and Ibn al-Rāwandī's *Barāhima*, and thus between Rāzī's freethinking and Ibn al-Rāwandī's. Both authors seem to have found in their cultural background accounts concerning pagan peoples, who acknowledge divine guidance to humanity while rejecting the Muslim idea of prophethood.

It thus seems that by the third Islamic century one could encounter several manifestations of a critical attitude towards various aspects of the monotheist prophetology. These manifestations had their origin in the pagan legacy of the Muslim philosophers: the Hellenic critique of the scriptures, the arguments raised during encounters and debates with representatives of Indian philosophy, the literary descriptions of the Sabeans. Ibn al-Rāwandī and Rāzī did not invent these groups or forge their arguments. Rather, they integrated these arguments into their system and used them as supporting blocks in the construction of their own freethinking.

---

[95] The question raised by the examples of Ibn al-Rāwandī and Rāzī is a general one: Is it really the case that, out of fear, dissidents preferred to attribute their own writings to pagans? Pseudepigraphy was, of course, a well-known phenomenon in the ancient and medieval world. But one must notice the recurring tendency among scholars in the first half of the present century to suggest literary fictions by Muslim heretics as an easy solution to difficult problems. Thus Ibn al-Muqaffaʿ was suspected of having forged Burzoe's introduction to *Kalīla wa-Dimna*, and Ibn al-Zayyāt was said to have forged Ibn Waḥšiyya's *al-Filāḥa al-nabaṭiyya*. Such suggestions are not necessarily wrong, but they must be reexamined with care.

[96] According to Ibn al-Nadīm (*Fihrist*, 259:20–21) Kindī wrote an epistle on the disagreements of peoples concerning monotheism. In this epistle he pointed out that the various nations oppose one another, despite their agreement on the belief in one God (*innahum mujmaʿūna ʿalā al-tawḥīd, wa-kullun qad ḫalafa ṣāḥibahu*). Rāzī's litany on the bloodshed caused by rival revealed religions could be regarded as taking Kindī's argument one step further.

CHAPTER SIX

THE IMPACT OF THE FREETHINKERS ON ISLAMIC PHILOSOPHY

1. The *Fortleben* of the Freethinkers

In his ground-breaking study of the *Barāhima*, Paul Kraus pointed out the pervasiveness of the description of *Barāhima* by later authors: Bāqillānī, Ibn Ḥazm, Ghazālī, Ṭūsī, Dhahābī and others, all seem to have been preoccupied with these putative Indian thinkers.[1] On examining the context of their discussions, however, it becomes clear that the preoccupation of these authors was not so much with the *Barāhima* as with the denial of prophecy. It would be no exaggeration to say that this preoccupation soon became a real fear, even an obsession, among both Sunnī and Shīʿī Muslim intellectuals. It is of course true that without Ibn al-Rāwandī, the *Barāhima* would probably not have retained much significance in the fourth and fifth Islamic centuries, for instance, in Spain. It is also clear, however, that the real issue was not the figure of Ibn al-Rāwandī, irritating as he may have been. Ibn al-Rāwandī did not establish a school of thought, he had no disciples, and there was no immediate reason why Muslim authors should continue to argue with him so vehemently. One would have expected the memory of a solitary heretic like Ibn al-Rāwandī to fade quickly, as indeed happened to the memory of many ninth-century heretics. The fact that it did not shows the extent to which his heterodox views hurt Islam and the Muslims.

As noted by Daniel de Smet, the fact that Muʾayyad chose to discuss the *K. al-Zumurrud* during official sessions of Fāṭimid propaganda, two centuries after Ibn al-Rāwandī's death, points to the importance of the topic, far above and beyond the colorful character of the freethinker himself.[2] The centrality of prophecy for their doctrine rendered the Ismāʿīlīs particularly sensitive to this kind of heresy, and triggered their responses to Ibn al-Rāwandī and to Rāzī.

---

[1] Kraus, "Ketzergeschichte," 344–45.
[2] See De Smet, "Polémique ismaélienne," 91–92.

Naturally, the most obvious reflection of the impact of the freethinkers on other intellectuals is found in polemical literature. Polemical books against Ibn al-Rāwandī were written by many theologians, most of them Muʿtazilites, such as Abū Sahl al-Nawbaḫtī, Abū Muḥammad al-Nawbaḫtī, al-Ḫayyāṭ, Abū ʿAlī al-Jubbāʾī and his son Abū Hāšim, Abū al-Qāsim al-Balḫī and Abū Zayd al-Balḫī. But theologians belonging to other schools, such as Abū al-Ḥasan al-Ašʿarī, and some philosophers, such as Kindī[3] and Fārābī also published polemical works against Ibn al-Rāwandī.[4] But the marks left by the freethinkers on Islamic thought can be detected beyond the literary genre of straightforward polemics. Both *mutakallimūn* and *falāsifa* found themselves obliged to formulate their thought in a way that would allow them to dodge the criticism of the freethinkers on the one hand and the orthodox accusation of heresy on the other. In such cases, we can sometimes detect the intellectual imprint of freethinking also in works that do not deal directly with this topic.

The attitude of the ṣūfīs, who were often suspected of harboring antinomian ideas, can offer an illustration of this fact. The Ḥanbalite author Ibn ʿAqīl associates the ṣūfīs with the *Barāhima*, while Ibn al-Jawzī (who often draws from Ibn ʿAqīl) simply accuses the ṣūfīs of denying prophecy.[5] In this case, Ibn al-Rāwandī has faded out of the argument, and the *Barāhima* have clearly become a strawman. What remains is the fear that someone might consider the guidance of prophets to have become redundant; this time, by mystical experience rather than by intellectual endeavor.

Another example of this phenomenon can be seen in a passage of Ghazālī (d. 1111). In his spiritual autobiography, *al-Munqidh min al-ḍalāl*, he describes a stage of his life in which he had become painfully aware of the fact that a person's religious identity is, to a large extent, the result of biographical chance. He then began to search for the true religion, the veracity of which could be proven decisively. In his description of the decisive religious proof which he was seeking, Ghazālī's wording is very reminiscent of that of the freethinkers. He evokes the familiar theme of the equivalence of religions (*takāfuʾ al-adyān*), and struggles to counter this dangerous notion with a belief that enjoys perfect certainty,

[3] See IAU, 292 (ed. Nizār Riḍā, Beirut n d.).

[4] Aʿsam, *History*, 8.

[5] *Talbīs*, 364–70; cf. Kraus, "Ketzergeschichte," 347–48. Bīrūnī may have been the first to point out the resemblance of the ṣūfīs with the *Barāhima*; cf. *Hind*, 5–6.

> so that, if someone were to challenge[6] my belief and refute it, for example, if someone would turn a stone into gold, or a cane into a serpent, this would cause me neither doubt nor disbelief. For, since I know that ten is more than three, therefore even if someone were to say: "On the contrary, three is more than ten; I can prove it by transforming the cane into a serpent," even if he were to perform this act, and even if I myself were a witness to it, all this would not cause me to doubt what I know with certainty. I will be impressed by this feat; but I will not doubt my knowledge.[7]

Ghazālī recycles the argumentation of the freethinkers against the prophets' miracles. Like the Meccans of old, his hypothetical opponent presents a challenge (*taḥaddī*); this time, however, the challenger is not the one who demands a miracle but rather he who offers to perform it. And like the freethinkers, Ghazālī refuses to accept the evidentiary power of miracles. A miracle, impressive as it may be, does not constitute a proof of the veracity of an idea or of the prophetic mission of anyone.

Ghazālī and the ṣūfīs are only two examples among many of the infiltration of the freethinkers into Muslim thought. Once admitted into Islamic discourse, the material—vocabulary, formulas, *topoi*, and arguments—was broken into small units and recycled. But it remains recognizable, even though it underwent considerable changes in both form and function. Furthermore, the general sensitivity concerning freethinking affected also the thought of non-Muslim thinkers.

In the remaining part of this chapter—and of this book—I propose to examine several cases where the image of the freethinkers has effected Islamic thought, as well as its repercussions beyond Islam.

2. The Physical Theory of the *Kalām*

The purpose of the present book is to show the importance of Ibn al-Rāwandī and Abū Bakr al-Rāzī for the development of Islamic thought. The overall argument would be stronger if one could show that these two figures were also responsible for significant or even major changes in the development of physical theories. Such a suggestion has been put forward by Alnoor Dhanani. It seems, how-

[6] *Taḥaddā*.

[7] Ghazālī, *Munqidh*, 82; compare McCarthy's translation, *Freedom and Fulfillment*, 63–64.

ever, that Dhanani's argument places too much responsibility on the freethinkers' threshold.

Dhanani, who regards Ibn al-Rāwandī and Rāzī as contemporaries, studies both of them in the context of *kalām*: the former, because of the books he wrote against the Muʿtazilites' arguments for the temporal creation of the world, his discussion of Divine justice and the Muʿtazilite polemic against Manicheism; the latter, because of his criticism of *kalām* doctrines and also because he became the target of *kalām* polemics against the *falāsifa*. According to Dhanani, Ibn al-Rāwandī's and Rāzī's critique of *kalām* doctrines herald the end of the first period of Muʿtazilism.[8] In the second phase of Muʿtazilism (called by van Ess "the scholastic phase"), Abū ʿAlī al-Jubbāʾī and his son Abū Hāšim "initiated the formation of systems with some measure of unity after the epistemological and terminological analysis of problems which had been discussed by their predecessors." Dhanani believes that in so doing, the Jubbāʾīs were reacting to the challenge of Ibn al-Rāwandī and the rise of the *falsafa*.[9]

Ibn al-Rāwandī attacked several tenets of Muʿtazilī cosmology. He is reported to have criticized doctrines relating to atomism, raising objections to the concepts of the finitude of matter and of time. Ibn al-Rāwandī tried to force the *mutakallimūn* to admit that some of their doctrines, which allowed for potential infinity but rejected actual infinity, went contrary to God's omnipotence, an essential component of their religious belief. Dhanani suggests that one result of Ibn al-Rāwandī's criticism was "to shut out the potential infinite, and to therefore make atomism the only theologically sound doctrine of the constitution of matter."[10] As it is, the suggestion suffers from some circularity: potential infinity did not exist for the *mutakallimūn*, but only for Ibn al-Rāwandī, and it is precisely because of the absence of this potential infinity that Ibn al-Rāwandī attacked them. Ibn al-Rāwandī's criticism exposed the gap between the logic of atomism and non-atomism, a gap which Dhanani rightly describes as "the incommensurability of the two geometries."[11] It is, however, undoubtedly true that Ibn al-Rāwandī's criticism of atomism encouraged the atomists to become entrenched in their positions. Although phrased

[8] Dhanani, *Physical Theory*, 9.
[9] Ibid., 10.
[10] Ibid., 171, 189.
[11] Ibid., 170.

in physical terms, Ibn al-Rāwandī's criticism was directed at the religious foundations of Muslim *kalām*. It thus highlighted for the *mutakallimūn* the religious implications of their physical theories, and encouraged them to add a religious, theological twist to their physical doctrines.

Dhanani argues convincingly that around this period the *mutakallimūn* became increasingly familiar with Greek and Syriac philosophical texts. At the same time, he suggests that the question of void was incorporated into the physical theory of the *kalām* towards the end of the third/tenth century because of Rāzī's acceptance of vacuum. According to Dhanani, the questions posed by Rāzī made atomist *mutakallimūn* realize the close relationship of vacuum and atomism in Greek atomism.[12] The chronology of the introduction of the concept of the void into *kalām* is compatible with this suggestion; but one could also argue that this introduction was contemporary with Rāzī's adoption of the vacuum, rather than dependent upon it. Although physical arguments crept into the discussion of the *kalām* with Ibn al-Rāwandī and Abū Bakr al-Rāzī, all in all, it seems that this was the result of the fact that arguments about physics were an integral part of intellectual and religious discussions. There is no reason to believe that it was at the core of the religious challenge to Islam in general and to the *kalām* in particular.

3. Ibn al-Rāwandī as a Dialectician

Unlike physics, the art of dialectics may seem at first sight to involve only the external form of the discussion and not its content. One can thus assume that its relevance to understanding the impact of freethinking on Islamic thought would be marginal. But the art of dialectics touches the very heart of *kalām*; as I shall now try to show, Ibn al-Rāwandī's dialectical style was intrinsically connected to his freethinking, and it left a lasting impression in both theological and philosophical circles.

Ibn al-Rāwandī excelled as a dialectician, and his fame as a dialectician is apparent even in his opponents' criticism. His *K. Adab al-jadal* is listed by Ibn al-Nadīm among those written before Ibn al-Rāwandī's apostasy (*kutub ṣalāḥihi*).[13] It was probably one of the

[12] Ibid., 190.

[13] *Fihrist*, 217:11; Cf. Gimaret, "Bibliographie," 247.

earliest books written in Islam about dialectics. It therefore stands to reason that it had some influence on subsequent literature. But in view of Ibn al-Rāwandī's notoriety as a heretic the nature of this influence must be evaluated with care. In the following pages I will first discuss the code of behavior in theological disputations and the ethos that underlined this code. I will then examine some anecdotes which offer us a glimpse of the harsher reality of the *majlis*, and analyze the references to this reality on the part of the Islamic theoreticians. Finally, I will concentrate on Ibn al-Rāwandī and on the role he played in the development of Muslim literature on the subject.

Manners

Like rhetoric in classical Greek education, dialectical disputation was in early medieval Islamic culture a discipline to be mastered by men of learning. Some people were considered to be gifted dialecticians by nature,[14] but the practice of the discipline was widespread, and could not be left to the intuition or talents of each individual. As Vajda aptly put it: "The discussions [of *kalām*] were controlled by procedures that were too regularly applied to be unconscious." In other words, they presuppose a technique, "a real eristic [art], an *ars disputandi*."[15] This "art" referred to the acceptable reasoning and to the methods of argumentation as well as to the manners of conduct. It instructed the participants on the kind of arguments to be presented, on the best reaction to any development in the discussion, on how to determine the winner and on general gentlemanly deportment. All these regulations of form, technique, and manners were known as *adab al-jadal* or *adab al-mujādala*.

But the same terms were also used in a narrower sense, to denote the manners alone. Occasionally, the etiquette of behavior in the debate is specifically designated by the term *ādāb al-jadal*, in the plural (as opposed to *adab al-jadal*, in the singular, which denotes the technical aspect of the disputation).[16] But quite often *adab* is used for

---

[14] E.g., Abū al-Hudhayl al-ʿAllāf, who is said to have won a debate when he was a mere youth of fifteen. Cf. van Ess, "Disputationspraxis," 35, citing *Ta'rīḫ Baghdād*, III, 367:5ff. This topos appears also concerning Abū ʿAlī al-Jubbā'ī, see Abū al-Qāsim al-Balḫī, *Ṭabaqāt al-muʿtazila*, ed. Fu'ād Sayyid, Tunis n.d., 287.

[15] Vajda, "Qirqisani," 7.

[16] Gimaret, "Document," 216 and note 147 bis. The tenth-century Jewish *mutakallim* Saʿadya lists both technique proper and good manners under the term *ḥadd al-munāẓara*, i.e., "the parameters (or 'the terms') of disputation." See *Kitāb ṭalab*

both. Van Ess, who studied "the ritual, the forms of courtesy . . . and the tricks with which one outdid one's opponents" has pointed out that, whereas the logical structure of the argumentation was registered in manuals, rules of behavior usually remained uncodified.[17] In fact, one cannot say that the rules of behavior were not codified altogether, but rather that no independent codes were dedicated to them. *Adab al-mujādala* in the narrow sense of "code of manners" was seen as an integral part of *adab al-jadal* in the wider sense of "principles of dialectic," and manners were codified as part of the codes of disputation. There are some extant examples of such "codes," for instance those preserved in Qirqisānī's *Kitāb al-Anwār*,[18] and in the *Naqd al-nathr*.[19]

Ādāb al-jadal included rules of fair play: one must grant the younger or weaker participant the privilege of asking the first question,[20] one must not presume to answer for one's opponent,[21] and so on. It also included rules of decorum: one must never yell or scream,[22] one should not boast of one's talents,[23] and one should always look straight at one's opponent.[24]

Good manners were regarded as both proper and advantageous. Self-control and composure, for instance, were considered not only a dignified behavior, but also a polemical asset, since they allowed the participant to remain in control of the scene and to navigate the discussion to his advantage. The tenth-century Karaite Qirqisānī, for example, cites an anonymous authority who instructs the polemicist as follows:

> When your opponent speaks rashly and yells, silence and abstaining from screams is a better response to him. It will silence him more effectively than the whip, and it will offer a more forceful demonstration of the weakness of his argument. For when he observes that

al-ḥikma—Saʿadia's Commentary on Proverbs, ed. J. Qāfiḥ, Jerusalem 1976, 197:23–198:24, esp. 197:25 and 198:3–4.

[17] Van Ess, "Disputationspraxis," 26.

[18] Qirqisānī, *Anwār*, 470ff.

[19] Cf. van Ess, "Disputationspraxis," 32, n. 4. This book, attributed by its first editors to Qudāma b. Jaʿfar, is in fact the *K. al-Burhān fī wujūh al-Bayān* of the Shiʿite Abū al-Ḥusayn Isḥāq b. Ibrāhīm al-Kātib (d. 946). It was edited anew by Aḥmad Maṭlūb and Ḫadīja al-Ḥadīti, Baghdad 1967. I was unable to see this edition.

[20] Van Ess, "Disputationspraxis," 37; Ibn Ḥazm, *Taqrīb*, 187:8–10.

[21] *Ḥayda*, 27.

[22] Qirqisānī, *Anwār*, 488:14–15.

[23] Ibn Ḥazm, *Taqrīb*, 196:19.

[24] *Mujarrad maqālāt al-Ašʿarī*, 318.

> you do not heed his screams, that you consider him a fool, that you hold him in contempt because he must rely on yelling, and that you reprove him for that, then he will calm down, he will be embarrassed and will retract his claims.[25]

Manners were thus endowed with more than a mere esthetic value. They were considered to be an integral part of the normative setting of the discussion, and as such closely linked to its content.

The code of polite behavior was, in a way, a reflection of the theoretical framework of the debate. A clear definition of both was considered essential for a successful discussion, and it was common practice to begin the discussion only after an explicit clarification of both the theoretical and the ethical parameters of the debate. This can be seen in the following two examples, taken from two of the better known accounts of *majālis*.

a) In the debate which supposedly took place in the *majlis* of al-Ma'mūn between ʿAbd al-ʿAzīz al-Kinānī and Bišr al-Mārisī, Kinānī was anxious to mark the common ground for the discussion: the Qur'ān and the Sunna. He also insisted, despite the Caliph's growing impatience, on delaying the beginning of the debate until he was given assurance that civilized procedure would be observed.[26]

b) In a famous description of the *majālis ahl al-kalām* in tenth-century Baghdād, Ḥumaydī recounts the impressions of a visitor from Spain. The witness describes two preparatory stages to the debate, the first ceremonial, the second more closely related to contents. As the participants gathered for the debate, the assembly rose respectfully to greet the dignitaries of each religious community. Then, at the outset of the debate, one of participants reminded all present that they should avoid preaching and citing their respective Holy Writs, and that they should bring forward only such arguments the premises of which were accepted by all parties.[27] As the outraged reaction of the orthodox witness makes clear, these two introductory stages stem from the same basic perception of the *majlis*'s nature and goal. The equally polite greeting of all participants foretell the tolerant theoretical parameters which were set as the common ground for the debate.

---

[25] Qirqisānī, *Anwār*, 488:16–19; see also *Naqd al-nathr*, 113:11–12.

[26] *Ḥayda*, 20–27. See Qāsim b. Ibrāhīm, *Rādd ʿalā al-mulḥid*, 17:7–9.

[27] Al-Ḥumaydī, *Jadwat al-muqtabas fī dhikr wulāt al-andalus*, ed. M. al-Tanjī, Cairo 1352H, 101–102.

The manners in the *majlis* thus reflected its basic principles. And similarly to a common theoretical frame of reference, good manners were considered a prerequisite for a fruitful discussion, inductive to Truth. Ibn Fūrak, for example, reports that the tenth-century theologian Aš'arī regarded good discussion manners as a religious obligation. According to him, through the disputation, a person should seek to fulfill the precept of "admonishing virtue and forbidding wickedness" (*al-amr bi'l-ma'rūf wa'l-nahy 'an al-munkar*). According to Aš'arī,

> In dialectical debates and disputations one should seek to get closer to God, the Exalted. They should serve as a way to worship Him and to fulfil His commands. Their motif should be the desire to achieve His reward and to avoid His punishment. When these are lacking, disputations have no reason except greed, obstinacy, or glee in defeating the opponent and overcoming him. Other animals, such as the stallions of camels, rams and roosters, share this drive to conquer.[28]

In other words, for Aš'arī, a debate represented the accomplishment of a religious duty. As such, it should be performed with the proper decorum, and should not be allowed to deteriorate into a cock fight.

Bad Manners

The extent to which manners were kept depended a great deal on the speakers' status. To judge from our texts, it would seem that members of religious minorities were often the target of discourteous behavior. This was especially true when high ranking officials were present.[29] An important Geniza fragment, published by Sasson Somekh and Mark Cohen, provides a good illustration of this fact. It recounts an incident in the court of the Fāṭimid vizier Ibn Killis. Ibn Killis used to convene sessions of theological discussion (*majlis al-kalām 'alā al-sharā'i'*), where philosophers, theologians, physicians, and other learned men from various religions were present.[30] The *majlis* was presented with an Arabic translation of the Jewish prayer book in Sa'adya's edition. The Jewish representatives, who were called upon

---

[28] *Mujarrad maqālāt al-Aš'arī*, 317:6–9. Cf. also *Naqd al-nathr*, 102:7–9.

[29] Van Ess, "Disputationspraxis," 29 and 34, cites the example of Timothy's disputation with the Caliph to demonstrate the delicate position of the Christians.

[30] For a description of these sessions, see Taqī al-Dīn al-Maqrīzī, *al-Mawā'iẓ wa'l-i'tibār bi-dhikr al-ḫiṭaṭ wa'l-āthār*, Cairo 1987, II, 6 and 341, cited in Cohen-Somekh, 294; Ibn Khallikān, *Wafayāt al-a'yān*, I, 34.

to defend the text in front of the assembly, were "vilified, ridiculed and scorned," and subjected to "ignominy and shame."[31]

To be sure, rudeness towards members of minority groups was not restricted to the courtly *majālis*. Many felt that they need not respect good manners when addressing a non-Muslim. ʿAbd Allāh ibn Ṭāhir al-Azdī recounts a disagreement he had with a Jew at Baghdād's market-place. Addressing his Jewish opponent, the words "You dog!" slipped from his mouth.[32] The mystic Abū Manṣūr al-Ḥallāj (d. 309/922), who happened to be passing by, rebuked him in strong terms. In all likelihood, Ḥallāj's standards of behavior were here, as in other matters, above those of most people.

It would be incorrect, however, to assume that bad manners were reserved to the treatment of minorities. Two famous ninth-century theologians, al-Ḥusayn al-Najjār and Ibrāhīm al-Naẓẓām, used to hold sessions of debate (*majālis wa-munāẓarāt*) with each other, until one of these debates proved fatal for al-Najjār. According to Ibn al-Nadīm, the debate had started calmly enough:

> Al-Najjār met al-Naẓẓām at the house of one of his colleagues. Ibrāhīm (al-Naẓẓām) greeted al-Ḥusayn (al-Najjār) and asked him: "Would you sit down, so that I may discuss theological matters with you?"[33] So al-Najjār sat down.

The formal agreement of the opponent was a prerequisite for the disputation.[34] Having received this agreement, al-Naẓẓām started to question al-Najjār on a problem related to free will. As one of the leaders of the *mujbira*, i.e., the predestinarians, al-Najjār claimed that human acts are created by God. Al-Naẓẓām tried to get him to "translate" this claim into terms which would make it easier for him to expose its absurdity. Al-Najjār refused to cooperate, obstinately sticking to his former formulas. Finally, al-Naẓẓām lost patience:

---

[31] Cohen-Somekh, 292–93.

[32] "*Kuntu uḫāṣimu yahūdiyyan fī sūqi Baghdād, wa-jarā ʿalā lafẓī an qultu lahu: yā kalbu;*" *Akhbār al-Ḥallāj*, ed. and tr. L. Massignon, Paris 1975³, #45, p. 133. The setting and the precise term (*uḫāṣimu*) suggest a chance disagreement. But a real *munāẓara* is not to be excluded, since the opponnent in a theological debate is often called the *ḫaṣm*; see, for instance, Qirqisānī, *Anwār*, 472:3.

[33] "*Tajlisu ḥattā ukallimaka.*" The expression as well as the presence of colleagues suggests that al-Naẓẓām was offering to hold a *majlis*. The formula "*lahu majālis wa-munāẓarāt maʿa fulān*" (which appears also here) seems to indicate that, like the term *munāẓara*, the term *majlis* could be used to indicate a disputation between only two people (in the presence of whoever happened to be there).

[34] Cf. van Ess, "Disputationspraxis," 34; *Naqd al-nathr*, 104:11–12.

> Then Ibrāhīm (al-Naẓẓām) kicked him and said: "Get up! May God disgrace whomever attributes to you any knowledge or understanding!" and he left, foaming with rage. This was the cause of a stroke from which al-Najjār died.[35]

As this anecdote vividly shows, rude behavior was part of the polemical reality. The participants in the debate held different views, and each regarded his own opinion as true and the other's as false. In their perception, the true opinion meant eternal salvation, while the erroneous opinion could entail eternal perdition. When the discussion threatened to have such grave consequences, few succeeded in keeping their calm, as in a purely intellectual excercise. It was only natural that even people who had begun the conversation courteously could forget their manners as the debate became heated.

The debate between al-Naẓẓām and al-Najjār had, perhaps, an unusually tragic outcome. But scenes of discourtesy and even violence must have been quite frequent. The Andalusian author Ibn Ḥazm counts the conditions for declaring an *inqiṭā*ʿ (i.e., the declaration that one of the participants has lost at one round of the debate). As a sixth condition he mentions bad manners, which he describes in the following words:

> [If one of the participants is] making insinuations by smiling to himself, or if he yells; if he imitates the other, or makes jest, or treats the other as a fool, or treats him rudely; or when he insults the other and calls him an infidel, curses and reviles him or makes foul accusations about his mother or father, let alone if this is accompanied by slapping and stamping the feet.

Ibn Ḥazm closes this long list with a grim observation:

> In the behavior of most people today, it is almost impossible to find any behavior except that which is listed above. May God help us![36]

In medieval disputations, however, bad manners were not only a fact of life, a reflection of human shortcomings and of the inability to abide by the rules. The codified rules of disputation, which set the standards of courtesy, were accepted by all participants. But another apparently accepted, though unwritten, code, allowed for a certain amount of discourtesy in the disputation.

---

[35] *Fihrist*, 179:12–22. Compare Dodge's translation, I, 446–47.
[36] Ibn Ḥazm, *Taqrīb*, 197.

The awareness of such acceptable rudeness is reflected in one of the earliest extant accounts of a theological debate. The ninth-century Jewish theologian al-Muqammaṣ recounts a disputation which he had held with a Muslim in Damascus. The topic of the discussion was the qualification of human acts. The Muslim, a certain Šabīb al-Baṣrī, claimed that all human acts are good. For him, "bad" is just a name we give to acts that cause us suffering, but this suffering may well be good for us, since we will be rewarded for it (presumably in the Hereafter). Muqammaṣ tried to demonstrate the absurdity of such a contention, which leaves no room for differentiation between various acts. According to him, Šabīb's claim would amount to condoning evil deeds like theft. It also may lead to the absurdity of blaming charitable deeds, since the receiver of charity is, as it were, robbed of the reward he could have got if he had suffered. Muqammaṣ described Šabīb in the highest laudatory terms, calling him "one of the best known ascetics of his community, and among the leading sages of his time." The disputation seems to have strictly followed the rules. In his argumentation against Šabīb, Muqammaṣ made no personal asides, did not lose his temper, and cannot be said to have been ill-mannered. And yet Muqammaṣ himself revealed the atmosphere of the debate (or at least, his own appreciation of this atmosphere) when he said about his opponent: "I have made fun (*laʿibtu*)" of him.[37] This boastful expression reflects first and foremost Muqammaṣ's pride at having defeated a particularly competent opponent. But it also shows that the rules for a civilized debate left ample room for carefully measured malice. The polemical techniques of exposing the absurdity of the opponent's claim allowed for scathing sarcasm. There were thus some kinds of humiliation and ridicule that were still acceptable within the code of manners. The atmosphere of the debate could thus become tense even when all formal rules were kept.

Maǧālis al-Ḫauf

The dialecticians themselves knew that rules alone were not enough to ensure a fair discussion. Qirqisānī's source instructs the disputants concerning a healthy setting for the debate, as follows:

[37] Muqammaṣ, *Twenty Chapters*, 249–51.

> Beware of theological discussions in the "terror session" (*majālis al-ḫauf*). Fear baffles the intellect, on which you must rely, and distracts it, so that instead of paying attention to the theological system, the intellect is preoccupied with anxiety.[38]
>
> Beware of anger and of theological discussions with people who make you angry. For anger kills ideas and prevents you from establishing them solidly.
>
> And beware of theological discussions with people you hate intensely. Hatred makes you impatient with your opponent and incites you to lose your temper for the slightest offense, and impatience oppresses the heart and weakens the intellectual faculties.[39]

In another passage, Qirqisānī describes yet another kind of *majlis* to be avoided, saying:

> Beware of *majālis* whose participants exercise no equity in their treatment, and do not grant you and your opponent equal sympathy and opportunity to be heard. They do not have the manners that will prevent them from hastening to judge against you, to show partiality (*ʿaṣabiyya*) to your opponent and to criticize you.[40]

In Qirqisānī's version, the last three instructions discuss the subjective feelings and frame of mind of the participants. But the first sentence describes an objective setting which induces this frame of mind. Qirqisānī does not say: "Beware of discussing theology with people from whom you are in awe." The feelings of fear against which he warns result from the character of the *majlis* itself rather than from the participant's psychological attitude. This kind of *majlis*, whose organisers are ill-intentioned from the outset, receives a name: *majlis al-ḫauf*. Presumably, this was not a term that the convener of the *majlis* would use. No one would receive a summons to a *majlis al-ḫauf*. And yet it is likely that everyone was familiar with the experience of participating in such sessions and knew what the name meant.[41] In theory at least, the experience was not connected to the frustration of failing to win the debate. As Qirqisānī's source reminds

---

[38] "*Wa-'ḥdhar al-kalām fī majālis al-ḫauf, fa-inna al-ḫauf yudhhilu al-ʿaqla alladhī minhu tastamiddu wa-tušghiluhu ḥirāsat al-rawʿ ʿan ḥirāsat al-madhhab.*" (Compare below, n. 45)

[39] Reading (with Vajda) "*quwā al-ʿuqūl*". Qirqisānī, *Anwār*, 492:10–14; French translation in Vajda, "Qirqisani," 48.

[40] Qirqisānī, *Anwār*, 492:15–17. *ʿAṣabiyya* is the loyalty, devotion and partiality that one owes one's family, tribe or nation. In the context of the debate, Qirqisānī probably intends first and foremost to warn against zealous loyalty to one's religious community. Cf. also *Naqd al-nathr*, 111:16, 113:16.

[41] A vivid description of the fear and the way in which it is instilled is given in *Ḥayda*, 12–15, esp. 12:9; cf. also Marcuzzo, *Dialogue*, 335.

us, such occasional frustration was inevitable.[42] A *majlis* was called "terrorizing" when intimidation was used as standard practice, in other words, when the manners of disputation were not respected.[43]

A very similar description appears in Ibn Fūrak's *Mujarrad Maqālāt al-Aš'arī*. In a chapter on *ādāb al-jadal*, rightly praised by Daniel Gimaret as "une belle leçon d'honnêteté et d'esprit scientifique,"[44] Ibn Fūrak says:

> One should beware of theological discussions in the terror sessions (*majālis al-ḫauf*), because it (i.e., the terror) baffles the intellect and distracts it, so that instead of paying attention to the theological system, the intellect is preoccupied with controling one's temper.[45]

Qirqisānī's other injunctions also appear in Ibn Fūrak, who combines the warning against *'aṣabiyya* with the demand to avoid anger. Ibn Fūrak has also a name for the sessions where partiality and frivolity are frequent occurrences. He calls them *majālis al-ṣudūr*,[46] an expression which may be translated as "sessions of high-officials." He says:

> One should also beware of discussing theology in the sessions of high-officials, where the participants seek to be diverted by what they hear, and do not grant the two opponents equal sympathy or equal opportunity to be heard. Participation in such sessions is either folly and subjection to humiliation or an incentive to anger and a source of depression.[47]

Like "terror sessions," "sessions of high-officials" are to be avoided because the participants are not likely to follow the code of manners. And as the context makes clear, both kinds of sessions must

---

[42] Qirqisānī, *Anwār*, 493:9–11. See also Ibn Ḥazm, *Taqrīb*, 188:5, who insists that losing one round (*inqiṭā'*) is not equivalent to defeat (*falj*).

[43] Cf. also Ibn Ḥazm, *Taqrīb*, 187:19–20: "We refuse to speak with him who instills fear, for the debate must be held in safety."

[44] Gimaret, "Document," 216.

[45] "*Wa-yajibu an yuḥdhara al-kalām fī majālis al-ḫauf, fa-innahu yudhhilu al-'aqla wa-yušghiluhu bi-ḥirāsat al-rūḥ* (or read: *raw'*) *'an ḥirāsat al-madhhab*" Cf. *Mujarrad maqālāt al-Aš'arī*, 317:19–20; comp. n. 38 above.

[46] The *ṣadr* "denotes an eminent or superior person or *primus inter pares*, whence its use for a chief, president or minister;" cf. "*Ṣadr*," *EI*², VIII, 748 (J. Calmard); cf. also E. W. Lane, *Arabic-English Lexicon*, New York 1956, IV, 1661: "*Ṣadr al-majlis*: The upper, or highest, part (of the sitting room); . . . *Ṣadr al-ṣudūr*: the chief of chiefs, . . . the chief judge." *Ṣadr al-majlis* then comes to mean the person who presides over the session (I am indebted to J. Sadan for drawing my attention to this meaning of the word). It thus seems that *majālis al-ṣudūr* are those sessions organized by or for the high officials, perhaps mainly for their entertainment.

[47] *Mujarrad maqālāt al-Aš'arī*, 317:20–23.

be shunned because they divert the participants from the true religious purpose of the disputation.

The similarity between the testimony of Qirqisānī and that of Ibn Fūrak points to a common source. Ibn Fūrak identifies his source: he claims to bring Ašʿarī's opinion on this subject, which he had compiled and edited from dispersed sayings in Ašʿarī's books.[48] It is therefore likely that Ašʿarī was also the source of Qirqisānī. Moreover, both the context and the content of the passage allow us to identify the book of Ašʿarī used by Qirqisānī and Ibn Fūrak. In the preceding chapters, Ibn Fūrak cites Ašʿarī's "refutation of Balḫī concerning the latter's objection to Ibn al-Rāwandī, (where) he (i.e., Balḫī) had claimed to correct Ibn al-Rāwandī's errors concerning the art of polemics."[49] It is plausible that Ibn Fūrak and Qirqisānī drew their respective chapters on *adab al-jadal* from this book.[50]

Bad Manners as Sabotage: The Case of Ibn al-Rāwandī

The identification of the source for these passages points to the theologians' reaction to Ibn al-Rāwandī as an important factor in the shaping of the literature on the subject of *adab al-jadal.*

In a short study on the art of disputation, Larry Miller argued that, although Ibn al-Rāwandī's *K. Adab al-jadal* is not extant, we can have a fairly good idea about its contents, as it apparently served as a model for subsequent works on *jadal.* As examples of such subsequent works, Miller names Kaʿbī, Ašʿarī, Qirqisānī, and Maqdisī.[51] Similarly, van Ess has suggested the possibility that Ibn al-Rāwandī was the source of both Qirqisānī and Ibn ʿAqīl.[52] A closer look at the available evidence confirms the suggestion that these authors were familiar with Ibn al-Rāwandī's book. It also shows, however, that none of these authors can be categorically pronounced to have used Ibn al-Rāwandī as a model for imitation. Abū al-Qāsim al-

[48] Ibid., 317:2–3.

[49] Ibid., 310; Cf. Gimaret, "Bibliographie," 247–49.

[50] The possibility that Qirqisānī was using Ašʿarī's response to al-Balḫī is further enhanced by the fact that both these sources refer to the anthropomorphists by the name *jismiyya,* rather than the more common *mujassima*; see *Mujarrad maqālāt al-Ašʿarī,* 315:6; Qirqisānī, 473:9.

[51] Miller, "Fārābī's Dispute," 185.

[52] Van Ess, "Disputationspraxis," 32; See further idem, *TG,* IV, 306–307.

Ka'bī al-Balḫī (d. 931) wrote a refutation of the *K. al-Jadal, K. Iṣlāḥ ghalaṭ Ibn al-Rāwandī.*[53] Aš'arī responded to the latter,[54] and, as we have seen above, it is this response which probably served as Qirqisānī's source. The same may well hold for Maqdisī and Ibn 'Aqīl.

In the fragments of Aš'arī's book cited by Ibn Fūrak, Aš'arī backed Ibn al-Rāwandī against al-Balḫī. But it is obvious that Aš'arī was more interested in exposing al-Balḫī's mistakes than in defending Ibn al-Rāwandī. Aš'arī's general opinion of Ibn al-Rāwandī, and in particular of his intellectual honesty, was rather low.[55] Nonetheless, he felt compelled to examine each of his arguments objectively, regardless of his opinion of their author. He could thus find himself supporting Ibn al-Rāwandī's technical argumentations while rejecting his underlying system.[56]

The same ambivalent attitude can be seen in Tawḥīdī's dispersed references to him. In his *K. al-Baṣā'ir*, Tawḥīdī says that Ibn al-Rāwandī

> makes no mistakes and his speech is free of dialectical barbarism. For he is a superb theologian, a discerning critic, an inquisitve dialectician,[57] and a persevering observer.[58]

This outpouring of praise on Ibn al-Rāwandī's linguistic excellence and erudition does not prevent Tawḥīdī from counting him among the heretics.[59] Such ambivalent attitude to Ibn al-Rāwandī, which, in the case of Aš'arī, can only be inferred from his writings, is made explicit in the tenth-century *K. Naqd al-nathr.* This book contains a chapter on dialectical disputation, where the etiquette of disputation is also discussed. Like Qirqisānī, the author of this chapter cautions his reader to attend only such meetings where the atmosphere is not hostile. But in his instructions, the injunction is given a religious meaning and is put in a historical context:

---

[53] Abū Rašīd al-Naysaburī, *Masā'il al-ḫilāf*, cited in M. Horten, *Die Philosophie des Abu Raschid*, Bonn 1910, 202; van Ess, "Disputationspraxis," 31, n. 2.

[54] Thus van Ess, "Disputationspraxis," 31, n. 3, citing McCarthy, *The Theology of al-Ash'ari*, Beirut 1953, 218, nr. 27. Compare Gimaret, note 49 above.

[55] Cf. Aš'arī, *Maqālat*, II, 222.

[56] A similar case of unbending intellectual honesty can be seen in Ibn al-Haytham, who criticized Jubbā'ī's refutation of Ibn al-Rāwandī; see above, Ch. 2, n. 133.

[57] *Baḥḥāth jadil*; lit.: "a disputative inquirer."

[58] *Li-annahu mutakallim bāri' wa-jahbadh nāqid wa-baḥḥāth jadil wa-naẓẓār ṣabūr; Baṣā'ir*, 183.

[59] Tawḥīdī, *Imtā'*, II, 20:9–14; and see above, Ch. 2, nn. 115–17

> The disputant should aim at the truth and his desire should be to achieve that which is right. If he perceives himself as endowed with strength of character, sound discernment, beautiful ideas, fine intuition, [capable of] clear presentation and solid argumentation, all this should not bring him to hasty affirmation and negation of anything. He must not begin to draw arguments for and against. Such haste is likely to tarnish the brightness of his science, to extinguish the light of his understanding and to make pious and devout people treat him as a heretic and as a skeptic.
>
> It is for this reason that people rejected al-Rāwandī and his ilk, despite their strength as dialecticians and their ability in speculative philosophy.[60]

While acknowledging Ibn al-Rāwandī's excellence in dialectic, the author of the *Naqd al-nathr* disqualifies him as a model. He warns the dialectician against using his skill for a "hasty affirmation and negation" and for drawing "arguments for and against."[61] As a prototype of a person who did misuse such skill he mentions Ibn al-Rāwandī.

The wording of this passage ("hasty affirmation and negation") is such that it can be understood as merely a condemnation of hastiness, whether in affirming or in negating. But another passage makes it clear that this is not the correct interpretation of the text here. A few pages later the author of *Naqd al-nathr* says:

> [The disputant] should further know that he will not be considered a skillful dialectician unless he is able to use his sound intuition, beautiful presentation and suave speech in order to make truth appear as falsehood and falsehood appear as truth whenever he decides to do so, and to make people take the one for the other.[62]

The words used here to describe the dialectician's talents are almost identical to those used in the previous quotation. And here too *Naqd al-nathr* strongly admonishes the disputant to use this skill only in good faith.[63] When one reads the two paragraphs together, it becomes obvious that the author of *Naqd al-nathr* chose his words carefully when he warned the skillful dialectician against the temptations of

---

60 *Naqd al-nathr*, 111; compare the French translation in Vajda, "Qirqisani," 65.

61 "*. . . an yusriʿa fī ithbāt al-šayʾa wa-naqḍihi, wa-yašraʿa fī al-iḥtijāj lahu wa-li-ḍiddihi*," *Naqd al-nathr*, 111:4.

62 "*Wa-an yaʿlama . . . annahu lā yuʿaddu fī al-mujādilīn al-ḥudhdhāq ḥattā yakūnu bi-ḥusni badīhatihi wa-jūdat ʿāriḍatihi wa-ḥalāwati manṭiqihi qādiran ʿalā taṣwīr al-ḥaqq fī ṣūrati al-bāṭil wa'l-bāṭil fī ṣūrati al-ḥaqq*." *Naqd al-nathr*, 114:10–13; compare Ibn Ḥazm, *Taqrīb*, 195:4–6.

63 *Naqd al-nathr*, 114:17–18.

his own talents. The danger he had in mind was the temptation to use the skill for both the affirmation *and* negation of anything, to draw arguments both for *and* against. As a prototype of the dialectician who succumbed to this temptation, he cited Ibn al-Rāwandī. And indeed, other sources tell us that Ibn al-Rāwandī was willing to use his excellent technique to prove both an argument and its opposite, and that he used to write books and then refute them.[64]

It is important to note that *Naqd al-nathr* mentions Ibn al-Rāwandī's shortcomings in the context of the etiquette of behavior and the moral requirements of the disputation rather than in the context of logic. The diagnosis of Ibn al-Rāwandī's weakness as related to manners is corroborated by still another source, Ibn al-Jawzī, who says:

> This cursed one (i.e., Ibn al-Rāwandī) mentioned other things in the same vein. He combined them with his bad manners (*sū' adab*) and with jokes in ill-taste, and he referred to the Creator, praised be He, in a way in which it is inappropriate to refer even to common people. We have never heard anyone defame the Creator and make jest about him as this cursed one did.... Had he denied the Creator altogether, it would have been better for him than to admit His existence, then to polemicize against Him and defame Him.... May God retaliate his deeds in the Hereafter in a more severe manner than He retaliates Iblīs. For Iblīs, although he contradicted God, at least retained a respectful form of address, as He said (Q. 38:82) "By Your Glory," and he did not answer back in such ill-mannered way as this cursed one.[65]

The sin of Ibn al-Rāwandī is identified in this paragraph not as breaking the rules of the *munāẓara*, but as blasphemous speech.[66] We know indeed that Ibn al-Rāwandī's Qur'ān criticism was characterized by its sarcasm (e.g., his suggestion that God is weak in arithmetic).[67] Understandably, the intellectual openness and the sense of humor of Muslim orthodoxy ended at this point. It is also not incidental that, instead of repeating the accusation of heresy, Ibn al-Jawzī prefers to criticize Ibn al-Rāwandī's ill-mannered polemics. For in the field of *jadal*, Ibn al-Rāwandī served as both a positive and a negative model: he was a master of the technique, and a master in abusing it.

---

[64] See above, Ch. 2, nn. 146, 149.

[65] Ibn al-Jawzī, *Muntaẓam*, 117:8–10.

[66] Blasphemous disrespect is mentioned as "*sū' adab*" also by Ibn Taymiyya, who regards disrespect towards Muslim sages as breaking the laws of disputation ("*innahu asā'a al-adab 'alā al-salaf*"); cf. *Tafsīr sūrat al-nūr*, Cairo 1343H, 113.

[67] Ibn al-Jawzī, *Muntaẓam*, 113:7–10; and see below, Ch. 7, apud n. 45.

Dialectics was the domain of the theologians, but Ibn al-Rāwandī's manipulations here also triggered a reaction on the part of the philosopher Fārābī (d. 339/950), who wrote a *K. al-Radd ʿalā Ibn al-Rāwandī fi'l-jadal.*[68] Larry Miller has suggested that for Fārābī "*jadal*" was identified as the subject of Aristotle's *Topica*, i.e., dialectics as a way to reach probability rather than certain knowledge. For Miller, Fārābī's criticism of Ibn al-Rāwandī reflects mainly the different approaches of the philosopher and the theologian. Miller therefore concludes that "disagreement over basic epistemological issues may have led Al-Fārābī to compose his refutation of Ibn al-Rīwandī's work on the *adab al-jadal.*"[69] Joseph van Ess too sees this book as a result of the tension between *kalām* and *falsafa.*[70] Van Ess associates the use the term *ādāb al-jadal* in the title of this book with the fact that it was directed against the Muʿtazilite *mutakallim* Ibn ar-Rāwandī, and that it was written "in the . . . desire of making Aristotelian logic acceptable for *kalām.*"

It is indeed probable that Fārābī's refutation of Ibn al-Rāwandī was phrased in terms of Aristotelian epistemology. It is also quite possible that Fārābī took this opportunity to express once more his basic disagreement with and contempt for the *mutakallimūn*. In fact, Fārābī's presentation of the arguments of the *mutakallimūn* reflects Ibn al-Rāwandī's arguments in the *K. al-Zumurrud.*[71] While drawing upon Ibn al-Rāwandī's work, Fārābī apparently saw his dialectic as typical of the *kalām*.

Nevertheless, it seems that on the issue of Ibn al-Rāwandī's dialectics, Fārābī may have been closer to the *mutakallimūn* than he would have liked to admit. For, as we have seen, not only philosophers, but also theologians like Kaʿbī and Ašʿarī took exception to Ibn al-Rāwandī's dialectics. For Fārābī, Ibn al-Rāwandī's dialectic may have presented a true portrait of the typical theologian, while for the theologians it was nothing but a nasty caricature. Yet for theologians and philosophers alike, there was something intrinsically

---

[68] IAU, II, 139:7. In Ibn al-Qifṭī, 279:15, this becomes two books: *Fī adab al-jadal* and *al-Radd ʿalā al-Rāwandī*; cf. van Ess, "Disputationspraxis," 31, n. 4.

[69] Miller, "Fārābī's Dispute," 188.

[70] "Logical Structure," 22, n. 6.

[71] Ibn al-Rāwandī's "intellectualist" argument against prophecy appears in Fārābī's *Iḥṣāʾ al-ʿulūm*, without Ibn al-Rāwandī's name and without the heretical conclusions; cf. Al-Fārābī, *Iḥṣāʾ al-ʿulūm*, ed. ʿUthmān Amīn, Cairo 1948, 107–111; compare van Ess, "Al-Fārābī," 394–98. On the philosophers' attitude to Ibn al-Rāwandī, see p. 69 above, and further below.

wrong with Ibn al-Rāwandī's *adab al-jadal*, even though his book on the subject was not counted among his heretical works. As *Naqd al-nathr* tells us, it was not Ibn al-Rāwandī's technique which was faulty. It was the underlying ethos which was lacking.

Ibn al-Rāwandī did not abide by the rules of the debate. The first direct source about him, al-Ḫayyāṭ, accused him of breaking basic rules of quotation: he attributed to his adversaries things they never said,[72] he quoted half-sentences out of context.[73]

While excelling in the techniques of dialectics, Ibn al-Rāwandī, like the sophists of old, questioned the efficacy of these techniques as a means for reaching truth with certainty. He rejected the notion of prophecy and preached reliance on reason. But it is not at all clear that he saw reason as an alternative way to achieve the same authoritative knowledge which was offered by prophetic religions. It seems that for Ibn al-Rāwandī knowledge was a dynamic, ever growing, ever changing process.[74] His provocative style and shifting positions, his manipulation of the rules and regulations of the debate and his insolence were for him the means to express his metaphysical stance. In other words, he used his mastery of the Art of Disputation in order to sabotage the *majlis* system and to reject its ethos. The theologians understood this well, and their reaction to his "bad disputation manners" did not reflect simple rebuke or disdain, as one would expect for bad manners. They took Ibn al-Rāwandī's bad manners for what they really were: a threat to the very core of their theological quest. His skeptical attitude was perceived by everyone else as permitting a cynical, mercenary use of the technique, and therefore the method itself was judged as sophistry. Ibn al-Rāwandī refused to abide by the rules of the *munāẓara*. He came to distrust the theologians and their system. But rather than walking out, he succeeded in manipulating the system, sabotaging it from within.

When Ibn al-Rāwandī became disillusioned with the theologians' wisdom, he also rejected their epistemology, logic, and their *majlis* system. But rather than entering the debate on the terms of his opponents, he transformed it into a farce. His rejection of disputation manners thus reflected a rebellion against the theologians and their claim to monopolize truth.

---

[72] *Intiṣār*, 11 (Aʿsam, *History*, 18).
[73] Ibid., 52 (Aʿsam, *History*, 27).
[74] A similar perception of knowledge [but without Ibn al-Rāwandī's cynical edge] is formulated clearly by Abū Bakr al-Rāzī; See Rāzī, *Šukūk*, 3; Pines, "Rāzī," esp. 482.

The purposefully ill-mannered behavior of a skeptic like Ibn al-Rāwandī contributed, in its turn, to the further development of the medieval codes of the art of disputation. Under attack, the theologians strove to reaffirm their convictions. As I have tried to show, Ibn Fūrak's "beautiful lesson in integrity and scientific spirit" was learnt the hard way, and Ibn al-Rāwandī was instrumental in teaching it. His manipulations of the system, his sarcasm, his out-of-context citations and his sneers, encouraged the theologians into formulating more clearly their code of manners and tying it more securely to their ideas.

4. Repercussions in Islamic Philosophy

Polemics is the art of theologians, and therefore, the influence of Ibn al-Rāwandī and Rāzī on the *mutakallimūn* is, in a way, expected. It was mostly theologians who responded to their books and tried to refute them.[75] In the preceding sections we have examined the impact of Ibn al-Rāwandī's heterodoxy on Muʿtazilite positions.[76] In the remaining part of this chapter, we shall investigate the possibility that the freethinkers also affected positions adopted by the philosophers.

We have seen above that philosophers like Kindī and Fārābī were aware of Ibn al-Rāwandī's writings and responded to it. Their responses did not survive, but one suspects that their reaction may have been less vociferous than that of the theologians. If, however, Kindī took the trouble to refute Ibn al-Rāwandī's views concerning God's unity, and Fārābī to refute his book on *jadal*, it is legitimate to ask whether the philosophers' encounter with the freethinkers had a lasting impact on their own thought.

In fact, Fārābī was familiar not only with Ibn al-Rāwandī's work, but also with Rāzī's, and more specifically, with his philosophical (rather than scientific) work. In Ibn Abī Uṣaybiʿa's bibliography of Fārābī we find twice "*K. al-Radd ʿalā al-Rāzī fi'l-ʿilm al-ilāhī.*"[77] As mentioned above, Kraus had suggested that in this book Rāzī discussed all the topics of his philosophy. We have also seen that both

[75] Cf. van Ess, *TG*, VI, 438–42 (list of refutations written against Ibn al-Rāwandī); Kraus, *Rasāʾil*, 167–70 (list of refutations written against Rāzī's *K. al-ʿilm al-ilāhī*).
[76] Cf. above, pp. 171–72.
[77] IAU, 608; cf. Ibn al-Qifṭī, 280.

Bīrūnī and Nāṣir-i Ḫusraw quote passages from this book which contain some of Rāzī's more virulent anti-prophetic sayings.[78] It follows that Fārābī was probably familiar with the anti-prophetic aspect of Rāzī's philosophy, although we do not know if it was this aspect of the *K. al-ʿIlm al-ilāhī* that he chose to refute.

Rāzī was no political philosopher. And yet Ibn Abī Uṣaybiʿa's list of his work contains one title that does not fit this generalization: "On the Perfect Conduct and on the Conduct of the People of the Perfect City" (*K. Fī'l-sīra al-fāḍila wa-sīrat ahl al-madīna al-fāḍila*).[79] This title is so out-of-line with Rāzī's thought, and so strikingly reminiscent of the title of Fārābī's famous book, that Kraus discarded it as a mere mistake, resulting from confusion. Alternatively, Kraus suggested that this book was identical with Rāzī's *K. al-Sīra al-falsafiyya*.[80] Of course, the possibility of confusion cannot be ruled out; but it is worth noting that nothing in Rāzī's *K. al-Sīra al-falsafiyya* could be naturally associated with the notion of *madīna fāḍila*.

Without trying to decide on the identity of the book intended by this puzzling title, its very existence forces us to consider the relations of Rāzī's philosophy to that of Fārābī. This was already suggested in a general way by Paul Walker, who said that "[Rāzī's] value as a negative voice against which an 'Islamic' political philosophy was to develop must be taken into account."[81] Beyond this point, Walker did not pursue the venue which he had suggested. This line of thought, however, is well worth pursuing. Walker's research led him to concentrate on the undeniabe lack of a political dimension in Rāzī's thought. He admonishes students of Islamic philosophy to remember that Rāzī preceded Fārābī, and that Rāzī's "failure to complete his philosophy by offering a place in it for both politics and religion becomes remarkable only in the face of Fārābī's achievement of precisely this combination."[82] This somewhat apologetic conclusion underlines the

---

[78] See above, Ch. 3.

[79] IAU, 426:3. Ibn al-Nadīm mentions only *K. Fī'l sīra al-fāḍila*, which is listed twice (*Firhist*, 301:12, 23). Incidentally, the next title in both sources is the puzzling *Fī wujūb al-duʿā' wa'l-daʿāwā* (or *Fī wujūb al-adʿiya*), which smacks of Ismāʿīlī vocabulary; see above, Ch. 3, n. 139. There is a remote possibility that this book reflected on some aspect of political or educational thought. But the sad fact is that we know nothing of its content. The order of the titles in Ibn al-Nadīm and IAU is, of course, often a mere chance.

[80] *Rasā'il*, 97.

[81] Walker, "Political Implications," p. 66.

[82] Ibid., 73–74.

novelty of Fārābī's political thought; if political thought is a new theme, introduced by the later philosopher, one cannot criticize the earlier one for ignoring it. The difference between the two philosophers is presented mostly in terms of chronology. Indeed, since Walker defines the topic of his investigation as political thought, which, it is agreed, Rāzī did not develop, it is hardly possible to speak of his influence.

If, however, we replace the broad modern term "political thought" by the more archaic and more specific "prophetology," the difference between the two philosophers receives its full significance. Rāzī is not known to have been an anarchist, who rejected the notion of a state (or a "city"), or the role of a leader. It was specifically the notion of prophecy and the role of the prophets in history which he denied. We have seen above that Rāzī's only political treatise, the *Epistle on the Marks of Political Ascendency* was probably a (sarcastic) response to the *Signs of Prophecy* genre, and in particular to the Ismāʿīlī authoritarian prophetology.[83] Disagreement on the notion of divinely inspired leadership was at the centre of Rāzī's discussion with the Ismāʿīlī Abū Ḥātim al-Rāzī, and his *K. Maḫārīq al-anbiyā'* was a sustained attempt to reject the Islamic view of the role of prophets. It is precisely concerning these notions that Fārābī made his most innovative contributions. The Greek philosophers who had developed a political theory did not need to relate it to a prophetic lore; with the exception of Philo, ancient Judaism did not develop a systematic philosophy before the ninth century; and the first Muslim philosopher, Kindī, was not particularly interested in political philosophy. The innovation occurred in the tenth century, in the work of Fārābī, who identified Plato's philosopher-king with a prophet, thus initiating the remarkable medieval tradition of political thought.[84]

The milieu in which Fārābī lived offers more than one possible source of influence in this respect. Ismāʿīlī thought, for instance, undoubtedly played some part in stressing the role of the prophet in society.[85] The concentration on prophecy points to Rāzī as to

[83] See above, Ch. 4, apud nn. 81–82.

[84] Cf. Pines, "Philosophy," 785.

[85] Cf. Daiber, "The Ismaili Background," esp. 144 and 148; Idem, "Political Philosophy," 848–49. But see S. Pines, "Some Problems of Islamic Philosophy," *Islamic Culture* 11 (1937), 71 (= *Collected Works*, III, 52), who points to Fārābī's influence on the political thought of *Iḫwān al-ṣafā'*. For an example of Fārābī's identification of *imām* with *faylasūf*, see *Taḥṣīl al-saʿāda*, ed. Jaʿfar Yāsīn, Beirut 1983, 93.

another likely source of influence on Fārābī (and perhaps also on the Ismā'īlīs). Prophecy played a major role in Fārābī's political philosophy, and it may well have been developed against Rāzī's negative voice.

The impact of this innovation was formidable. After Rāzī, rationalists in general and philosophers in particular stood the risk of being automatically suspected of freethinking. By introducing the prophets into the philosophical hierarchy, Fārābī succeeded in diverting the fire from the philosophers.

This does not mean that orthodox thinkers accepted Greek philosophy. It only means that they no longer could accuse the philosophers of plainly rejecting the notion of prophecy. Thus, when Ghazālī listed the philosophical ideas for which, according to him, the philosophers should be branded as infidels, he listed their denial of three doctrines: creation *ex nihilo*, God's knowledge of the particulars, and resurrection of the dead. Prophecy does not appear in this list, nor in Ghazālī's list of the philosophers' seventeen less grievous offenses. Nevertheless, since Ghazālī wants to bring home to his reader the seriousness of the philosophers' heterodoxy, he immediately continues the list of three heretical doctrines by saying:

> These three doctrines are in no way compatible with Islam; anyone believing in them, believes that the prophets lie, and that they have said what they did [only] as a stratagem, speaking in symbols to facilitate the multitudes' understanding; this is sheer disbelief, which none of the Muslim sects upholds.[86]

Ghazālī's words demonstrate that he still considered the impingement on prophecy to be the worst possible disbelief. The fact that the philosophers' doctrines may lead in this direction represents, for him, a serious offense. But Ghazālī also exemplifies, perhaps better than most theologians, that as far as the declared belief in prophecy was concerned, the *falāsifa* were not suspect.

Another example can be seen in the work of another pillar of orthodox Islam, Ibn Taymiyya. His book on "The Distinction between the Friends of the Merciful God and the Friends of Satan" was written as a response to ideas circulated by the ṣūfīs, and in particular by Ibn 'Arabī. Ibn Taymiyya takes this opportunity to fight other heretical thinkers, including philosophers like Avicenna. He begins

[86] Abū Ḥāmid al-Ghazālī, *Tahāfut al-falāsifa*, ed. M. Bouyges, Beirut 1927, 378.

by saying that the "core of disbelief and hypocricy is disbelief in the prophets and their message."[87] Like Ghazālī, Ibn Taymiyya accuses latter-day philosophers, like Avicenna, of interpreting prophecy as a divine ruse meant for the benefit of the multitudes. In the final analysis, he argues, what they say amounts to a repudiation of the principles of faith.[88] But he has to admit that all their false ideas are part of their attempt "to establish the concept of prophecy on their false principles."[89]

It thus seems that, from the time of Fārābī on, the prophetology of the *falāsifa* immunized them to the suspicion of freethinking. Once the *falāsifa*, at the forefront of rationalism, had integrated prophecy into their thought, the accusation of freethinking would become increasingly similar to that of atheism: the philosophers' doctrine was said to be *like* denying prophecy, or it *amounted* to denying prophecy. The *falāsifa* did not profess real, open and blatant freethinking, and they were not accused of professing it.

[87] *Al-Furqān bayna awliyā' al-raḥmān wa-awliyā' al-šayṭān*, Damascus 1962, 37.
[88] Ibid., 93: *wa-ghāyat ḥaqīqat hā'ulā'—inkār uṣūl al-īmān.*
[89] Ibid., 86: *an yuthbita amra al-nubuwwāt 'alā uṣūlihim al-fāsida.*

CHAPTER SEVEN

FROM MUSLIM HERESY TO INTER-RELIGIOUS POLEMICS

Islamic theology developed as the art of the polemicists, and the *mutakallimūn* were also known as *ahl al-jadal*. The examination of one theological theme, the "signs of prophecy," has allowed us to see how inter-religious encounters contributed to the crystallization of Islamic prophetology during the first centuries of Islam. The freethinkers reacted to this prophetology, and their rebellion can thus be considered a typically Islamic phenomenon. The theme of freethinking, however, did not remain contained within the arena of inner-Islamic polemics. The same theologians who took it upon themselves to preserve the integrity of Islam by fighting Muslim heretics also felt called upon to defend Islam against other religions; their activity in one domain naturally affected the other. The theme of freethinking, and the arguments connected to it, were thus integrated quite early into interreligious polemics. The present chapter will investigate two cases in which Ibn al-Rāwandī's arguments against Islam seem to have found their way into Islamic polemics with Christianity and Judaism.

1. Christian anti-Muslim Polemics: The *Risāla* of ʿAbd al-Masīḥ al-Kindī

The *Epistle* (*Risāla*) of ʿAbd al-Masīḥ b. Isḥāq al-Kindī is presented as the second part of a theological correspondence between courtiers and men of letters. ʿAbdallāh al-Hāšimī, a pious Muslim from the family of the Caliph al-Maʾmūn (813–833), had written to his colleague Kindī, a Nestorian Christian dignitary from the court, inviting him to convert to Islam. In response to this invitation Kindī wrote an outspoken defense of Christianity and attack on Islam.[1] The *Risāla* was first published in the West by William Muir at the

[1] The earliest manuscripts, written in Karshūnī, are copies of manuscripts dated 1173 C.E; cf. Tartar, *Dialogue*, 22.

end of the nineteenth century.[2] Muir accepted the dating offered in the text itself, as did Armand Abel and, much more recently, Bruce Lawrence and Georges Tartar, all of whom treated the *Risāla* as an early ninth-century document.[3] Louis Massignon and Paul Kraus, on the other hand, opted for a later date.[4] Kraus pointed out that the earliest evidence for the existence of the text is found in Bīrūnī. He therefore suggested that the *Risāla* was composed around the beginning of the tenth century. This relatively late date of composition served Kraus as important supporting evidence for his theory that Ibn al-Rāwandī was the earliest thinker to refer to the *Barāhima*, and thus the source of all later authors. Kraus's arguments for the later dating are, I believe, convincing. If we accept his dating, then several of the Christian arguments in the *Risāla* would go back to Ibn al-Rāwandī's *K. al-Zumurrud*.

Related to the question of dating the *Risāla* is the question of its authenticity. For both Muir and Tartar the text records the authentic exchange of letters between two friends belonging to two different religions.[5] Kraus, on the other hand, regards the *Risāla* as a literary fiction in which the names of both protagonists do not reflect historical personalities, but stand for typological representatives of the two religions. Kraus also pointed out the sharp tone of the *Risāla*, a sharpness unequalled in Christian polemical literature. Kraus assumed that no Christian would have dared to speak so derisively of Islam in a real debate.

Indeed, although Christian-Muslim debates and dialogues were probably quite frequent, this *Risāla* bears the marks of a wholly Christian composition.[6] This is particularly evident in the epistle which is presented as composed by the Muslim protagonist. ʿAbdallāh al-Hāšimī appears to be not only thoroughly acquainted with Christianity, but also very attached to it. He seems to relish the opportu-

[2] W. Muir, *The Apology of Al Kindy*, London 1882; *Risālat ʿAbd Allāh bn. Ismāʿīl al-Hašimī ilā ʿAbd al-Masīḥ bn. Isḥāq al-Kindī*, London 1912.

[3] Muir, *Apology*, 24–36; Abel, *Apologie*, 523 (quoted in Tartar, *Dialogue*, 44); Lawrence, *Shahrastānī*, 76; Tartar, *Dialogue*, 59–66.

[4] Kraus, "Ketzergeschichte," 335–41; art. "al-Kindī, ʿAbd al-Masīḥ," *EI*, II, 1021 (L. Massignon).

[5] Cf. Tartar, *Dialogue*, 54.

[6] Compare Tartar, *Dialogue*, 48–54. Tartar believes the two epistles which constitute the text to have been written by two different authors, a Muslim and a Christian. According to him, the analysis of the two epistles proves the different religious denomination of their respective authors.

nity to indulge in detailed descriptions of Christianity and its traditions, regardless of whether or not these descriptions are needed to prove his point. His insistence on Christian piety is much more lengthy than what seems to be required for the sake of politeness or even as calculated flattery.[7]

The *Risāla* contains arguments that are strikingly similar to those in the *K. al-Zumurrud.* In response to the doctrine of *iʿjāz al-Qurʾān*, the *Risāla* says that Muḥammad's challenge to the Arabs to produce verses similar to those of the Qurʾān was left unanswered only because the Qurʾān "was given to ignorant people, the vile, barbarian, and uncouth Nabateans."[8] It attempts to ridicule Muslim tradition by pointing out its irrational aspects: if the angels were willing to help Muḥammad at Badr, where were they at Uḥud, when their help was so badly needed?[9] The *Risāla* mentions some of the miracles attributed by Muslim tradition to Muḥammad: Umm Maʿbad's sheep, the intoxicated sheep who talked, and the wolf who talked.[10] It ridicules the same elements of Muslim ritual, and in particular the rites of the *ḥajj*: throwing stones, circumambulating a house that cannot respond to prayers, running between stones that can neither help nor harm.[11] These rites are compared by the author to the cult of the Buddhists (*Sumaniyya*)[12] and the *Barāhima.* Kindī's concludes his tirade, saying:

> What is blameworthy with a person who shaves his head, undresses himself, runs and throws stones is that only someone who has lost his reason and denied his intellect acts in that way, someone who is possessed of demons.

---

[7] Cf. Kindī, *Risāla*, 12–15.

[8] Kindī, *Risāla*, 123: "*qawm ummiyyīn, anbāṭ suqqāṭ ʿajam ʿulūj.*" A similar comparison of the Arabs with the Nabateans was made by Ibn al-Rāwandī in his *K. al-Dāmigh.* According to Ibn al-Jawzī, Ibn al-Rāwandī said that the brocade which the Qurʾān promises for the righteous in paradise is of the coarse kind, suitable for drapery rather than for clothes. Any person imagining himself in paradise, wearing such coarse garments and drinking milk and ginger, would resemble the brides of the barbarian Kurds and Nabateans; see *Muntaẓam*, 115 (Aʿsam, *History*, 163). Rāzī, *Rasāʾil*, 42, mentioned the "uncouth Bedouins and Kurds and the vulgar Nabateans" as part of his attack on the *udabāʾ* and their idealization of love. In so doing, he insinuates that the supposedly refined Arab culture is identical to the coarse life of the desert dwellers.

[9] Kindī, *Risāla*, 73; comp. above, Ch. 2, n. 190.

[10] Ibid., 88–89; comp. above, Ch. 2, n. 192.

[11] Ibid., 134–45.

[12] *Šamsiyya* in the text (*Risāla*, 140), corrected already by Kraus.

> ... God is wise and thus He cannot command His creatures to worship Him according to detestable and harmful rules, which are repulsive to human nature, and which horrify the intellect. Rather, God commands that He be worshipped only according to rules which agree with the human intellect.[13]

As in Ibn al-Rāwandī's *Zumurrud*, the rejection of ritual is based on appreciation of the intellect. The intellect is given a place of honor, and is hailed as the only suitable basis for the commandments of a wise God.

The similarity of argumentation in the *Risāla* and the *Zumurrud* does not in itself prove the dependence of the former on the latter. In fact, Sidney Griffith has argued that "given the state of the development of Christian Arabic Apologetics in the first Abbasid century, it seems more reasonable to suppose that Ibn al-Rāwandī was influenced by the Christians." Griffith points out that Ibn al-Rāwandī's arguments "have about them the ring of the Christian anti-Muslim polemical pamphlets."[14] Indeed, it is quite possible that earlier Christian polemics were among Ibn al-Rāwandī's ultimate sources. But, as noted by Griffith, the *Risāla* is unique among Arabic and Syriac apologies of this period in its defiant, impudent tone.[15] In other words, there is no obvious precedent for the style of the *Risāla* in the Christian contemporaneous apologetic tradition. It therefore seems likely that its author, rather than drawing solely from his own tradition, had used Ibn al-Rāwandī's bolder reworking of this tradition. The practice of improving on earlier polemical arguments by taking into consideration the opponents' version of these arguments was, after all, commonplace.

Among medieval polemical Christian compositions, the *Risāla* is unusually outspoken in its criticism of Islam, and is, as far as I know, unique in its clear similarity to the *K. al-Zumurrud.* Its unique character is highlighted by comparison to another treatise, which clearly belongs to the same genre of Christian apologetics. This was apparently also the perception of the Christian scribes, who often copied both tracts in the same codex.[16] Like the *Risāla*, this treatise was also

---

[13] Kindī, *Risāla*, 141; compare Lawrence, *Shahrastānī*, 77.

[14] S. H. Griffith, "The Prophet Muhammad, His Scripture and His Message According to the Christian Apologies in Arabic and Syriac from the First Abbasid Century," *La vie du Prophète Mohamet, Colloque de Strasbourg, Oct. 1980*, Paris 1983, 99–146, esp. 107, n. 1.

[15] Ibid., 108.

[16] See Kh. Samir, *Correspondance islamo-chrétienne*, 526.

presented in the form of a theological correspondence, but more specifically between Christian and Muslim philosophers. In response to the Muslim Ibn al-Munajjim, the Christian Qusṭā b. Lūqā tried to refute the doctrine of *iʿjāz al-Qurʾān*, arguing that not every deed which surpasses previous human achievement could qualify as an evidentiary miracle. Thus a person who succeeds in using mirrors more cleverly than his predecessors is not necessarily a prophet.[17] In response to Ibn al-Munajjim's claim that Muḥammad was accepted by all humankind, Qusṭā b. Lūqā says that the Greek have higher esteem for their philosophers than for Muḥammad.[18] Although these arguments are somewhat similar to those found propounded by Ibn al-Rāwandī,[19] Qusṭā's courteous, mild and rather general argument is markedly different from the specific and insolent arguments used by Ibn al-Rāwandī. The author of Kindī's *Risāla*, on the other hand, seems to enjoy the detailed and outspoken attack on the Qurʾān and on Muslim tradition. The difference between Kindī's *Risāla* and Ibn Lūqā's may reflect the difference between a fictitious correspondence, intended for Christian internal use (and therefore often written in Karshūnī) on the one hand, and a more authentic record of real correspondence with Muslims on the other hand.

Assuming now that the *Risāla* is indeed dependent on the *Zumurrud*, and that (like most polemical compositions) it was indeed meant for internal use, we should underline the difference between the source and its Christian reworking. Unlike the Muslim freethinker, the Christian apologist did not see the necessity to combine his arguments with a rejection of prophecy. In fact, nowhere in the whole bulky text of the *Risāla* do we find a juxtaposition of the rational divine governance of the world with the notion of prophetic religion.

The *Risāla* was quite popular among Christians in the east, as the number of manuscripts (in both Arabic and Karshūnī) attests. It also reached Spain, where it was translated into Latin,[20] and this translation played a significant role in introducing the critique of irrational

---

[17] Ibid., 674.

[18] Ibid., 608.

[19] Cf. van Ess, *TG*, IV, 341, who points to the similarity with Ibn al-Rāwandī's *K. al-Farīd.*

[20] Cf. Sj. van Koningsveld, "La apologia de al-Kindi en la España del siglo XII. huellas Toledanas de un 'animal disputax'," *Actas del II Congreso Internacional de Estudios Mozárabes*, Toledo 1989, III, 107–29; José Muñoz Sendino, "Al-Kindī; Apologia del Christianismo," *Miscelanea Comillas'*, 11–12 (1949), 339–460; J. Kritzeck, *Peter the Venerable and Islam*, Princeton 1964, 101–107 (both quoted in S. Griffith, "The Monk in the Emir's Majlis").

religiosity into European literature. It has been suggested that the translator of the *Risāla*, Peter of Toledo, is the same person as Petrus Alphonsi (d. 1146), a Jew who had converted to Christianity. Petrus was familiar not only with Judaism, but also with Islam, and among his Muslim acquaintances was the famous jurist Ibn Ḥazm,[21] who plays a leading role in the following episode.

2. Ibn al-Rāwandī's *K. al-Dāmigh* and its Afterlife in Ibn Ḥazm's Polemics

The particular case which will be our concern in the following pages involves Ibn al-Rāwandī's connection with Judaism. According to one account, his father was a Jewish convert to Islam. The same source tells us that the Jews warned the Muslims against Ibn al-Rāwandī, saying that just as the father had used his knowledge of Judaism in order to undermine it, so the son was liable in a similar way to use his knowledge of Islam.[22]

In another anecdote we are told that Ibn al-Rāwandī advised the Jews on the most effective ways to carry on their theological debates with the Muslims.[23]

It is important to note that none of the sources relative to Ibn al-Rāwandī's Jewish connections implies that he actually became a Jew at any point. For Muslim writers, the anecdotes about Ibn al-Rāwandī's relationship with the Jews—and it need not concern us here whether these anecdotes contain any element of truth—only served to show his pathological hatred of Islam and his mercenary irreligiosity. The most persistent and detailed of these anecdotes relates the story of Ibn al-Rāwandī's last days:

> Abū ʿAlī al-Jubbāʾī said: the Sultan was searching for Abū ʿĪsā al-Warrāq and for Ibn al-Rāwandī. Al-Warrāq was caught and imprisoned, and died in jail. Ibn al-Rāwandī escaped and was hidden by Ibn Lāwī the Jew. He wrote for him the *Book of Fatal Blow* (*K. al-*

[21] Cf. A. H. and H. E. Cutler, *The Jew As Ally of the Muslim: Medieval Roots of Anti-Semitism*, Notre Dame, Ind. 1986, 56–57; Lazarus-Yafeh, *Intertwined Worlds*, 140, and n. 29.

[22] *Muntaẓam*, 108:5–6 (= Ritter, *Philologica*, 2); Ibn Kathīr, *Bidāya*, 2, 112 (who depends on Ibn al-Jawzī) improves on the story by saying that the father only dissimulated conversion to Islam.

[23] *Muntaẓam*, 108:6–7.

> *Dāmigh*),[24] which is a calumny (*ṭaʿn*) of Muḥammad, may God bless him and give him peace, and of the Qurʾān. A few days after [composing the work] he fell ill and died.[25]

This Jewish connection of Ibn al-Rāwandī is also mentioned by ʿAbd al-Jabbār, who says that the enemies of the Apostle of God rely on Q. 34:24 and Q. 28:55, and say that during his Meccan period, Muḥammad adopted an appeasing attitude towards the idolaters. Only in Madīna, when he had a supporting crowd, did he declare war upon them:

> This is what Ibn al-Rāwandī said, when he got together with Lāwī the Jew, and other wretched people collaborated with them . . .; In their interpretations [of the Qurʾānic verses], they abandoned the self-evident criteria,[26] and they named the book in which they included this interpretation and others like it: "*Kitāb al-Dāmigh.*"[27]

Although none of Ibn al-Rāwandī's heretical books was well-received by the Muslim community, it was the *K. al-Dāmigh*, more than any other work, which infuriated his former coreligionists. This book, which included "such blasphemous sayings as to make one's hair stand on end,"[28] was the subject of at least two refutations, one by al-Ḫayyāṭ, the other by Abū ʿAlī al-Jubbāʾī.[29] These works, like the book they meant to attack, are no longer extant. But passages from Abū ʿAlī's book survive in works by ʿAbd al-Jabbār[30] and Ibn al-Jawzī,[31] and these passages include a few quotations from the *K. al-Dāmigh* itself. From them we can see something of the nature of the *K. al-Dāmigh*, a biting and detailed criticism which presents the Qurʾān

[24] Lit. "The skull-smasher." The "skull-smashing blow" was directed to the Qurʾān, against which the book attempted to produce an irrefutable proof (*ḥujja dāmigha*). In Ibn Kathīr, *Bidāya*, 113, it thus becomes *al-Dāmigh li'l-Qurʾān*. On the titles of Ibn al-Rāwandī's books, see above, Ch. 2, apud n. 168.

[25] *Muntaẓam*, 112:7–10; Ibn Kathīr, *Bidāya*, 113. The statement that Ibn al-Rāwandī wrote a book *for* Ibn Lāwī (*ṣanafa lahu*) or for the Jews and Christians (*waḍaʿa kitāban li'l-yahūd wa'l-naṣārā*) is intended to point out his unprincipled behavior of renting out his dialectical skills. A similar behavior is recorded in a story concerning the *K. al-Baṣīra*, see above, Ch. 2, n. 149.

[26] *Fa-inṣarafū ʿan al-ḍarūrāt bi'l-taʾwīlāt*. Or: "By these [far-fetched] interpretations, they departed from self-evident understanding"; see van Ess, *TG*, VI, 487.

[27] *Tathbīt*, II, 356:17–357:7 (translated by van Ess, *TG*, VI, 486–87).

[28] *Taqšaʿirru minhā al-julūd; Muntaẓam*, 112:12.

[29] *Muntaẓam*, 109:1–2; al-Balḫī (quoted in Nyberg's introduction to his edition of *Intiṣār*, 36), adds that Ibn al-Rāwandī himself also wrote a refutation of this book of his. On *K. al-Dāmigh*, see van Ess, *TG*, IV, 337–39; IV, 483–87.

[30] *Mughnī*, XVI, 389:11–394:8.

[31] *Muntaẓam*, 112:7–117:12.

as an absurd composition, riddled with internal contradictions. Moreover, as Ibn al-Jawzī puts it, to the exposition of the so-called contradictions in the Qur'ān, Ibn al-Rāwandī

> added his ill manners and jokes in ill-taste, and he referred to the Creator, praised be He, in a way in which it is inappropriate to refer even to common people. We have never heard anyone defame the Creator and make jest about him as this cursed one did.[32]

In other words, what characterized Ibn al-Rāwandī was not simply his pointing out flaws in the logic of the Qur'ān; this was apparently done also by others and is dealt with, to some extent, in the context of *mutašābihāt al-Qur'ān*. This kind of literature consisted of presenting problems (*masā'il*) that arise while reading or discussing the Qur'ān, and of solving them through various hermeneutical techniques.[33] Indeed, some of these questions in the fragments of the *K. al-Dāmigh* are also found in respectable Muslim sources.[34] And yet, by changing or adding even a few words, the *mas'ala* could be turned into a *ṭa'n*: an argument intended to rock and shatter the Muslim faith in the

---

[32] *Muntaẓam*, 117:4–6.

[33] On this branch of literature, see Ḥājjī Ḫalīfa, *Kašf al-ẓunūn 'an asāmī al-kutub wa'l-funūn* (1943), 1584. In modern scholarship the *mutašābihāt* are usually discussed in the context of the *tafsīr*, see J. Wansbrough, *Quranic Studies: Sources and Methods of Scriptural Interpretation*, Oxford 1977, 148–70; A. Rippin, "The Present State of *Tafsīr* Studies," *Muslim World* 72 (1982), 227. Their role in the development of Islamic theology deserves further study. Here it need only be mentioned that both in his *Mutašābih al-Qur'ān* (ed. 'Adnān Muḥammad Zarzūr, Cairo 1969) and in his *Tanzīh al-Qur'ān 'an al-maṭā'in* (Beirut n.d.), 'Abd al-Jabbār includes only a few questions which we can identify with certainty as deriving from the *K. al-Dāmigh* (e.g., *Tanzīh*, 99, in reference to Q. 4:56, corresponds to *Muntaẓam*, 114:10–11; *Tanzīh*, 369, in reference to Q. 41:9, corresponds to *Muntaẓam*, 113:7–10). More frequently, even when he refers to the verses in question, 'Abd al-Jabbār does not address the problems raised by Ibn al-Rāwandī (for example, *Mutašābih*, 604, deals with Q. 42:44, but does not note the contradiction with Q. 16:63, as Ibn al-Rāwandī had done according to the *Mughnī*, XVI, 391). It would seem that when, in the context of the *mutašābihāt al-Qur'ān*, 'Abd al-Jabbār addressed questions originally raised by Ibn al-Rāwandī, he rephrased these questions in a much less pointed form so as to make them fit for publication in an exegetical work. On the other hand, Ibn Ḥanbal's polemical treatise *al-Radd 'alā al-zanādiqa wa'l-Jahmiyya* (*'Aqā'id al-salaf*, ed. 'Alī Sāmī al-Naššār and 'Ammār Jam'ī al-Ṭālibī, Alexandria 1971) begins with a list of twenty-two *mutašābihāt*. According to Ibn Ḥanbal, the *zanādiqa* use these verses to show that the Qur'ān contains contradictions, and hence to cast doubt on its divine nature. Although only one of these alleged contradictions (in Q. 4:59) can be identified with assurance as deriving from the *K. al-Dāmigh*, it is quite possible that some, if not all, of the others were also drawn from this book. But since they lack the typical Rāwandian sneer, we cannot, in the present state of knowledge, establish their origin with any certainty.

[34] See the examples adduced by van Ess, *TG*, IV, 337–38.

Qur'ān. Ibn al-Rāwandī was distinctive in his jeering and personal attack, as it were, on God: He presented God as vindictive, quarrelsome, weak in arithmetic, etc.[35] Such concentrated abuse did not accompany the scholarly discussions of the *mutašābihāt*, and indeed, as we have just seen, Ibn al-Jawzī saw this style of argument as peculiar to Ibn al-Rāwandī.[36] If we accept Ibn al-Jawzī's judgment (and I see no reason to reject it), and if we look closely at the hitherto identified quotations from the *K. al-Dāmigh*, then we can, I believe, identify other fragments of the same book in a most unlikely place.

The famous Ẓāhirite scholar Ibn Ḥazm dedicated many pages to the refutation of Judaism. From his major heresiographical work, *al-Fiṣal fī'l-milal wa'l-ahwā' wa'l-niḥal*, we learn that he was also involved in oral disputations with Jews. One of his opponents in these disputations, Ismā'īl Ibn al-Naghrīla[37] (or, in Hebrew, Shmu'el, i.e., Samuel ben Levi; d. 1056), was later to become the head (*Nagid*) of the Jewish community in Granada, and the vizier of Bādīs, Granada's Berber ruler.[38] In the *Fiṣal*, Ibn Ḥazm speaks about his opponent in laudatory terms, calling him "the most learned and skilled in disputation of all [the Jews]."[39] Other Muslim writers follow Ibn Ḥazm and speak of Samuel ha-Nagid with great respect.[40] It is evident that Samuel ha-Nagid achieved his prominent position (and managed to keep it) because of his keen sensitivity to human whims and his ability to navigate prudently between the opposing factions that constituted the complex political scene of Granada.

It is therefore somewhat surprising to find, years later, the same Ibn Ḥazm pouring abuse on Samuel. In the epistle entitled *al-Radd 'alā Ibn al-Naghrīla al-Yahūdī*, Ibn Ḥazm calls him "a fool" and "an ignoramus."[41] It is true that things had changed in the years that

[35] Cf. *Muntaẓam*, 112:14, 113:7, 116:6–7.

[36] On the general Muslim sensitivity to Ibn al-Rāwandī's style, see above, Ch. 6.

[37] = Ibn Negrella. On this name see S. M. Stern, "Life of Shmu'el ha'Nagid," *Zion* 15 (1950), 135, n. 2 (in Hebrew); idem, "Les vers finaux en espagnol dans les *muwaššaḥs* hispano-hébraïques," *Al-Andalus* 13 (1948), 330–31, n. 3.

[38] See Wasserstein, *The Party-Kings*, 197–99; idem, "Samuel ibn Naghrīla."

[39] "*A'lamuhum wa-ajdaluhum*;" Ibn Ḥazm, *Fiṣal*, I, 245 (where Ibn Ḥazm dates the conversation to the year 404/1013), and see also 225.

[40] Ibn al-Ḫaṭīb, *al-Iḥāṭa fī aḫbār gharnāṭa*, ed. Muḥammad 'Abd Allāh 'Inān, Cairo 1973, I, 438:10–439:5; Ibn Ṣā'id, *Ṭabaqāt*, 115–16; Ibn Bassām, *al-Dhaḫīra fī maḥāsin al-jazīra*, Cairo 1942, I, 265–67; Ibn 'Idhārī al-Marākušī, *al-Bayān al-mughrib fī aḫbār al-Maghrib*, ed. E. Lévi-Provençal, Beirut 1967, III, 264:16–17.

[41] These and the other abuses were conscientiously collected by García Gómez,

had passed between the meetings recorded in the *Fiṣal* and the composition of the epistle. At the time of their first meetings the two men were both young and promising scholars, but in the years that followed Ibn Ḥazm suffered many misfortunes, while Ibn al-Naghrīla rose to the highest rank at the court of Granada. Yet if envy and frustration were the true reasons for Ibn Ḥazm's anger, he hid them behind less personal motives.

According to Ibn Ḥazm, what triggered the outburst of his wrath was a book which Ibn al-Naghrīla had written in refutation of the Qur'ān. Such insolence on the part of a Jew led Ibn Ḥazm to vindicate the Holy Book. Since, he says, he was unable to procure a copy of Ibn al-Naghrīla's book, he had to make do with a refutation of it written by another Muslim.

On the basis of that Muslim refutation of Ibn al-Naghrīla's polemical work, Ibn Ḥazm quotes eight samples of the anti-Qur'ānic arguments of his opponent. The first five refer to apparent contradictions between Qur'ānic verses:[42]

1. Q. 4:78 reproaches people who, "if a good thing visits them, they say: 'This is from God,' but if an evil thing visits them, they say, 'This is from thee.'" Q. 4:79 says "Whatever good visits thee, it is of God; whatever evil visits thee is of thyself." Ibn Ḥazm then quotes the Qur'ān refutation, and says: "So the ignorant *zindīq* said: 'He is thus found justifying their words and contradicting what He said at the beginning of the verse.'"[43]

2. Q. 79:27–32 says that God spread out the earth after the creation of the heavenly vault. Q. 2:29 says that He created all that is in the earth, then He lifted Himself to heaven. The author of the refutation said that the two verses presented opposing versions concerning the order of creation.[44]

3. Q. 41:10, Q. 41:11–12 and Q. 50:38 give the numbers of days in which God created the earth as two, four and six. From Ibn

"Polémiqua," 3, n. 7. Because of the inordinate violence of this treatise's language, García Gómez views the *Radd 'alā Ibn al-Naghrīla* not merely as an ordinary refutation of Judaism, but rather as a "document . . . for the study of antisemitism in Muslim Spain" (ibid., 5). David Wasserstein is preparing an English translation of this epistle.

[42] The verses are quoted in the translation of A. J. Arberry, *The Koran Interpreted*, New York 1955.

[43] Ibn Ḥazm, *Radd*, 43.

[44] Ibid., 46.

Ḥazm's response it appears that the author of the refutation saw these numbers as irreconcilable.[45]

4. Q. 77:35 says that on Judgment day the sinners "shall not speak, neither be given leave to excuse themselves." Q. 16:111 says that "every soul shall come disputing in its own behalf." The author of the refutation pointed to the apparent contradiction between these two verses.[46]

5. The ignorant *zindīq* pointed to an apparent contradiction between Q. 55:39, which says that "on that day none shall be questioned about his sin," and between Q. 7:6, which says: "We shall question the envoys."[47]

These "contradictions" are presented as serious scholarly questions, but the feigned seriousness hardly hides the sneer. The exposition of these "contradictions" is clearly nitpicking for its own sake, and this irreverent attitude to the Qur'ān is even more apparent in the last three samples:

6. Q. 10:94 says: "If thou art in doubt regarding what We have sent down to thee." According to Ibn Ḥazm, "this posessed madman" (*majnūn*) said: "This Muḥammad was skeptical about the things he was preaching."[48]

7. Q. 16:71, says that the honey is "healing for men," but the author of the refutation pointed out that honey is detrimental to people who suffer from certain fevers.[49]

8. Q. 50:9 says: "We sent down out of heaven water blessed." The "ignorant *zindīq*" objected to this view of rain, saying: "How can it be blessed, when it demolishes houses and brings perdition on many living things?"[50]

Modern scholars have always found it difficult to reconcile the provocative sarcasm of Ibn Ḥazm's opponent with Samuel ha-Nagid's famous political wisdom and prudence. While some scholars marvel

---

[45] Ibid., 47.

[46] Ibid., 49.

[47] Ibid., 50–51.

[48] Ibid., 53.

[49] Ibid., 55. The disparity between God's endorsment of honey and its actual therapeutic merits is also evoked in a *ḥadīth* quoted by Ibn Rušd, but there seems to be no connection between Ibn Rušd and Ibn al-Rāwandī's criticism of the Qur'ān; cf. *Faṣl al-Maqāl wa-taqrīr mā bayna al-šarī'a wa'l-ḥikma min al-ittiṣāl*, ed. A. Nādir, Beirut 1973, 34; G. F. Hourani, *Averroes on the Harmony of Religion and Philosophy*, London 1976, 48, and n. 50.

[50] Ibid., 56.

at the audacity of Samuel in writing such a virulent attack on Islam,[51] others adopt a cautious reserve or express skepticism about the attribution of the refutation of the Qur'ān to Samuel ha-Nagid.[52] Still others suggest that Ibn Ḥazm's opponent should be identified with Samuel's arrogant and indiscreet son, Joseph Ibn al-Naghrīla.[53]

Scholars who accept the authorship of a Jewish Ibn al-Naghrīla (father or son) of a Qur'ān refutation, search for a Jewish tradition from which such a refutation could stem. Joseph van Ess has suggested Ḥīwī al-Balḫī's *Questions* concerning the Bible as a source of inspiration for a Jewish polemicist.[54] These *Questions* prove indeed that criticism of the scriptures was not foreign to the Medieval Jewish tradition. But it is evident that Ḥīwī, who was only interested in the Bible, could not have been the source of specific questions against the Qur'ān. If at all, his questions might have found their way to Ibn Ḥazm's critique of the Bible rather than to the book Ibn Ḥazm attacked.

This scholarly unease concerning the attribution of a Qur'ān refutation to Ibn al-Naghrīla has been based only on what we know of al-Naghrīla's personality. But a closer look may allow us to go one

---

[51] See for example, E. Ashtor, *The Jews of Moslem Spain*, II, trans. A. Klein and J. Machlowitz Klein, Philadelphia 1979, 122–24.

[52] See M. Perlmann, "The Medieval Polemics Between Islam and Judaism," in S. D. Goitein, ed., *Religion in a Religious Age*, Los Angeles 1974, 109; idem, "Eleventh Century Andalusian Authors on the Jews of Granada," *Proceedings of the American Academy for Jewish Research* 18 (1948–49), 280ff.; B. Lewis, *The Jews of Islam*, Princeton 1984, 87; Powers, "Ibn Ḥazm," 109–11 and n. 11, withholds judgment; Wasserstein, *The Party-Kings*, 201–204, weighs the arguments for and against the authenticity of the attribution, and concludes that "on balance, it seems perhaps less than likely that the Nagid did write such a tract."

[53] I. ʿAbbās in his introduction to Ibn Ḥazm, *Radd*, 7ff.; P. B. Fenton, "Jewish Attitudes to Islam: Israel heeds Ishmael," *Jerusalem Quarterly* 29 (1983), 91; and see already R. Dozy, *Histoire des Musulmans d'Espagne*, III, in E. Lévi-Provençal's revised edition, Leiden 1932, 70. R. Arnaldez rejects this suggestion and believes that Samuel was the author ("Controverse d'Ibn Ḥazm contre Ibn Naghrila le Juif," *Revue de l'Occident Musulman et de la Méditerranée* 13–14 [1973], 40). Arnaldez suggests that Samuel's book was a retort to Ibn Ḥazm's refutation of the Bible in the *Fiṣal*, for he thought it unlikely that Samuel would have dared to initiate a debate in this style (ibid., 42). The question remains, of course, whether Samuel ha-Nagid would have dared even to respond in this style.

On Joseph Ibn al-Naghrīla, see Wasserstein, *The Party-Kings*, 206; idem, "Samuel ibn Naghrīla," 113–14.

[54] Van Ess, *TG*, IV, 339. Powers, who is cited by van Ess, does not mention Ḥīwī as proof for a Jewish tradition which Ibn al-Naghrīla could continue, but rather as a possible source for Ibn Ḥazm; cf. Powers, "Ibn Ḥazm," 118.

step further, and identify the author of the book attacked by Ibn Ḥazm. Indeed, a comparison of the various fragments involved shows a striking similarity between the book quoted by Ibn Ḥazm and Ibn al-Rāwandī's *K. al-Dāmigh*.

In the list of anti-Qur'ānic arguments summarized above, the second and the third items correspond to two of 'Abd al-Jabbār's six quotations from the *K. al-Dāmigh*,[55] while the third and the seventh items correspond to two of the fourteen quotations from the *K. al-Dāmigh* given by Ibn al-Jawzī.[56] Thus we know that of the eight anti-Qur'ānic arguments quoted by Ibn Ḥazm, three were to be found in the *K. al-Dāmigh*. Whether or not the other five arguments appeared in the full text of the *K. al-Dāmigh* remains an open question. To my mind, the fact that three out of eight arguments of the anti-Qur'ānic tract appear in the *Dāmigh*, suggests the possibility, and even the likelihood, that the *Dāmigh* was also the ultimate source of the other five. This hypothesis in strengthened by the unmistakably Rāwandian character of the other five arguments. They too bear the mark of the feigned seriousness typical of Ibn al-Rāwandī, a seriousness which only highlights his sarcasm and insolence, and were in all probability part of the same book.

How did it happen that fragments of the *K. al-Dāmigh* are ascribed to Ibn al-Naghrīla in the work of Ibn Ḥazm? There seem to be four possible answers to this question:

The first solution would be to claim that, inspired by the *K. al-Dāmigh*, Ibn al-Naghrīla wrote a refutation of the Qur'ān, and that it is this book that Ibn Ḥazm refutes. But this is most unlikely. We have already mentioned the long-standing doubts as to whether a man as prudent as Ibn al-Naghrīla would engage in such folly. Moreover, Ibn Ḥazm himself says that he was unable to obtain a copy of the book he attacks, and was therefore compelled to rely on quotations in a Muslim refutation of it. We are thus asked to believe

[55] *Mughnī*, XVI, 393:6–11 = Ibn Ḥazm, *Radd*, 47; *Mughnī*, XVI, 394:1–8 = *Radd*, 46. The quotations in the *Mughnī* are more extensive and present Ibn al-Rāwandī's arguments in a more consequential way. Cf. also 'Abd al-Jabbār, *Tathbīt*, II, 359:2 (translated by van Ess, *TG*, VI, 487), on Q. 46:9, which presents the same logic as the sixth argument in Ibn Ḥazm's *Radd*.

[56] *Muntaẓam*, 113:7–10 = Ibn Ḥazm, *Radd*, 47; *Muntaẓam*, 115:5–9 is probably the fuller version of *Radd*, 55. We should also note that *Mughnī*, XVI, 391:12–14 = *Muntaẓam*, 113:3–4; and *Mughnī*, XVI, 390:10 corresponds (in part only) to *Muntaẓam*, 113:14–114:1 (and, therefore, we do have two arguments which appear both in the *Mughnī* and in *Muntaẓam*); compare Vajda, "Ibn al-Rāwandī," 905, col. 2.

that Ibn Ḥazm was unable to procure a copy of a book written in his own time and place; nor can it be argued that the book remained unpublished, since, according to Ibn Ḥazm, another Muslim author saw fit to write a refutation of it.

Special attention should be drawn to the last item on Ibn Ḥazm's list of quotation from this supposedly Jewish Qur'ān refutation. At closer look, this argument turns out to be, rather than an attack on the Qur'ān, an attack on the Creator of the universe, whose so-called "blessed rain" works havoc. This argument betrays the identity of its author as someone who rejects the basic monotheist approach to theodicy. Such an accusation against God could hardly have come from the pen of a Jewish author, one of the pillars of the Jewish community. In particular, this specific argument could not have been invented or adopted by Samuel ha-Nagid, in whose poems rain is described as a blessing to humanity, and the world is called "a wisely constructed creation" (*yeṣira meḥukama be-maʿaséhā*). Ibn al-Rāwandī, on the other hand, dedicated a whole book, the *K. al-Taʿdīl wa'l-tajwīr*, to the exposition of God's injustice and the futility of His creation (*ʿabath al-ḥikma*).[57]

To be sure, the fragmentary state of our knowledge does not allow for any unequivocal conclusion. Nevertheless, until clear evidence to the contrary is adduced, it is reasonable to assume that Ibn al-Naghrīla never wrote a book against the Qur'ān, and that the "other refutation of it by a Muslim" referred to in Ibn Ḥazm's work, is none other than Abū ʿAlī's refutation of the *K. al-Dāmigh.*

It must be mentioned, however, that later Muslim historians take it for granted that Ibn al-Naghrīla did write a Qur'ān refutation. Ibn Ṣāʿid (d. 685/1286) even quotes some Arabic verses in which Ibn al-Naghrīla brags about his achievement.[58] But Ibn Ṣāʿid clearly confuses Samuel with his son Joseph. Moreover, Arabic-speaking Jews rarely wrote poetry in any language but Hebrew,[59] and to this rule Samuel ha-Nagid was no exception.[60] Although educated Jews also

---

[57] Cf. above, Ch. 4, apud n. 50.

[58] *Al-Mughrib fī ḥulā al-Maghrib*, ed. Shawqī Ḍayf, Cairo 1964, II, 134. Ibn Bassām, *Dhaḥīra*, 269:9–10, is confused, and clearly dependent on Ibn Ḥazm.

[59] See J. Blau, *The Emergence and Linguistic Background of Judaeo-Arabic*, Jerusalem 1981², 22–24, 223; A. Halkin, "Judaeo-Arabic Literature," in L. Finkelstein, *The Jews, Their History, Culture and Religion*, New York 1949, II, 791–92.

[60] See *Divān Shmu'el ha-Nagid*, ed. D. Yarden, Jerusalem 1966.

learned Arabic poetry, and knew how to compose it, they made little use of this knowledge. In the rare occasions when they did write Arabic poetry, it was restricted to a limited number of subjects, excluding anything pertaining to religion.[61] A poet like Judah al-Ḥarīzī, who composed some Arabic poems, was quite unusual, and it seems that he carefully avoided anything from which a Muslim could take offense.[62] Thus an Arabic poem mocking the Qur'ān is even less likely to have been written by Samuel (or by his son) than a book to the same effect.

Assuming now that Ibn al-Naghrīla is not the author of the book attacked by Ibn Ḥazm, there remain three other possible explanations for Ibn Ḥazm's ascription of fragments from the *K. al-Dāmigh* to Ibn al-Naghrīla.

It could be argued that Ibn Ḥazm wrote a *Radd ʿalā Ibn al-Rāwandī*, for which he drew his material from Abū ʿAlī's book of the same name, and through some error or confusion Ibn Ḥazm's book eventually received the title *al-Radd ʿalā Ibn al-Naghrīla.*

The very contents of Ibn Ḥazm's book speaks against such a solution. Ibn Ḥazm's *Risāla* is clearly an anti-Jewish tract. Nor can we hypothesize that Ibn Ḥazm included anti-Jewish elements in his attack on Ibn al-Rāwandī simply because he was led (by Abū ʿAlī al-Jubbāʾī) to believe that Ibn al-Rāwandī was a Jew. The target of Ibn Ḥazm's *Radd* is clearly a contemporary Jewish dignitary, not a long-dead Muslim who had become a dubious convert to Judaism. *Al-Radd ʿalā Ibn al-Naghrīla al-yahūdī* was, in all likelihood, the original title given by Ibn Ḥazm to his tirade.

In sum, Ibn Ḥazm really did write against Ibn al-Naghrīla, even though the Qur'ān refutation he attacks was the work of Ibn al-Rāwandī. In this case, there are only two possibilities left: either Ibn Ḥazm himself was the innocent victim of some bibliographical

[61] The only Arabic verses plausibly attributed to Samuel ha-Nagid are "a straightforward love poem," see M. Stern, "Arabic Poems by Spanish-Hebrew Poets," in *Romanica et Occidentalia, Etudes dédiées à la mémoire de Hiram Peri (Pflaum)*, ed. M. Lazar, Jerusalem 1963, 255, and the lines that he wrote as part of a poetic contest in wit, in which he engaged with his daughter; see J. M. Nichols, "The Arabic verses of Qasmūna bint Ismāʿīl Ibn Baghdallah," *IJMES* 13 (1981), 155–58: J. A. Bellamy, "Qasmūna the Poetess: Who Was She?," *JAOS* 103 (1983), 423–24; Wasserstein, "Samuel ibn Naghrīla," 120–22.

[62] See J. Sadan, "Judah Alḥarizi as a Cultural Junction—An Arabic Biography of a Jewish Writer as Perceived by an Orientalist," *Peʿamim* 68 (1996), 16–67 (in Hebrew), esp. 36–38.

confusion; or he was aware of the real identity of the author of the Qur'ān refutation.

It would have been easier to decide on this matter if we knew the precise nature of Ibn Ḥazm's familiarity with the *K. al-Dāmigh*. Ibn Ḥazm wrote a book entitled *K. al-Taršīd*, a refutation of Ibn al-Rāwandī's attack on prophecy in his *K. al-Farīd*.[63] This shows that Ibn Ḥazm was familiar with Ibn al-Rāwandī's name and thought, and that he was—predictably—annoyed by his denial of prophecy. It does not, unfortunately, tell us which of Ibn al-Rāwandī's books he actually read, or which of them was available in Spain at that time: the lot of the *K. al-Dāmigh* may well have been different from that of the *K. al-Farīd*. Furthermore, the fact that Ibn Ḥazm wrote a refutation of *K. al-Farīd* does not necessarily prove that he had direct access to this book. The case of Ibn Ḥazm may not be different from that of Ibn al-Jawzī, who says that he came across Ibn al-Rāwandī' books (*waqaʿat ʿalayya kutubuhu*), of which he lists an impressive number: *K. Naʿt al-ḥikma, K. Qaḍīb al-dhahab, K. al-Zumurrud, K. al-Tāj, K. al-Dāmigh, K. al-Farīd*, and *K. Imāmat al-mafḍūl*. Kraus considered it improbable that Ibn al-Jawzī had actually seen any of these books. As noted by Kraus, for each and every quotation from these books, Ibn al-Jawzī referred to an indirect source in which he had found it.[64]

Of course, there is no way in which we can fathom Ibn Ḥazm's innermost thoughts, but a close reading of the *al-Radd ʿalā Ibn al-Naghrīla al-yahūdī* is highly suggestive. One curious feature of the epistle is what we may call its shifting target. In the opening paragraphs, as well as throughout the second half of the epistle, the opponent is the vile *yahūdī*, the attack is on the Bible and on the Jews, and the Qur'ān refutation is never mentioned. On the other hand, in the first half of the epistle, where the Qur'ān refutation is quoted and refuted, no mention is made of a Jew, nor of an infidel (*kāfir*) as one would expect; the reference is only to "the heretic who *hides behind* Judaism" (*al-zindīq al-mustasirr bi'l-yahūdiyya*)[65] or to the "heretic who, in his innermost faith, adheres to the views of the materialists, but outwardly seeks refuge in the Jewish ark."[66] These descriptions—

---

[63] The refutation is mentioned in Dhahabī, *Siyar aʿlām al-nubalā'*, Beirut 1984, XVIII, 195; cf. van Ess, *TG*, IV, 339; VI, 442.

[64] Cf. *Muntaẓam*, 108 (Aʿsam, *History*, 154); "Ketzergeschichte," 364.

[65] Ibn Ḥazm, *Radd*, 43:10.

[66] Ibid., 42:18–19: "*al-zindīq al-mustabṭin madhhaba al-dahriyya fī bāṭinihi, al-mutakaffin*

zindīq, dahrī—would fit Ibn al-Rāwandī perfectly, but they would not fit Ibn al-Naghrīla. One cannot avoid noticing that the *Radd* is not made of one cloth; the author, who pieced the two parts of the *Radd* together, did not bother to remove the basting stitches.

All the above considerations put together lead to the inevitable conclusion that the Qur'ān refutation which Ibn Ḥazm attributes to Ibn al-Naghrīla is, in fact, the *K. al-Dāmigh* as quoted by Abū ʿAlī's refutation of the *K. al-Dāmigh*, and that Ibn Ḥazm was well aware of the real identity of his source.

This conclusion, which I have presented in a previous publication, was rejected by van Ess, who objected that arguments quoted by Ibn Ḥazm appear in sources that do not depend on the *K. al-Dāmigh.*[67] Van Ess's judicious objection, however, presents only a partial picture of the facts, and should be rephrased: some of the arguments that appear in the *Radd* can indeed be found scattered in sources that do not include the handful of fragments identified so far with the *K. al-Dāmigh*. Whether or not these sources depend in any way on the *K. al-Dāmigh* is, at the present, an unanswerable question. For, until the full text of the *K. al-Dāmigh* is retrieved, we have no way of knowing whether or not a text depends on it, unless it includes at least one of the identified fragments. Ibn Ḥazm's *Radd*, as we have seen, includes three such arguments.

Since none of the three relevant compositions (scil. the Qur'ān-refutation attributed to Ibn al-Naghrīla, the *K. al-Dāmigh*, and Abū ʿAlī's refutation of the *K. al-Dāmigh*) is extant, my claim (that the first composition is actually the second as quoted in the third) must remain an unproven hypothesis. It is, however, the only hypothesis suggested so far that takes into consideration the concentration of the arguments from the *K. al-Dāmigh* in *Radd*, the style of all the anti-Qur'ānic arguments, and their attribution to the *zindīq* or the *dahrī* (rather than to the *kāfir* or the *yahūdī*).

Maribel Fierro accepted my argument concerning the ultimate dependence of the *Radd* on the *K. al-Dāmigh*. She argued, however,

---

bi-tābūt al-yahūdiyya fī ẓāhirihi." See also ibid., 43:7 (*al-zindīq al-laʿīn*). The use of the very vague term *zindīq* (when we would expect something like *kāfir*) and the strange fact that Ibn Ḥazm never calls his opponent by name, were noted also by Arnaldez, "Controverse d'Ibn Ḥazm," 41. See also *Radd*, 70:8–9, where Ibn Ḥazm sums up his *risāla* by saying that the information he had quoted about the Jews includes things "which makes one's hair stand on end" (*taqšaʿirru minhā al-julūd*); compare above, n. 28.

[67] Van Ess, *TG*, IV, 339; and see Stroumsa, "From Muslim Heresy."

that this fact does not exclude a Jewish participation in the composition of the Qur'ān refutation quoted in the *Radd*.[68] According to her, in the intellectual atmosphere of eleventh-century Andalus, it was quite possible for a Jew to point out the contradictions of the Qur'ān, using arguments that were previously used by Muslims. Assuming that one must search for an Andalusian Jew who would fit the description given by Ibn Ḥazm (a *zindīq*, a *dahrī*), she points out to the existence of two other Jewish interlocutors of Ibn Ḥazm, the physicians Ismāʿīl b. al-Qarrād and Ismāʿīl b. Yūnus al-Aʿwar.[69] In the *Fiṣal* Ibn Ḥazm identifies them as "*min al-dahriyya*" and they are said to have argued for the equivalence of religions (*takāfuʾ al-adyān*). Fierro thus suggests that Ibn Ḥazm was aware of these Jewish dahrites, and, having heard of a Jewish dignitary who used arguments from the *K. al-Dāmigh*, he responded to this book (which he had not seen) as he would respond to the *Dāmigh* (which he had not seen either). According to Fierro, Ibn Ḥazm indeed used Abū ʿAlī's refutation of the *Dāmigh*, but because of his hostility to the Muʿtazila, he was reluctant to mention Abū ʿAlī as his source.[70]

Fierro's description of the atmosphere in al-Andalus, in which polemical compositions circulated freely between intellectuals of the three monotheistic religions, is very convincing. It does not, however, solve the riddle of the *Radd*. The "dahrite" physicians mentioned by Ibn Ḥazm say nothing that would identify them as "materialists"; Ibn Ḥazm seems to have used the term *dahriyya* rather freely, because of the Jewish physicians' argument for the equivalence of religions. As suggested by van Ess, Ibn Ḥazm's Jewish interlocutors probably reverted to this argument as a convenient defense in an uneasy polemical situation.[71] Their use of this argument thus corroborates the prevalent impression, that in face-to-face confrontations with Muslims, Jews preferred various forms of prudent non-committal arguments to direct attack. This is not to say that an Andalusian Jew could not have expressed relativist opinions: to the extent that religious non-conformism and relativism existed among Muslim intellectuals, it must have existed also among their Jewish and Christian counter-

---

[68] Fierro, "Ibn Ḥazm," 84.

[69] Cf. M. Perlmann, "Ibn Ḥazm on the Equivalence of Proofs," *Jewish Quarterly Review* 40 (1949–50), 281–82.

[70] Fierro, "Ibn Ḥazm," 84–86; see also C. Adang, *Islam frente a Judaísmo: La polémica de Ibn Hazm de Córdoba*, Madrid 1994, 40–41.

[71] Van Ess, "Disputationspraxis," 47.

parts. But Ibn Ḥazm's debate with these two Jews can tell us little of the extent and the nature of such non-conformism.

Furthermore, polemical material, which indeed circulated freely in al-Andalus as in the east, was not automatically transferred unedited to open interreligious debates. The comparison with an anti-Muslim Jewish book of the same period is illuminating. This book was composed by the Karaite Yūsuf al-Baṣīr, and it was meant to prepare Jews for disputations with Muslims in the *majālis*. As with most Jewish compositions, it was written in Judeo-Arabic, that is to say, in Hebrew characters. This was no guarantee that the material would never reach Muslims, but the book was nevertheless intended for internal use. We can thus assume that the author wrote as freely as a Jew ever could. The book contained detailed criticism of the Qur'ān, of the doctrine of *iʿjāz al-Qur'ān*, and of Muslim historiography. At the same time, it also praised Muḥammad for having introduced the pagan Arabs to monotheism.[72] The arguments used by al-Baṣīr resemble, in many ways, the anti-Muslim arguments which we have seen in a Christian apologetic composition, the *Risāla* of ʿAbd al-Masīḥ al-Kindī.[73] The polemical intention of al-Baṣīr is unmistakable, and the possibility that, like Kindī's *Risāla*, he used material from the *Zumurrud*, cannot be excluded. And yet nothing in al-Baṣīr's book resembles the polemical style of Ibn al-Rāwandī, the mocking and jeering tone of his arguments, or his constant jibes at the God of Scriptures. It also does not resemble the polemics found in Ibn Ḥazm's *Radd*, with its mock seriousness and its criticism of divine governance of the world.

Having taken the facts adduced by Fierro into consideration, I come back to the same conclusions that I have reached earlier: it is most unlikely that a Jewish dignitary (Ibn al-Naghrīla or anyone else) would have published a Qur'ān refutation based on *K. al-Dāmigh* without some major re-editing. Ibn Ḥazm accused Ibn al-Naghrīla of writing precisely such a book. He seems to have been fully aware of the ultimate origin of the arguments, and yet he failed to mention not only the Muʿtazilite Abū ʿAlī, but, more significantly, Ibn al-Rāwandī or the *K. al-Dāmigh*.

I am unaware of any case in the history of medieval polemical

---

[72] Cf. D. Sklare, *Judaeo-Arabic Manuscripts in the Firkovitch Collections: The Works of Yusuf al-Basir*, Jerusalem 1997, 101–102 and 137–39 (in Hebrew).

[73] See above.

literature in Arabic in which the book of a heretic was ascribed to an infidel. Nevertheless we seem to be brought to the conclusion that Ibn Ḥazm ascribed to Ibn al-Naghrīla what he knew to be the *K. al-Dāmigh*. This unusual procedure might have been prompted by the similarity of names of the people involved: Abū ʿAlī al-Jubbāʾī's claim that the *K. al-Dāmigh* was written for an Ibn Lāwī, might have suggested to Ibn Ḥazm the idea of using this book as a weapon against Ibn al-Naghrīla who, as Ibn Ḥazm well knew, also happened to be called Ibn Lāwī.[74]

By ascribing to Ibn al-Naghrīla a refutation of the Qurʾān, Ibn Ḥazm hoped (as he himself says), to strip the Jew of the protection granted him by Islamic law (*barāʾatihi min al-dhimma al-ḥāqina damahu*).[75] His professed intention was to make sure that his opponent would receive

> what he deserves by law, that is to say, the shedding of his blood and the confiscation of his property and captivity of his women and children.[76]

Samuel Ibn al-Naghrīla's career was, apparently, unaffected by Ibn Ḥazm's accusations. But in 1066 the Berbers of Granada rose against their Jewish neighbors, killing thousands and sacking their houses. Both Ibn Ḥazm and Ibn al-Naghrīla were no longer alive, but among the victims of this pogrom was Samuel's son, Joseph Ibn al-Naghrīla, also called *ha-Nagid*.[77] Whether the *Radd ʿalā Ibn al-Naghrīla*, published some ten years earlier, contributed in any way to these riots is a question that deserves further study.

The mobilization of arguments from the *K. al-Zumurrud* or from the *K. al-Dāmigh* for Muslim anti-Jewish polemics, as well as of arguments from the *K. al-Zumurrud* for Christian anti-Muslim polemics, are two examples of the same phenomenon. In both cases, an attempt

---

[74] See *Fiṣal*, I, 225, where Ibn Ḥazm records an Arabicized form of his interlocutor's Hebrew name, "Shmūʾl Ibn Yūsuf al-Lāwī."

[75] Ibn Ḥazm, *Radd*, 42:20–43:1. On the conditions under which the status of *dhimma* may be withdrawn, see A. Fattal, *Le statut légal des non-mulmans en pays d'Islam*, Beirut 1958, 81–84; and see Wasserstein, *The Party-Kings*, 200–206.

[76] Ibn Ḥazm, *Radd*, 42:19–20.

[77] Cf. Ibn ʿIdhārī, *al-Bayān*, 264: 17–266:6; Ibn Bassām, *al-Dhaḫīra*, 268–72; Ibn al-Ḫaṭīb, *Aʿmāl al-aʿlām*, ed. E. Lévi-Provençal, Beirut 1956, 230–33; *al-Iḥāṭa*, 434. See Ashtor, *The Jews of Moslem Spain*, 204; Dozy, *Histoire des Musulmans d'Espagne*, III, 70, 73; Lewis, *The Jews of Islam*, 158–65, 320–23; Wasserstein, *The Party-Kings*, 208–11.

is made to strip the arguments of their irreligious character before recycling and introducing them into the inter-religious debate.[78] Neither ʿAbd al-Masīḥ al-Kindī nor Yūsuf al-Baṣīr is a freethinker; and although Ibn Ḥazm calls Ibn al-Naghrīla a *zindīq*, he still intends to stress the fact that he is a Jew. The polemicists appear as faithful adherents of a specific scriptural religion, who may use available arguments to refute Islam, but who do not intend to deny the very notion of revelation.

Ibn al-Rāwandī's books were apparently used by polemicists, who were looking for exceptionally offensive anti-Islamic arguments. Such arguments precisely were found in Ibn al-Rāwandī's books, but they were not likely to be presented by a *dhimmī* in a public debate or in an independent Arabic composition. Their introduction into the inter-religious debate was thus only a superficial borrowing from Islamic freethinking. While affecting the relations of Islam with other religions, in such cases the phenomenon of freethinking itself cannot be said to have penetrated Judaism or Christianity. The following chapter will examine the possibility of such a penetration, which affected the way in which both Jews and Christians reflected about their own religion.

[78] As we have seen above, it is the failure to remove profoundly irreligious arguments that most clearly betrays Ibn Ḥazm's source.

CHAPTER EIGHT

REPERCUSSIONS OF ISLAMIC FREETHINKING BEYOND ISLAM

1. Repercussions in Christianity

The crucial role played by Christian Arabs in the development of Arabic thought is well known. Arabic-speaking Christian communities existed long before the rise of Islam, and for Christian intellectuals in the areas conquered by Islam, the linguistic transition, from predominently Greek and Syriac to predominanty Arabic philosophical literature, was relatively easy.[1] Despite their position as a tolerated minority, Christian Arabs living under Islam felt confident and proud in their Arabic heritage.[2] As custodians of the Greek philosophical tradition, they were the most important factor in the stupendous achievement called "the translation movement." Thanks to their centuries-old experience in dialectical theology and apologetics, they were able to contribute greatly to the shaping and formation of Muslim *kalām*. In other fields, such as the development of exegetical literature in Islam, the Christian influence, which has been hardly investigated, must have been as formidable.

During the "Renaissance of Islam" in the tenth century, Christian authors went on translating and revising previous translations of Greek philosophical and scientific literature,[3] and they fully participated in the intellectual and theological discussions in the *majālis*. Nevertheless it seems—although this is as yet only an impression, which calls for further investigation—that while fully participating in the nascent Islamic culture, the Christian Arabs maintained a clear topical independence in their theological literature. The theme of "the signs of prophecy" may serve as an illustration of this independence. As noted above, this theme was discussed at length by the-

[1] See, for instance, S. H. Griffith, "The Gospel in Arabic: An Inquiry into its Appearance in the First Abbasid Century," *OC* 69 (1985), 126–67.

[2] See, for example, Kindī, *Risāla*, 123 (Tartar, *Dialogue*, 196).

[3] Cf. A. Mez, *Die Renaissance des Islāms*, Heidelberg 1922, esp. 34–55; J. L. Kraemer, *Humanism in the Renaissance of Islam*, Leiden 1986, 104–39.

ologians of all three monotheistic religions. But, as opposed to both Muslims and Jews, who developed this theme on almost identical lines, the Christians gave it a twist of their own.[4] Similarly, although the penetration of some Qur'ānic terminology can be detected in Christian writings (as in the use of *al-asmā' al-ḥusnā* for God's epithets), both the theological vocabulary and the semantics of Christian theological compositions in Arabic appear to be quite distinguishable from their Muslim (and Jewish) parallels. It is possible that the existence of a well-established philosophical and theological Eastern Christian tradition (both in Greek and in Syriac) before the rise of Islam, and its continuation in the Islamic period (in Syriac as well as in Arabic) had rendered the Christians less prone to cultural and literary assimilation.[5] While adopting the Arabic *lingua franca*, the Christians retained their own theological voice: to a large extent, their choice of topics and the manner in which they were treated remained dictated by previously established traditions.

This phenomenon, which we may dub "the Christian Arab theological aloofness," obtains also in regard to the topic of the present book. Prophetology had played a relatively marginal role in Christian theology before the rise of Islam,[6] and this marginality was largely preserved in Christian theological literature in Arabic. The Christian Arabic tradition shows very little interest in the freethinkers or in the questions associated with them. One example of this lack of interest is the already noted puzzling absence of the *Barāhima* from later, post-Islamic Christian Arabic texts.[7] To the extent that such interest exists, it usually appears in anti-Muslim polemical context (as, for example, in Kindī's *Risāla*) or in the works of Christian converts to Islām. But when no polemical background exists, the impact of the Islamic freethinkers on the Christian Arabic tradition seems to be negligeable.

This topic deserves further research, which cannot be attempted here. The question of a possible impact of the Islamic freethinkers on Christianity beyond the Arabic-speaking world has been studied even less. Several instances in the history of ideas in Europe present tantalizing similarities to the freethinkers of Islam, and encourage

---

[4] See above, Ch. 1.
[5] See below, n. 20.
[6] See Ch. 1, n. 16 and after n. 27.
[7] Cf. Ch. 1, n. 25.

speculation on possible influences. Abélard (d. 1142), for example, wrote a *Dialogue* between a Christian, a Jew, and a Philosopher whose depiction is clearly meant to identify him as belonging to the Muslim cultural world. This philosopher, however, does not attempt to defend Islam, but rather advocates a natural, rational religion which is not derived from a prophetic revelation. According to him, the laws which were supposedly revealed to the prophets did not bring humanity closer to God than it had been in the times of Adam; such revealed laws are, therefore, superfluous.[8] It has been suggested that Abélard's philosopher may have been modeled after the image of Abélard's contemporary, the Spanish philosopher Ibn Bājja (d. 533/1139); not his image as it transpires from his own works, but as it circulated in Arabic literature.[9] If we accept this plausible suggestion, it would mean that the (sometimes slanderous) image of the Islamic philosopher as a freethinker was transmitted, as early as the twelfth century, to Latin Christianity.

In the fifteenth century, the Byzantine Georgios Gemistos Plethon adopted what he believed to be a wholly Platonist philosophy, rejecting Christianity and all revealed religions. Plethon's teacher was a certain Jew named Elissaeus (Elisha) who had come to Byzantium from the East. Like Rāzī, Plethon regarded Moses as a sophist and not a proper law-giver; unlike Rāzī, Plethon read Plato as a political philosopher. Plethon took his inspiration from various sources which included, besides Plato, the Chaldean Oracles and Zoroastrian wisdom, and he hailed the Brahmans as the wisest of sages. None of this is enough to prove a direct influence of the Islamic freethinkers, but an indirect influence (perhaps through the writings of Suhrawardī) cannot be ruled out.[10]

---

[8] Cf. "Dialogue entre un philosophe, un juif et un chrétien," *Œuvres choisies d'Abélard*, trans. Maurice de Gandillac, Paris 1945, esp. 230–31.

[9] Cf. J. Jolivet, *Abélard ou la philosophie dans le langage*, Paris 1969, 90–91; Cf. also R. Brague, "L'entrée d'Aristote en Europe: l'intermédiaire arabe," *Aristote, l'École de Chartres et la cathédrale*, Chartres 1997, 77.

[10] Cf. F. Taeschner, "Georgios Gemistos Plethon. Ein Beitrag zur Frage der Übertragung von islamischem Geistesgut nach dem Abenland," *Der Islam* 18 (1929), 236–40; F. Masai, *Pléthon et le platonisme de Mistra*, Paris 1956, esp. 270ff. and 308ff.; H. Corbin, *En Islam iranien: Aspects spirituels et philosophiques*, II: *Sohrawardī et les platoniciens de Perse*, Paris 1971, 34; J. Nicolet and M. Tardieu, "Pletho arabicus, Identification et contenu du manuscrit arabe d'Istanbul, *Topakapi Serāi, Ahmet III 1896*," *JA* 268 (1980), 35–57; C. M. Woodhouse, *George Gemistos Plethon, The Last of the Hellenes*, Oxford 1986, esp. 71–72; B. Tambrun-Krasker, "Le prologue du 'Traité des Lois'

Since around the sixteenth century, an anonymous book called *De Tribus Impostoribus* circulated in Europe.[11] The book attempted to expose the founders of the three monotheistic religions—Moses, Jesus, and Muḥammad—as impostors. The striking similarity of this book to the arguments propagated by Ibn al-Rāwandī and Rāzī has been noted for some time.[12] It has been suggested that the influence of Ibn al-Rāwandī or Rāzī was behind the composition of this book, although no obvious chain of transmision is known to us.[13] The notion of "the three impostors" was first mentioned in Europe at the court of Friedrich II (1212–1250).[14] This could point to a Sicilian connection, through some thirteenth-century translation from the Arabic, which would have transmitted to the Latin-speaking world some ideas of Ibn al-Rāwandī or of Rāzī (although not necessarily their books). The *De Tribus Impostoribus* itself, however, did not surface before the sixteenth century. Friedrich Niewöhner, who has studied the peregrinations of the notion of the equivalence of religions throughout the Middle Ages and the early modern times, has suggested that the author of the *De Tribus Impostoribus* may have been a sixteenth-century Jewish Marrano from the Iberian peninsula.[15] If this suggestion is accepted, the book (or its ideas) would be one of many transmitted, through translations or adaptations, from Muslims, via Jews, to Christian Europe.

A thorough examination of these intriguing cases is beyond my competence as it is beyond the scope of this book. But it is hoped that greater awareness of the nature of freethinking in medieval Islam and its importance may allow a better appreciation of its significance for the development of ideas in European history.

---

de Pléthon et le regain d'intérêt pour le scepticisme aux XIVe et XVe siècles," *Entrer en matière: les prologues*, eds. J.-D. Dubois and B. Roussel, Paris 1998, 271–91; M. Stausberg, *Faszination Zarathushtra: Zoroaster und die europäische Religionsgeschichte der früher Neuzeit*, Berlin and New York 1998, I, 37–44, 57–82.

[11] Cf. W. Gericke, *Das Buch "De tribus impostoribus,"* Berlin 1982, 23–57; *Anonymus Traktat über die drei Betrüger: Traité des trois imposteurs (l'esprit de Mr. Benoit de Spinosa)*, ed. W. Schröder, Hamburg 1992, vii–xliii.

[12] Cf., for instance, Walzer, *Greek into Arabic*, 15; Niewöhner, *Veritas*, esp. 189, 246; van Ess, *TG*, IV, 333.

[13] Kraus and Pines, *EI*; Corbin, introduction to *Jāmiʿ al-ḥikmatayn*, 139; S. W. Baron, *A Social and Religious History of the Jews*, Philadelphia 1957², V, 103–104.

[14] Cf. F. Meier, "Zum Vorrang des Glaubens und des guten Denkens vor dem Wahrheitseifer bei den Muslimen," *Oriens* 32 (1990), 1–49, esp. 10.

[15] "Are the Founders of Religions Impostors?" esp. 237.

2. Repercussions in Judaism

Very little is known about the Jewish communities during the first two centuries of Islamic rule. But from the beginning of the third/ninth century a vigorous Jewish culture surfaces, which would soon transform Jewish society in the East. Only the language of liturgical poetry remained what it had been before the rise of Islam, i.e., Hebrew. In all other fields of intellectual endeavor, Arabic replaced Aramaic as the language in which Jews normally expressed themselves.[16] The common Arabic language reflects the almost total immersion of Jews in the Islamic civilization that surrounded them, an immersion that was aptly characterized by Shlomo Dov Goitein as "symbiosis."[17] In the sciences, in philosophy, and in theology, the Jews became an integral part of Islamic culture.[18] Judeo-Arabic works on these subjects are often indistinguishable from Muslim or Christian compositions on the same topics. In some extreme cases, nothing in the contents or form of a book betrays the Jewish identity of its author.[19] More often, however, Jewish philosophical and theological works in Arabic apply the common philosophical language and ideas to the particular Jewish frame of reference. The prevalent Arabic terminology, the *topoi*, the concerns and the mentality of their non-Jewish neighbors, all became totally internalized by Jewish authors. These new ingredients were merged with older parts of the Jewish cultural heritage, to produce a Judeo-Arabic philosophy and theology.[20] The whole-

---

[16] On the existence of Arabic-speaking Jewish communities even before the rise of Islam, cf. H. Ben-Shammai "Observations on the Beginnings of Judeo-Arabic Literature" (forthcoming).

[17] S. D. Goitein, *Jews and Arabs: Their Contacts Through the Ages*, New York 1964, 11, 127. Goitein's magisterial study dealt mostly with social and economic aspects of this symbiosis, and less with the history of ideas; cf. Wasserstrom, *Between Muslim and Jew*, 3–12.

[18] The active participation of Jews and Christians in this culture, and their significant role in its shaping, account for the common practice of avoiding the term "Muslim philosophy," and the preference of either "Islamic" or "Arabic" as a more apt epithet; see, for instance, O. Leaman's Introduction to O. Leaman and S. H. Nasr, eds. *History of Islamic Philosophy*, London and New York 1996, I, 1–10.

[19] Two examples are *al-Kitāb al-muʿtabar* of the twelfth century Abū al-Barakāt al-Baghdādī, and Ibn Gabirol's *Fons Vitae*, which in the Latin Middle Ages was actually believed to have been written by a Muslim Avencebron.

[20] Rémi Brague proposes to distinguish two models of assimilation, "inclusion" and "digestion"; cf. "L'entrée d'Aristote en Europe: l'intermédiaire arabe," (see n. 10 above) 77. The Jewish medieval philosophical tradition definitely fits Brague's "digestion model," i.e., "appropriation in which the foreign elements are assimilated

hearted absorption of the "signs of prophecy" genre, which we have seen above, is one example of this phenomenon.[21]

Although Jews remained usually indifferent to internal strife among Muslim theologians, they were generally alive to intellectual developments among Muslims.[22] In the case of the freethinkers, the discussions of the *Barāhima* in Jewish *kalām* works, such as the works of al-Muqammaṣ, Saʿadya, Yūsuf al-Baṣīr, and Yefet ben ʿAlī, testify to the fact that freethinking, i.e., the denial of prophecy, became as much a concern for Jewish intellectuals as it was for their Muslim counterparts. Occasional remarks also allow us to see that the identity of the freethinkers of Islam was known to Jewish authors.[23]

Usually, however Ibn al-Rāwandī and Rāzī were perceived as a "Muslim problem." When Judeo-Arabic authors searched for a relevant example of a *mulḥid*, they often refered to Ḥīwī al-Balḫī (fl. between 850 and 875, that is to say, a contemporary of Ibn al-Rāwandī).[24] Saʿadya, for example, counted his refutation of Ḥīwī as part of his defense against "what the heretics (*mulḥidūn)* have used to argue (*ṭaʿana*) against us." In Saʿadya's long list of possible opponents, these

---

to the point of losing their independent character." The Christian theological tradition in Arabic may require a third model.

[21] See above, Ch. 1, esp. nn. 13, 44, and apud nn. 77–83.

[22] See, for instance, H. Ben-Shammai "Kalām in Medieval Jewish Philosophy," in D. H. Frank and O. Leaman, eds., *History of Jewish Philosophy* (*Routledge History of World Philosophies*, II), London and New York 1997, 129. Ben-Shammai notes Saʿadya's use of Muʿtazilite terminology on the sinner's status; cf. *Amānāt*, 181 (Rosenblatt, 218–19). The penetration of such typical terminology reflects the extent of the Jewish adaptation to Islamic theological discourse, to include problems which are hardly present in previous Jewish thought. For another example, see G. Vajda, "La parole créée de Dieu d'après le théologien karaïte Yûsuf al-Baṣîr," *SI*, 1974, 59–76.

[23] Rāzī's name and his religious ideas appear quite frequently in Jewish works; see for example, Maimonides, *Guide*, III, 12; and see Pines, "Nouvelles Études," 142–43 (on Ibn Ezra's *Maqālat al-ḥadīqa*) Ibn al-Rāwandī's name, on the other hand, is rarely mentioned. But see, for example the Geniza fragment T-S Ar. 27.9 (possibly a fragment of a Commentary on Ecclesiastes by the tenth-century Karaite Salmon b. Yerūḥīm), which mentions the book of a certain Rīwandī as an example of "the books of the Gentiles;" cf. Poznanski, "Miszellen," 722, 731. Another Karaite author, Yeshuʿa ben Yehuda, mentions in his *Kitāb al-Tawriya* a book called *al-Radd ʿalā Ibn al-Rāwandī* (Russian National Library, Ms. Yevr.-Arab. I:4816, fol. 44v). I am indebted to David Sklare for these references.

Furthermore, the presence of Muslim theological works in Jewish manuscript-collections testify to the latter's familiarity with Ibn al-Rāwandī's name; for instance, a copy of ʿAbd al-Jabbār's *K. al-Muḥīṭ*, which mentions a refutation of Ibn al-Rāwandī's *K. al-Ḫāṭir* (Russian National Library, Ms. Yevr.-Arab. I:3097, fol. 51v). I owe this reference to Bruno Chiesa.

[24] On Ḥīwī, see Urvoy, *Penseurs libres*, 133–40.

heretics were clearly distinguished from the *kuffār*, infidels like the pagan king of Ashur.[25]

In an attempt to establish a typology of heretics, the Jewish *mutakallim* Shmuel ben Ḥofni (d. 1013) distinguished between "dissolute, insolent people" and "rational deviators." The latter, he said, may hold a wrong opinion, but they do so in good faith, following rational speculation, whereas the former's attitude is characterized by insolent scoffing and jeering.[26] In this context, Shmuel ben Ḥofni quoted II Kings 19:22 (= Is. 37:23), thus equating the heretics' disrespectful and dissolute insolence (*ḫalāʿa* wa-*mujūn*) with Sennacherib's "vilification and blasphemy."[27] As the typical example of this kind of heresy, Shmuel ben Ḥofni named Ḥīwī al-Balḫī. The specific characteristics of Ḥīwī's hersey were his blasphemous Bible criticism and his insolent tone. In these, we can easily recognize the traits which Muslim authors found so offensive in Ibn al-Rāwandī.[28]

Modern scholars are thus tempted to compare Ḥīwī to Ibn al-Rāwandī.[29] The similarities between Ibn al-Rāwandī's Qur'ān criticism and Ḥīwī's Torah criticism brought Kraus to give serious consideration to the possibility that Ibn al-Rāwandī's father was indeed a Jewish heretic, as claimed by Tanūḫī.[30] Ḥīwī had written a Hebrew poem containing "questions" which aim at exposing the absurdity and contradictions of the Bible.[31] This biting Bible criticism was indeed perceived by Jews as an abomination that puts him outside the pale of Judaism, much like Muslims felt about Ibn al-Rāwandī's criticism of the Qur'ān.[32] Our sources, however, give no indication whatsoever that Ḥīwī was a freethinker and denier of

---

[25] Cf. Saʿadya's Arabic introduction to *Sefer ha-Galuy*, 165 and 177. See also Poznanski, "Miszellen," 732.

[26] *Inna hādha kalām ahl al-ḫalāʿa wa'l-mujūn, wa-laysa min kalām al-mulḥidīn min ṭarīq al-naẓar alladhīna yataʿallaqūna bi-šubha, wa-in kānū muḫṭi'īn.*

[27] See H. Ben-Shammai, "New Findings in a Forgotten Manuscript: Samuel b. Hofni's Commentary on *Ha'azinu* and Saʿadya's 'Commentary on the Ten Songs,'" *Qiryat Sefer* 61 (1986), 316–17 (in Hebrew); comp. Davidson, *Hiwi*, 96–97.

[28] See, for instance, above, Ch. 6, apud nn. 60 and 65.

[29] See, for instance; van Ess, "Image," 12, 19. Islamic influence on Ḥīwī is suggested also by J. Rosenthal, who mentions Ibn al-Rāwandī in this context; cf. Rosenthal, "Ḥīwī," 429.

[30] "Ketzergeschichte," 365, n. 3.

[31] Only one fragment was discovered so far from the original of Ḥīwī's Hebrew poem; see E. Fleischer, "A Fragment of Hīwī al-Balḫī's Bible Criticism," *Tarbiz* 51 (1984), 49–57 (in Hebrew); The bulk of the poem was reconstructed from its refutations by later Jewish authors, especially Saʿadya; see Rosenthal, "Ḥiwi"; Davidson, *Hiwi.*

[32] Ḥīwī al-Balḫī can thus be said to have remained a Jew only nominally, in a

prophecy. As M. Stein has convincingly shown more than half a century ago, Ḥīwī was probably a Marcionite.[33] It would thus be more appropriate, perhaps, to compare him to Abū ʿĪsā al-Warrāq.

Nevertheless, the Islamic images of Ibn al-Rāwandī and of Rāzī did infiltrate Jewish consciousness, and they contributed to the shaping of a stereotyped Jewish heretic. In the remaining part of this chapter I will explore the manifestations of this stereotype in the writings of Moses Maimonides (d. 1204).

3. Muslim Heretics in Maimonides' Writings

Without any doubt, the dubious honor of being the most notorious arch-heretic in Jewish tradition goes to the first-century Elisha ben Avuya. Before his apostasy Elisha was considered to be one of the great sages of his day. But after his apostasy he became anathema to his former colleagues. His sin was considered to be so great that, according to a Talmudic saying, a divine decree singled him out as the only person for whom no repentance was possible. His sobriquet *aḥer* came to indicate his position as someone who is completely beyond the pale of Judaism.[34] Talmudic literature recounts various stories concerning the reasons and the nature of Elisha's apostasy. He is said to have been appalled by the presence of unjustified and inexplicable suffering in this world; he is described as having been attracted to Greek learning; he is claimed to have reached the conclusion that "there are two powers in heaven." Perhaps the most ancient anecdote about his apostasy associates it with his unsuccessful mystical experience:

> Four [scholars] entered paradise (i.e., engaged in esoteric learning), Ben Azzai, Ben Zoma, *Aḥer* and R. Akiva . . . Elisha looked [into the divine secrets] and he uprooted the plants (*qizez ba-neṭiʿot*, i.e., became a heretic).[35]

---

narrow halakhic sense that does not allow any way out of Judaism, not even by conversion to another religion. Compare van Ess, "Image," 19.

[33] M. Stein, "Ḥīwī al-Balḫī, the Jewish Marcion," *Klausner Volume*, Tel-Aviv 1937, 210–55 (in Hebrew). On the presence of Marcionism in Khorasan as late as the tenth century, see Madelung, "Abū ʿĪsā," 320; cf also Wasserstrom, *Between Muslim and Jew*, 151.

[34] On the meaning of this sobriquet, see G. G. Stroumsa, "Aḥer: A Gnostic," in B. Layton, ed., *The Rediscovery of Gnosticism*, II, Leiden 1981, 808–18.

[35] Cf. *Tosefta Ḥagiga* 2:3; BT *Ḥagiga* 14b–15a. For an English translation of the relevant talmudic passages compare *The Babylonian Talmud*, London 1938, pt. 2,

The figure of Elisha has fascinated Jewish intellectuals through the ages, and they tried to interpret his sin in a way that would fit their religious ideas and philosophical approach. In Maimonides's writings Elisha appears in several places. I shall attempt to demonstrate that Maimonides's presentation of the figure of Elisha is not based upon the Jewish sources, and that certain features in the personality of the Maimonidean Elisha, which have thus far been unexplained, can be understood on the basis of the assumption that Maimonides was familiar with the freethinkers as they are depicted in Muslim theological literature.

In *Guide* I, 32. Maimonides explicitly mentions Elisha and his sin:

> For if you stay your progress because of a dubious point; if you do not deceive yourself into believing that there is a demonstration with regard to matters that have not been demonstrated; if you do not hasten to reject and categorically to pronounce false any assertions whose contradictories have not been demonstrated; if, finally, you do not aspire to apprehend that which you are unable to apprehend—you will have achieved human perfection and attained the rank of Rabbi Aqiva, peace be on him, who "entered in peace and went out in peace" [BT, *Ḥagiga* 14b] when engaged in the theoretical study of these metaphysical matters. If, on the other hand, you aspire to apprehend things that are beyond your apprehension; or if you hasten to pronounce false, assertions the contradictories of which have not been demonstrated or that are possible, though very remotely so—you will have joined Elisha *Aḥer*. That is, you will not only not be perfect, but will be the most deficient among the deficient.... In this regard it is said: "Hast thou found honey? Eat so much as is sufficient for thee, lest thou be filled therewith and vomit it" [Prov. 25:16]. In a similar way, the Sages, may their memory be blessed, used this verse as a parable that they applied to Elisha *Aḥer*.[36]

Several striking features of the Maimonidean Elisha emerge in this passage: he attempted to reach beyond his powers of apprehension; he deluded himself into thinking that there is a proof for that which in fact cannot be proved; and he pronounced false certain matters "the contradictories of which have not been demonstrated." Shortly after his remarks concerning Elisha in that same chapter, Maimonides returns to this subject and says:

---

vol. 8, pp. 30ff. On the sin of Elisha in the Jewish tradition, see Liebes, *Elisha's Sin*, 29–51 and bibliography there; and see art. "Elisha ben Avuyah" *Encyclopedia Judaica*, VI, col. 668–70 (Sh. Safrai).

[36] *Guide* I, 32, pp. 68–69 (= *Dalāla*, 46).

> This is what the Sages intended to signify by their dictum, "whoever considers four things," and so on, [BT *Ḥagiga* 13b], completing the dictum by saying, "he who does not have regard for the honor of his Creator;" whereby they indicated what we have already made clear: namely, that man should not press forward to engage in speculative study of corrupt imaginings. When points appearing as dubious occur to him or the thing he seeks does not seem to him to be demonstrated, he should not deny and reject it, hastening to pronounce it false, but rather should persevere and thereby "have regard for the honor of his Creator." He should refrain and hold back.[37]

Maimonides repeats here what he already stated several lines earlier concerning the need to refrain from reaching conclusive decisions about matters for which there is no demonstrated proof. This time, however, he mentions a further characteristic of those who act in the manner which was described as the behavior of Elisha—he who hastens to deny those things the contradictories of which have not been proved: namely, that such a man has no regard for the honor of his Creator. Maimonides alludes here to what is stated in the *Mishna*:

> One does not expound upon matters of illicit sexual relations before three, nor upon the Account of the Creation before two, nor upon the Account of the Chariot before a single person, unless he is wise and understands matters by himself. Whoever contemplates four things, it is a pity that he was ever born: What is above, what is below, what is before, and what is after. And whoever has no regard for the honor of his Creator—it is a pity that he was ever born.[38]

In his *Commentary on the Mishna, Ḥagiga* 2, Maimonides expounds this entire *Mishna* as one unit: he first explains the significance of the limitations which must be imposed on the discussion of the account of the Creation and of the Divine Chariot, concerning which the *Mishna* warns that one ought not to involve oneself without suitable preparation. He connects this to the condemnation of anyone involving himself with the question of "what is before and what is after"—that is, of a person who attempts to know "what there was before the heavens were created, and what will be after they are destroyed." All these things—the rash involvement with the mysteries of the Chariot and of Creation, and the condemnation of one who looks at what is before and what is after—are connected by Maimonides

[37] *Guide* I, 32, p. 70 (= *Dalāla*, 47).
[38] BT *Ḥagiga* 11b.

in his *Commentary on the Mishna* to, "Whoever has no regard for the honor of his Creator." He writes:

> And he states, regarding the restraint of a person who imposes his mentality (*dhihn*) also on the contemplation of metaphysics by way of the imagination alone, without graduated study of the sciences, "Whoever has no regard for the honor of his Creator."[39]

As we have seen, in *Guide* I, 32 Maimonides also mentions the contemplation of these four questions and the unwillingness to take heed of the honor due to one's Creator as two interrelated issues:

> This is what the Sages intended to signify by their dictum, "whoever considers four things," and so on, [BT *Ḥagiga* 13b], completing the dictum by saying, "He who does not have regard for the honor of his Creator" . . .

The person who reflects upon the acts of Creation ("what is before and what is after") is the same one who has no regard for the honor of his Creator, and is also he who is too hasty in rejecting those matters the contradictories of which have not been demonstrated—like Elisha-*Aḥer*.

As might be anticipated, against Elisha Maimonides poses R. Akiva, who "entered in peace and went out in peace." In this same context, Maimonides also defines the realm of speculation within which R. Akiva succeeded and Elisha failed as "the divine matters." As is known, the divine matters are first of all the science of the Godhead, metaphysics or "the Account of the Chariot" (*Maʿaseh Merkavah*). But Maimonides also thought, regarding the science of physics or "the Account of the Creation" (*Maʿaseh Bereshit*), that "there is a close connection between these matters and divine science, and they too are secrets of that divine science."[40]

Indeed, in his interpretation of the chapter concerning the four

---

[39] Maimonides, *Commentary, Ḥagiga* 2.1 (*Moʿed*, 378).

[40] *Guide*, Introduction, 7 (*Dalāla*, 3). On the significance of the Account of the Beginning and the Account of the Chariot, see Maimonides, *Commentary, Ḥagiga* 2.1 (*Moʿed*, 377): "By the name 'The Account of the Beginning' [the Sages] designate physics and the preoccupation with the origins of Being. By 'The Account of the Chariot' they intend metaphysics, which is the discussion of the entirety of Being, the existence of the Creator, His knowledge and His attributes, the necessitating of existent things from Him, the angels, the soul, the intellect which is conjoined with the human being, and the destiny after death." See also Maimonides, *Guide*, Introduction, 9 (*Dalāla*, 5): "The Account of the Beginning which . . . is natural science."

who "entered *pardes*" (i.e., engaged in esoteric speculation), Maimonides would seem to have accentuated the tendency to see involvement in physics as a part of the "divine sciences." In *Guide* II, 30, he distinguishes between the terms "first" (*al-awwal*) and "beginning" (*mabda'*); in that same chapter, in his interpretation of the terms "water" and "firmament" in the Account of the Creation, Maimonides chooses to cite R. Akiva:

> With regard to the fact that that which is above the firmament is called water in name only and that it is not the specific water known to us, a statement setting this forth has also been made by the Sages, may their memory be blessed. They made it in the following passage: "Four entered Paradise," and so on. [BT *Ḥagiga* 14b] . . . Rabbi Aqiva said to them: When you come to the stones of pure marble, do not say, Water, Water, for it is written: "He that speaketh falsehood shall not be established before mine eyes" [Ps. 101:7].[41]

Chapter 30 of Book II belongs to the complex of chapters dealing with the question of whether the world is eternal or created in time.[42] Indeed, Maimonides associates the path of the four who entered into *pardes*, and hence the path of Elisha, not only with the question of creation in a general way, but specifically with that of the eternity of the world.

This understanding enables us to re-examine the figure of Elisha. Let us return to Maimonides's remarks in I, 32, and to the stumbling blocks which may cause a person to behave like Elisha. As we have seen, this danger confronts one who "aspires to apprehend things that are beyond his apprehension" or who "hastens to pronounce false, assertions the contradictories of which have not been demonstrated." The specific question concerning which Maimonides takes the trouble to argue that there is not, and cannot be, a proof, is indeed that of the eternity of the universe. In *Guide* II, 15, he states:

> As for me, I have no doubt that the opinions mentioned by Aristotle with regard to these subjects—I mean such opinions as that concerning the eternity of the world . . . are none of them provided with a demonstration. And Aristotle never at any time had the fantasy that

[41] *Guide* II, 30, p. 353 (*Dalāla*, 247–48).

[42] The meaning of the statement of R. Akiva in this connection has already been widely discussed. See, e.g., S. Klein-Braslavy, *Maimonides' Interpretation of the Story of Creation*, Jerusalem 1987 (in Hebrew), 163–74.

> what he said in this connection constituted a demonstration. On the contrary, he thought, as he says, that the gates of the ways to inferential reasoning on these matters are closed before us and that we have at our disposition no principle pertaining to them from which to start to draw inferences.[43]

In a more explicit and specific way, the danger that one may come to resemble Elisha awaits him who "hastens to pronounce false, assertions the contradictories of which have not been demonstrated or that are possible, though very remotely so." Here, too, the specific issue that Maimonides tries to keep within the realm of the possible, and to argue that no proof has been brought disproving it, is the belief in the creation of the world in time. Thus, he says in II, 23:

> I have drawn your attention to this in order that you should not be deceived. For someone may some day lead you into vain imaginings through setting forth a doubt concerning the creation of the world in time, and you may be very quick to let yourself be deceived. For in this opinion is contained the destruction of the foundation of the law and a presumptuous assertion (*iftiyāt*) with regard to the deity.[44]

In Chapter 29 of Book II, he again explains that:

> Aristotle, on the other hand, thinks that just as the world is eternal *a parte post* and will not pass away, it is also eternal *a parte ante* and has not been produced. Now we have already said and explained that this doctrine can be arranged in a coherent way only through recourse to the law of necessity and that necessity contains a presumptuous assertion with regard to the deity, as we have explained.[45]

And shortly thereafter:

> Not everything mentioned in the *Torah* concerning the *Account of the Beginning* is to be taken in its external sense as the vulgar imagine . . . And [the Sages] say explicitly: "As from the beginning of the book up to here, 'the honor of God [requires] to conceal the thing.'" [Gen. R. 9:1, ref. to Prov. 25:2].[46]

In these last passages, an additional element appears in Maimonides's remarks. We initially saw his logical and methodological claim that there is no proof and cannot be any proof regarding the question of the creation of the world. But he then adds a claim that might

---

43 *Guide*, II, 15, p. 292 (*Dalāla*, 203); cf. also *Guide*, II, 22, p. 320 (*Dalāla*, 223).
44 *Guide*, II, 23, p. 321 (*Dalāla*, 224).
45 *Guide*, II, 29, p. 346 (*Dalāla*, 243).
46 *Guide*, II, 29, p. 346–47 (*Dalāla*, 243–44).

be described as an ethical one: casting doubts upon the doctrine of the creation of the world in time is "a presumptuous assertion with regard to the deity." The Account of the Creation and its hidden aspects belong to "the honor of God"; this is the meaning of the command, "to have regard for the honor of his Creator."[47] It is in this light that we ought to reexamine what we saw above, in I, 32:

> This is what the Sages intended to signify by their dictum, "whoever considers four things," and so on, [*BT Ḥagiga* 13b], completing the dictum by saying, "he who has no regard for the honor of his Creator"; whereby they indicated what we have already made clear: namely, that man should not press forward to engage in speculative study of corrupt imaginings. When points appearing as dubious occur to him or the thing he seeks does not seem to him to be demonstrated, he should not deny and reject it, hastening to pronounce it false, but rather should persevere and thereby "have regard for the honor of his Creator". He should refrain and hold back.

Let us summarize the argument to this point by stating that the central elements comprising the image of the sin of Elisha in the *Guide* are: excessive effort to attain that which is above human comprehension, the bold curiosity for "what is before and what is after"; the hasty denial of things which have not been logically disproved; and lack of regard for the honor of the Creator. The very same elements appear in Maimonides's discussion of the question of the eternity of the world. Maimonides nowhere explicitly says that Elisha sinned in denying the creation of the world in time. But the use of the same key expressions in both cases is striking, and it seems to me that we would not be far wrong if we said that Elisha ben Avuya is portrayed by Maimonides in the *Guide* as a man who stumbled into the pitfall of the Aristotelian philosophers (to be distinguished from Aristotle himself), and decided, regarding the "Account of the Creation" in favor of the belief in the eternity of the universe.

This is a wholly original interpretation of the figure of Elisha and his sin. While there had been earlier attempts to connect the question of the eternity of the world to the story of the "four who entered Paradise," such attempts were all related to the statement of Ben

[47] Maimonides also applies this expression to the discussion of the divine attributes in *Guide*, I, 59, p. 142 (*Dalāla*, 97). Elsewhere he interprets this dictum by identifying "the honor of God" with the intellect ("*wa'l-ʿaql huwa kvod ha-shem*"); cf. *Commentary*, *Ḥagiga* 2.1 (*Moʿed*, 378).

Zoma and the warning of R. Akiva.[48] It is self-evident that one cannot find the roots of the Aristotelian philosopher Elisha in the Talmudic description of Ben Avuya. Fundamental elements in the image of Elisha in the Talmud—such as his confrontation with Metatron and the matter of "two powers in heaven," or the concern for the apparent arbitrary suffering in the world, are completely absent in the image of the Maimonidean Elisha. But what most stands out in Maimonides's interpretation is primarily the absence of the Talmudic formulation of the nature of Elisha's sin: "*Aḥer* uprooted the plants."[49] In its place, added stress is placed by Maimonides upon the *Mishna*, which discusses those who "it is a pity that they were ever born": i.e., he who contemplates that which was before, and he who does not have regard for the honor of his Creator.

Moreover: the emphasis upon the excessive effort in order to attain that of which it is impossible to achieve comprehension causes Maimonides to apply to Elisha the verse, "Hast thou found honey? Eat so much as is sufficient for thee" (which in the Talmudic source is applied by the Sages to Ben Zoma), rather than the verse in Ecclesiastes 5:5, which the Sages expounded with regard to Elisha ("Do not cause your mouth to cause thy flesh to sin").[50] One may of course simply assume that Maimonides committed an error, and confused the Talmudic text concerning Elisha ben Avuya with that concerning Ben Zoma.[51] But such an error is not at all characteristic of Maimonides, and seems particularly unlikely in light of the numerous times that Maimonides alludes to the second chapter of *Ḥagiga* in the *Guide*.[52] This is without doubt a chapter which Maimonides studied a great deal and knew well.

The possibility of an ordinary error is weakened further, in my opinion, in light of the fact that this is not his only "slip of the pen" concerning this matter, but part of a systematic commentary, in which Maimonides frees the images of the four who entered *pardes* from the bonds of the Talmudic text, using them as he wishes. We would have thought that, of the four, the figure to whom it would be most appropriate to attribute involvement with the eternity of the world

---

[48] See Liebes, *Elisha's Sin*, 111–29, and the bibliography there.

[49] See above, n. 35.

[50] *Ḥagiga* 15a.

[51] As suggested by Liebes, *Elisha's Sin*, 125.

[52] For example, in *Guide*, Introduction, 7 (*Dalāla*, 3); I, 32, pp. 69–70 (*Dalāla*, 47); I, 33, pp. 71–72 (*Dalāla*, 48); I, 34, p. 78 (*Dalāla*, 53), and I, 35, p. 81 (*Dalāla*, 54).

would be Ben Zoma, about whom the Talmud says that "he glimpsed and was harmed," and whose reflection upon what was between the upper waters and the lower waters led Rabbi Joshua to remark that, "Ben Zoma is presently outside."[53] We might have thought that Maimonides, loyal to the above-mentioned approach (*Guide* II, 30), would be happy with this version, and would state that Ben Zoma dealt with the Account of the Creation. Instead, Maimonides completely ignores the Talmudic context. The saying of R. Joshua, "Ben Zoma is presently outside," is taken by him to indicate that Ben Zoma only engaged in the preliminary disciplines of mathematics and logic, and did not at all engage in natural sciences.[54] Maimonides prefers to attribute involvement in physics as "a presumptuous assertion with regard to the deity" to the great rebel, Elisha-*Aḥer*. We thus find here a clear departure from the Talmudic images of Elisha and of Ben Zoma. It seems to me very unlikely that this departure was performed without deliberate thought.

Neither can the source of Maimonides's interpretation of Elisha be found in the writings of Jewish philosophers who preceded him. None of the peculiarities of Maimonides's Elisha can be found in Judah Halevi's treatment of this figure (as, for example, in the *Kuzari*, III, 65).

Analysis of the passages pertaining to Elisha has bearing upon the famous and perplexing question as to what Maimonides really thought about the issue of the creation or eternity of the world. The formulations used by Maimonides in order to describe the sin of Elisha in the *Guide* emphasize the theological-educational aspect (as opposed to the philosophical-scientific aspect) of the demand to preserve the possibility of temporal creation of the world, "even if very remotely so." This demand follows from the need to "have regard for the honor of the Creator," and not from considerations of scientific rigor.

[53] *Ḥagiga* 15a; cf. Liebes, *Elisha's Sin*, 111–28.

[54] *Guide*, III, 51, p. 619 (*Dalāla*, 455–56): "Know, my son, that as long as you are engaged in studying the mathematical sciences and the art of logic, you are one of those who walk around the house searching for its gate, as [the Sages], may their memory be blessed, have said resorting to a parable: 'Ben Zoma is still outside.' If however, you have understood the natural things, you have entered the habitation and are walking in the antechambers...." Ralph Lerner has drawn my attention to the fact that, whereas Elisha is accused of doing too much, Ben Zoma's fault is that he did too little. Maimonides thus seems to be using Ben Zoma in order to counterbalance his demand for restraint by pointing to the necessity of ambition and audacity in the search for knowledge.

In addition, the blatant "error" of attributing an inappropriate verse to Elisha is intended, in my opinion, to serve as a "pointer," arousing the attention of the reader to the fact that Maimonides's statements here are not intended to be understood literally.[55] It may well be that these statements are correct only on one level, and that one needs to investigate the possibility of their being understood on another level. But it is not with this matter that I wish to deal here, but with the uniqueness of the Maimonidean Elisha.

The original interpretation of the *Mishna* offered by Maimonides is not an unusual phenomenon in the *Guide*. If only this one interpretation were available to us, we could point toward Elisha as just one more example of the fact that Maimonides translates the words of the Rabbis in a manner suitable to the problems with which he is dealing, but we would not have much success in identifying the source of his inspiration for this interpretation. This is made possible by another text, in which Maimonides mentions Elisha ben Avuya, in an entirely different context and with a different, but no less original, interpretation. In his *Commentary on the Mishna, Ḥullin* 1.2, Maimonides deals with the question of who is to be considered as an apostate Jew (*yisra'el meshumad*), who may validly perform ritual slaughter, and who is considered a heretic (*min*), who may not. He states in this connection:

> "And he may not be a *min*": In the language of the Sages, *minim* are the heretics among Israel (*zanādiqat yisra'el*) . . . And these are people whose minds were dulled by ignorance, and whose souls were darkened by lusts; therefore, out of ignorance, they defame (*yaṭʿanū*) the Torah and the prophets, of blessed memory, and give the lie to (or: deny) the prophets concerning matters about which they have no knowledge (*yukadhdhibūnā al-anbiyā' bi-mā lā ʿilma lahum bi-hi*), and they abandon the commandments in contempt. They are of the congregation of Jesus the Nazarene, Doeg, Aḥitophel, Geḥazi, and Elisha-*Aḥer*, and those who follow in their ways—may the name of the wicked rot. It is possible to know that a person belongs to this congregation when one sees that he shakes off any one of the commandments contemptuously, without deriving any benefit from this act.[56]

[55] See Maimonides' explicit remarks concerning the function of the problematic places in this book: *Guide*, Introduction, 18–20 (*Dalāla*, 12–13). On another famous "error" of Maimonides (in *Guide*, I, 71), see S. Stroumsa, "Al-Fārābī and Maimonides on the Christian Philosophical Tradition: A Re-evaluation," *Der Islam* 68 (1991), 284–85.

[56] Maimonides, *Commentary, Ḥullin* 1.2 (*Qodashim*, 175). Maimonides's description

The people included here as heretics are those whom one might expect to see in such a list: Jesus and Elisha, along with three Biblical figures whom the *Mishna* (*Sanhedrin* 11:2) enumerates as "laymen" (i.e., not kings) who have no portion in the World to Come: namely, Doeg, Aḥitophel, and Geḥazi. These figures are mentioned in the Talmudic discussion in *Ḥagiga* (in connection with Elisha's involvement with Greek songs and heretical books), as having been remarkably learned.[57] In his *Commentary on the Mishna* in *Sanhedrin*, Maimonides likewise stresses the fact that these same three laymen were renowned for their great erudition (*ʿiẓm darajatihim fī al-ʿilm*).[58]

Spiteful contempt for the commandments—contempt which, according to this passage, characterizes a "heretic"—fits the Elisha of the Talmud. One may also note a certain connection between the image of Elisha here and that in the *Guide*: the contempt towards the commandments and the prophets exhibited by heretics such as Elisha is consistent with the portrait of Elisha as one who has no regard for the honor of his Creator.

But the description of Elisha as one of those who "give the lie to the prophets concerning matters about which they have no knowledge" (a description formulated, incidentally, in Qur'ānic language),[59] includes a certain feature which we have not encountered in the *Guide*—one which, as far as I know, also has no basis in the description of Elisha in the Jewish tradition. The fact that in the *Commentary on the Mishna* Maimonides does not concern himself with the question of Physics and the creation is not surprising: this was not a proper subject for the audience intended in this book. However, his interpretation of the figure of Elisha remains as original as his interpretation in the *Guide*.

If we combine the description given in the *Guide of the Perplexed* with that in the *Commentary on the Mishna*, we are confronted with the image of a heretic, the paradigmatic Jewish *zindīq*, of whom Elisha is the archetype. He is a person of great erudition and learning, with a great appetite to know beyond what it is possible to

---

of Elisha as abandoning the commandments contemptuously is perhaps influenced by Judah Halevi. Halevi depicts Elisha as a person for whom the commandments are mere means to a goal, and who, having attained a glimpse of the intelligibles, abandons the commandments; cf. *Kuzari*, III, 65, pp. 140–41.

[57] *Ḥagiga* 15b.

[58] Maimonides, *Commentary*, *Sanhedrin* 10, 2 (*Neziqin*, 217).

[59] See, for example, Q. 22:71; 24:15; 40:42.

know. This combination of craving for knowledge and of erudition causes him to speak with "presumptuous assertion with regard to the Deity," not to have regard for His honor, to declare the falsity of that which has not been proven false, and to decide in favor of the belief in the eternity of the world. Such a person contemptuously and openly denigrates the commandments, without any motivation of personal benefit. He does not believe in the prophets, and in declaring this reveals the limits of his own knowledge. As we have already mentioned, several of the primary features of this image have no source in the Jewish literature concerning Elisha-*Aḥer*; however, they all appear in the figure of the Muslim archetype of the *zindīq*, Ibn al-Rāwandī.

It is perhaps not superfluous to recall the relevant traits of his history. At the beginning of his career, Ibn al-Rāwandī was an ordinary *mutakallim*, and was even thought of as a respected figure among the Muʿtazila of Baghdād. But, due to reasons which are not clear (the Muslim sources speak of some blow to his pride),[60] he turned his back upon his Muʿtazilite colleagues and began to direct verbal barbs against them—and not only against them. Notwithstanding the haziness of his historical image and of the true nature of his apostasy, one can reconstruct a rather clear, consistent picture of what one might call "the polemical image of Ibn al-Rāwandī"—that is, the image he assumes in polemical literature. He is designated as a *zindīq*. He is connected to the *dahriyya*—that is, those who assert the eternity of the world—and it is related that one of his books, *K. al-Tāj*, was devoted to a presentation of the proofs for the eternity of the universe.[61] He is described as an educated man whose erudition led him astray, causing him to underestimate the limitations of his own wisdom (in this connection, the Muslim sources apply to him the formula used by orthodox intellectuals to discount heretical intellectuals: that his erudition outstripped his intelligence).[62]

But Ibn al-Rāwandī is primarily known for two other characteristics: the one concerning his character, the other concerning his ideas. In terms of character, he is described as a haughty individual who, with a sharp tongue and arrogance, enjoyed satirizing and

---

[60] This is a *topos* in heresiographical literature: Marcion is said to have become a heretic after he was turned down as bishop of Rome, and Anan is said to have established Karaism because of his disappointed hopes to become Exilarch.

[61] See above, Ch. 2.

[62] See below, Conclusion, nn. 4 and 6.

mocking his rivals and all that was holy to them. In terms of his ideas, Ibn al-Rāwandī became a byword for one who denied prophecy. According to him, the notion that God sends prophets is opposed to common sense. As God has given human beings intellect, there are only two possibilities: if the words of the prophets correspond to common sense, they are unnecessary; if they are opposed to the command of the intellect, one ought not to heed them. Ibn al-Rāwandī's *K. al-Zumurrud* is devoted to proving the falsehood of prophecy in general, but he went even further and directed this argument explicitly against Islam: in the *K. al-Firind* he disproved the prophecy of Muḥammad. In another book, the *K. al-Dāmigh*, he pointed out, in impudent language, the internal contradictions within the Qur'ān and the laughable and absurd arguments found there. And his father is said to have made a similar criticism of the Bible.[63]

Thus, the features which mark Ibn al-Rāwandī as a *zindīq* – i.e., denial of the prophets and attachment to the *dahriyya* – are the same ones which we find emphasized in the Maimonidean Elisha, and which are not anticipated anywhere else in the Jewish tradition. Hence, we may legitimately ask the question whether Maimonides's portrayal of Elisha ben Avuya may not have been influenced by the figure of Ibn al-Rāwandī in Islamic literature.

One first needs to ask whether Maimonides may have himself read the numerous writings of al-Rāwandī. While such a possibility may not be ruled out entirely, it is rather unlikely. The Muslims persecuted Ibn al-Rāwandī while he was alive, and did not give him rest even after his death; his books were in effect banned, and there is reason to suspect that already during the eleventh century even Muslims found it difficult to find any manuscripts of his books.[64]

On the other hand, it is highly improbable that Maimonides did not know anything at all about Ibn al-Rāwandī. True, Maimonides does not mention Ibn al-Rāwandī by name, nor does his name appear prominently in the writings of the Muslim philosophers whom he was accustomed to read: i.e., Fārābī, Ibn Bājja, and Averroes.[65] But the image of Ibn al-Rāwandī, the debate with him, and quotations from his works are extremely common in the works of Muslim theologians, whether in their polemical writings or in *kalām* writings.

---

[63] See above, Ch. 2.
[64] See above, Ch. 7, apud n. 64.
[65] On the philosophers' attitude to Ibn al-Rāwandī, see above, Ch. 6.

Ibn al-Rāwandī's words are occasionally quoted when he expresses his opinion concerning some inoffensive issue, but he is primarily mentioned as one who argues against prophecy. One ought to mention that, even though Muslim theological and heresiographical literature contains a plethora of heretics, Ibn al-Rāwandī occupies a central place in this literature as the arch-heretic, one who is so out of spite—a role parallel to that occupied by Elisha ben Avuya in the rolls of Jewish heretics. We know that Maimonides met with Muslim theologians on various occasions, and even participated with them in discussions of *kalām*.[66] Shlomo Pines has convincingly argued that Maimonides often reacted in the *Guide* to thinkers, both philosophers and theologians, whose name he does not mention.[67] If Maimonides was familiar with the writings of Muslim theologians, and not only with those of Muslim philosophers, we must assume that he encountered the image of al-Rāwandī in one way or another. As mentioned above, Ibn al-Rāwandī's name is cited by Jewish authors and appears in Muslim books which were in the posession of Jews.[68] If Maimonides encountered any mention of al-Rāwandī, we may reasonably assume that this included his description as a *zindīq*, a Dahrite, and as one who denied the prophets. This being so, it seems quite possible that this image influenced him when he shaped the image of the Jewish *zindīq*, Elisha ben Avuya.

Maimonides behaves extremely freely in harnessing Ibn al-Rāwandī to his own purposes. He "breaks down" the Talmudic figure to its components, drawing from it those elements which comprise the stereotype of the *zindīq*, the dahrite or the "denier of prophecy." But Maimonides does not see any need to relate to Ibn al-Rāwandī himself or to mention him anywhere by name. All of the elements characteristic of Ibn al-Rāwandī enter in Maimonides's writings under the name of Elisha-*Aḥer*.

The same method, of taking the characteristic elements of a Muslim heretic and applying them to a Jewish one, is used by Maimonides with regard to Rāzī, whom he specifically attacks by name.

---

[66] See F. Rosenthal, "Maimonides and a Discussion of Muslim Speculative Theology," in M. Maswari Caspi, ed. *Jewish Tradition in the Diaspora*, Berkeley 1981, 109–12; cf. P. Fenton, "A Meeting with Maimonides," *BSOAS* 45 (1982), 1, n. 1.

[67] Cf. Pines, "Translator's Introduction: The Philosophic Sources of the *Guide of the Perplexed*," Maimonides, *Guide*, lvii–cxxxvii; see also Appendix vii to Pines, "Shī'ite Terms," 248–51.

[68] See above, n. 23.

In his introduction to *Pereq Ḥeleq*, Maimonides enumerates three classes of people in relation to the interpretation of the aggadic words of the Sages: those who understand them literally and accept them, who are numerous; those who read them literally and reject them, who are also a large group; and finally, the small number of people who understand the words of the Sages to contain hidden wisdom, concealed within riddles and parables. The second group—those who understand the words of the Sages in a literal manner, and who therefore reject them—is marked by contempt and scorn for the words of the Rabbis, rooted in excessive arrogance. Maimonides says of this group:

> They repeatedly mock the sayings of the Sages; they claim to be more intelligent and brighter than the Sages, and that the Sages, peace upon them, were simpletons who suffered from inferior understanding, that they were ignorant of the entirety of Being, and understood nothing whatever. Most of those who have stumbled into this error are those who claim [to know] medicine and those who rave about the decree of the stars. They claim to be cultivated men, physicians and philosophers. How remote they are from true humanity in the eyes of real philosophers![69]

One may ask: who are these people, who claim to be physicians, who deal with astrological delusions and claim to engage in philosophy, and who hold to scorn the words of the Sages and the Talmudic midrashim? Offhand, one would seek here Jewish doctors, and more than one, as Maimonides speaks here of a group of people that "are also numerous."

In his letter to Ibn Tibbon, Maimonides refers critically to the tenth-century Jewish philosopher Isaac Israeli, who is "only a physician" (and not a philosopher). But Isaac Israeli is far from meeting the specifications given here by Maimonides, and it is generally difficult to know where Maimonides would find a group of Jewish physicians to fit this description.

On the other hand, Abū Bakr al-Rāzī, who is also mentioned in that same letter to Ibn Tibbon as "only a physician," is very suitable for this description. Like Ibn al-Rāwandī, Rāzī was noted for

---

[69] Maimonides, *Commentary*, *Sanhedrin* 10, Introduction (*Neziqin*: 202): "*wa-akthar mā yaqaʿu fī hādha al-iʿtiqād muddaʿiyī al-ṭibb, wa'l-hadha'iyyīn bi-qaḍāyā al-nujūm, li-annahum bi-zaʿmihim ḥudhdhāq ḥukamā' falāsifa, wa-mā abʿadahum min al-insāniyya ʿinda al-falāsifa ʿalā al-ḥaqīqa.*" Disrespect to the Sages and to religious authorities is mentioned again by Maimonides shortly thereafter, as part of his definition of *epikoros*; cf. ibid., 209.

his claim that prophecy is opposed to Divine wisdom, and for the contemptuous remarks he made against the prophets. Like Ibn al-Rāwandī, Rāzī became an archetype of the heretic, but his reputation was more widespread in philosophic circles. Rāzī of course did not write against the exegeses of the Talmudic sages, but we do know of contemptuous remarks which he directed against religious sages in general, whom he calls "the bearded goats," and his criticism of the *ḥadīth* literature.[70] It is possible that, in his portrayal of those who hold the words of the Sages in contempt because they understand them literally, Maimonides incorporated the image of Rāzī, as a stereotype of one who mocks the tradition with the arrogance of the man of science and philosopher. It is not that Maimonides had any reason to think that Rāzī's words or his approach had any hold among Jewish physicians; rather, as Maimonides was engaging in typologies, and attempted to portray clear-cut types in relation to their understanding of the words of the Sages, he tried to build the image of one who has contempt for their words in the sharpest and most extreme way possible. The image of the physician-heretic Rāzī served him as a model for this purpose.

A similar use of Rāzī as the archetypal physician-heretic may also be present in the *Guide*. In *Guide*, III, 12, Maimonides attacks at length Rāzī's "ravings" (*hadhayānāt*), and quotes Rāzī's book *al-ilāhiyyāt*. According to Maimonides, Rāzī argues there that "there is more evil than good in what exists."[71] Maimonides goes on from there to discuss the words of the Aristotelian philosophers who, noting the injustice in the division of good and evil in the world, challenge the notion of Divine providence (III, 16). Then, in III, 19, while still within the broad context of the question of providence and God's knowledge of individuals, Maimonides cites the verse from Psalm 94:8–9, "Consider, ye brutish among the people; and ye fools, when will you understand? He that planted the ear, shall He not hear? He that formed the eye, shall He not see?" Maimonides then goes on to say:

> I shall now explain to you the meaning of this argument after I have mentioned to you the way in which those who assault the prophets' discourse[72] misunderstand this discourse. Several years ago some dis-

[70] See above, Chs. 3 and 4.

[71] Maimonides, *Guide*, III, 12, p. 441 (*Dalāla*, 318); see above, Ch. 3.

[72] *Al-mutahāfitīn li-kalām al-anbiyā'*. Pines's translation (p. 478) "Those who are

> tinguished individuals of our religious community, who were physicians, told me[73] that they wondered at this dictum of David. They said: According to his way of reasoning, it would follow necessarily that the creator of the mouth eats and the creator of the lung shouts and the same would hold for all the other members.[74]

The question cited here by Maimonides challenges the logic of this verse, on the assumption that the argument made in the verse is that whoever makes any sort of vessel, acts in accordance with the qualities of that vessel. Maimonides replies by explaining that the logic of the verse is in fact different: one who makes any vessel must certainly understand its way of functioning. At first glance, this passage does not present any difficulty: the question as such is a plausible exegetical one. Maimonides's answer to the question, intended to show the providence and wisdom inherent in creation, is related to the subject of the chapter: since the issue at hand is the wisdom revealed in the creation of the organs of the human body, one can understand why physicians would specifically be interested in this.

Nevertheless, the reader is left with a certain feeling of unease: the question itself is a logical one, which has nothing to do with medicine. One can understand why the one *answering* it might be a physician, who constantly encounters the complexity of the human body; but his interlocutor could be anyone. One cannot understand Maimonides's emphasis on the fact that those referred to are "some distinguished individuals of our religious community, who were physicians," in the plural, as if this was a typical question of physicians. Even less clear is why Maimonides formulates this exegetical question in the words: "I have mentioned to you the way in which those who assault the prophets' discourse, misunderstand this discourse." It follows from this that those selfsame "distinguished individuals of our religious community, who were physicians," who ask a question stemming from their misunderstanding of the verse in Psalms, are defined as those who "assault the prophets' discourse"—surely a severe accusation.

Since the accusation placed in the mouth of the physicians is not specifically a medical one, and its wording has a provocative ring,

---

overhasty in their interpretation of the prophets' discourse" does not seem to reflect the belligerence of the expression *muʿtahāfitīn*.

[73] *Qāla lī mundhu sinīn ašḫāṣ min nubalāʾ millatinā al-aṭibbāʾ*.

[74] *Guide*, III, 19, p. 478 (*Dalāla*, 346).

pointing out the absurd and ridiculous in the verse[75] – we have reason to suspect that here, too, Maimonides has mixed into the framework of the discussion elements which do not properly pertain to it in the historical sense. Because of the proximity of this discussion to that concerning the philosopher who "assaulted the prophets' discourse," Abū-Bakr al-Rāzī, it is possible that these elements were borrowed from the Muslim struggle against this freethinking philosopher.

These suggestions concerning Maimonides's use of the Muslim freethinkers are primarily based upon literary analysis, which by its nature is never unequivocal. But when combined together, they accumulate and strengthen one another in a manner that reveals a certain tendency within Maimonides's exegesis. Maimonides lived within an Arabic environment, predominantly Muslim, and breathed its culture. The fundamental problems which engaged him – prophecy and its authority, Divine providence, and strengthening the status, the "honor" of God as Creator of the world, within an Aristotelian philosophical position—these problems were common to him and to the Muslim thinkers of his time. But it is for the Jewish community that Maimonides wrote, and the authority which he wished to strengthen was that of the Jewish scriptures. Ibn al-Rāwandī did not interest him as such – he was the "problem" of the Muslims. Neither did Maimonides engage in a systematic debate with Rāzī, except to the extent that his name was known to the Jewish intellectuals for whom the *Guide* was written. Insofar as the problems represented by these two figures were relevant to Maimonides, he "judaized" them, translated them into the language of the Jewish reader, and applied them to a real historical figure – that of Elisha ben Avuya-*Aḥer*, or to a fictitious collective figure – that of the Jewish physicians.

---

[75] Compare the manner in which Maimonides presents a question which is not provocative, but legitimate, in *Guide* III, 32, p. 527 (*Dalāla*, 385).

CONCLUSION

The main purpose of the present book was to draw attention to the existence of the phenomenon of freethinking in the medieval Islamic world, and to its importance in the development of Islamic thought. As clearest examples of freethinkers I have chosen Ibn al-Rāwandī and Rāzī, although, on the one hand, they were very different from each other, and on the other hand, other thinkers shared some of their characteristics.[1] Nevertheless, the term "freethinking" focuses on a central and typical aspect of their thought: the belief in the sufficiency of the intellect and the rejection of all religions based on revelation. Ibn al-Rāwandī was "a veritable precursor" of Rāzī.[2] Although Rāzī never mentions him, it is plausible that Rāzī was aware of his existence, and may have been influenced by his anti-prophetic teaching. The differences between them on every other aspect of their thought does not allow us to regard them as establishing "a school of freethinking."[3] But their similar approach to knowledge and to authority, and the fact that this approach surfaces in the thought of other thinkers, forces us to regard freethinking as a recurring phenomenon.

The identification of the phenomenon, however, is only a first step. Many questions, which I have chosen to ignore in the attempt to isolate the notion of freethinking and its implications for religious history, are now pressing for answers. One would need to investigate the presence of freethinking in the thought of other Muslim thinkers, to draw the contours of the phenomenon, its appearance in varying degrees of intensity and consistency, in various intellectual circles: among scientists, mystics, and poets, and to study the social and political ramifications of its spread.

In a book dedicated to the freethinkers and to their ruthless attack on Islam, it seems appropriate to balance the picture by giving the final word to their opponents. Rather than arguing for the limitations of the human intellect and the need for prophecy, Muslim thinkers often resort to an *ad hominem* dismissal of the freethinker's pretence

[1] See above, Introduction.
[2] Pines, "Philosophy," 794.
[3] Cf. Niewöhner, *Veritas*, 245, n. 34.

to intellectual astuteness: "His knowledge was greater than his intelligence" (*kāna ʿilmuhu akthara min ʿaqlihi*).[4] This caustic retort aims at disqualifying the freethinkers on the basis of their own criteria of excellence. A similar method is used by Ibn Ṣāʿid, who says:

> [Rāzī], however, did not penetrate the divine science (*al-ʿilm al-ilāhī*), nor did he achieve knowledge of its profound intention. Therefore, his [own] opinion (*raʾy*) became confused; he accepted blindly (*taqallada*) weak opinions and adopted the ideas of weak schools of thought. He associated with people whom he did not understand, and whose system provides no guidance.[5]

Ibn Ṣāʿīd dismisses Rāzī's pretensions by showing that he failed precisely where he aspired to excel. He, who wrote a book on divine science, never really grasped it. Having boasted of relying solely on his independent opinion, he turned out to have poor judgment, which confused him. And having scoffed at those who accept blindly the authority of scriptures and clergy, he had in fact practiced *taqlīd* of much weaker and less trustworthy authorities, who led him astray.

Historically, indeed, Islamic orthodoxy had the final word. One cannot ignore the fact that this typical freethinking of Islam appears in its full-fledged, outspoken form, only on a very limited scale, and for a relatively short period: the ninth and tenth centuries. In the eleventh century Abū'l-ʿAlāʾ al-Maʿarrī professed very similar opinions, but his epistles usually remain within the boundaries of the accepted Muslim tradition. In his *Risālat al-ghufrān*, al-Maʿarrī refers to Ibn al-Rāwandī, using the expression that good Muslim intellectuals reserved to sneer at the freethinkers: "His knowledge was greater than his intelligence."[6] Al-Maʿarrī's own freethinking, his rejection of prophecy, where he stands closer to al-Rāwandī, emerges mostly in

---

[4] See Kraus, "Ketzergeschichte," 117, n. 3; *Fihrist*, 261; Rosenthal, *Saraḫsī*, 33 and n. 98; Tawḥīdī, *Imtāʿ*, II, 19–20; *Maʿāhid al-tanṣīṣ*, I, 156. This *bon mot*, which gained a certain popularity, is found in various contexts. A similar formula was used, for example, by Ibn al-Muqaffaʿ concerning al-Ḫalīl; see Chokr, *Zandaqa et zindīqs*, 195; Murtaḍā, *Amālī*, I, 135; and compare J. H. Haywood, *Arab Lexicograpy*, Leiden 1960, 21, who quotes al-Zubaydī and al-Qifṭī. A different phrasing of the same idea was used concerning *šayḫ al-išrāq* al-Suhrawardī: *kāna ʿindahu naqṣu ʿaqlin, lā ʿilmin*; cf. Cl. Cahen, "La 'Chronique des Ayyoubides' d'al-Makin b. al-ʿAmīd," *BEO* 15 (1955–57), 150–51.

[5] *Ṭabaqāt*, 52–53.

[6] *Risālat al-Ghufrān*, Beirut 1980, 232; cf. also Kraus, "Ketzergeschichte," 117, n. 3.

his poetry, perhaps because of the possibilities offered by poetic license.

It appears that after the tenth century, blunt prose expression of freethinking was no longer possible. The preoccupation of intellectuals with prophecy then found very different expressions. Philosophical parables like Ibn Sinā's, or poetry like al-Maʿarrī's and Jālāl al-Dīn al-Rūmī's, offered ways for discussing this preoccupation that were deemed safer for the writers, and perhaps also intellectually more rewarding. For, rather than forcing these thinkers into a headlong collision with the notion of prophetic religion, these new ways made it possible to integrate transformed echoes of freethinking into the Islamic legacy.

BIBLIOGRAPHICAL ABBREVIATIONS

(Abbreviations of Periodicals follow *EI*[2])

Primary Sources

'Abd al-Jabbār, *Tathbīt* = 'Abd al-Jabbār b. Aḥmad al-Hamadhānī, *Tathbīt dalā'il al-nubuwwa*, ed. 'Abd al-Karīm 'Uthmān, Beirut 1966–67.

——, *Mughnī* = ——, *Al-Mughnī fī abwāb al-tawḥīd wa'l-'adl*, various editors, Cairo 1958–66.

Abū-Qurra, *Mayāmir* = *Mayāmir Thaudūrus Abī Qurra*, ed. Q. Basha, Beirut n.d.

Abū Ḥātim, *A'lām al-nubuwwa* = Abū Ḥātim al-Rāzī, *A'lām al-nubuwwa* (*The Peaks of Prophecy*), ed. Ṣalāḥ al-Sāwy, Tehran 1977.

Abū Rā'iṭa, *Rasā'il* = *Die Schriften des Jacobiten Ḥabīb ibn Ḫidma Abū Rā'iṭa*, ed. G. Graf, Louvain 1951.

'Ammār al-Baṣrī, *Burhān* = 'Ammār al-Baṣrī, *Kitāb al-burhān*, in M. Hayek, *'Ammār al-Baṣrī: Apologie et controverses*, Beirut 1986.

'Ammār al-Baṣrī, *Masā'il* = ——, Ibid.

'Askarī, *Awā'il* = Abū Hilāl al-'Askarī, *Kitāb al-awā'il*, ed. Muḥammad al-Miṣrī and Walīd Qaṣṣāb, Damascus 1975.

Aš'arī, *Maqālat* = al-Aš'arī, *Maqālāt al-islāmiyyīn wa-iḫtilāf al-muṣallīn*, ed. M. M. 'Abd al-Ḥamīd, Cairo 1950.

'Aynī, *'Umdat al-Qārī'* = Badr al-dīn Abū Muḥammad al-'Aynī, *'Umdat al-Qārī' fī šarḥ al-Buḫārī*, Cairo 1308H.

Baghdādī, *Farq* = 'Abd al-Qāhir b. Ṭāhir al-Baghdādī, *Kitāb al-Farq bayna al-firaq wa-bayān al-firqa al-nājiya minhum*, ed. Muḥammad Badr, Cairo n.d (English translation in *Moslem Schisms and Sects*, pt. I, trans K. C. Seeyle, New York 1919; pt. II, tr. A. S. Halkin, Tel Aviv 1935).

Bāqillānī, *Bayān* = Abū Bakr Muḥammad b. al-Ṭayyib b. al-Bāqillānī, *Al-bayān 'an al-farq bayna al-mu'jizāt wa'l-karāmāt wa'l-ḥiyal wa'l-kahāna wa'l-siḥr wa'l-nāranjāt*, ed. R. J. McCarthy, Beirut 1958.

Bīrūnī, *Risāla* = *Risālat al-Bīrūnī fī fihrist kutub Muḥammb bn. Zakariyā' al-Rāzī—Epître de Bērūnī contenant le répertoire des ouvrages de Muḥammad b. Zakarīyā ar-Rāzī*, ed. P. Kraus, Paris 1936.

——, *Hind* = ——, Abū al-Rayḥān al-Bīrūnī, *K. Fī taḥqīq mā li'l-hind min maqūla maqbūla fī'l-'aql aw mardhūla*, Hyderabad 1958 (translated in E. C. Sachau, *Alberuni's India*, London 1910).

Correspondance Islamo-chrétienne = *Une Correspondance Islamo-chrétienne entre Ibn al-Munaǧǧim, Ḥunayn Ibn Isḥāq et Qusṭā ibn Lūqā*, ed. Khalil Samir, tr. P. Nwyia (PO 40/4) Turnhout 1981, 525–723.

Fihrist = Ibn al-Nadīm, *Kitāb al-Fihrist*, ed. G. Flügel, Leipzig 1872, reprinted Beirut 1964 (English translation by B. Dodge, *The Fihrist of al-Nadīm*, New York and London 1970).

Ghazālī, *Fayṣal* = Abū Ḥāmid al-Ghazālī, *Fayṣal al-tafriqa bayna al-Islām wa'l-zandaqa*, ed. Sulaymān Dunyā, Cairo 1961 (English translation in R. J. McCarthy, *Freedom and Fulfillment*, Boston 1980, 145–74).

——, *Munqidh* = ——, *al-Munqidh min al-ḍalāl wa'l-mūṣil ilā dhī al-'izza wa'l-jalāl*, ed. Jamīl Ṣalība and Kāmil 'Iyyād, Cairo 1981 (English translation in R. J. McCarthy, *Freedom and Fulfillment*, Boston 1980, 61–114).

K. al-Ḥayda = (Pseudo) ʿAbd al-ʿAzīz al-Kinānī, *Kitāb al-ḥayda*, ed. Jamīl Ṣalība, Damascus 1964.
IAU = Ibn Abī Uṣaybiʿa, *ʿUyūn al-anbāʾ fī ṭabaqāt al-aṭibbāʾ*, Cairo 1882.
Ibn Ḥazm, *Fiṣal* = Ibn Ḥazm, *al-Fiṣal fīʾl-milal waʾl-ahwāʾ waʾl-niḥal*, ed. Muḥammad Ibrāhīm Naṣr and ʿAbd al-Raḥmān ʿUmayra, Beirut 1985.
——, *Radd* = ——, *Risāla fī al-radd ʿalā Ibn al-Naghrīla al-yahūdī*, in *Rasāʾil Ibn Ḥazm al-Andalusī*, ed. *Iḥsān ʿAbbās*, Beirut 1981, 41–70.
——, *Taqrīb* = ——, *al-Taqrīb li-ḥadd al-manṭiq waʾl-madḫal ilayhi biʾl-alfāẓ al-ʿāmmiyya waʾl-amthila al-fiqhiyya*, ed. Iḥsān ʿAbbās, Beirut 1959.
Ibn al-Jawzī = (Ritter).
Ibn Kathīr, *Bidāya* = Ibn Kathīr, *al-Bidāya waʾl-nihāya*, Cairo 1932 .
——, *Tafsīr* = *Tafsīr al-qurʾān al-ʿaẓīm liʾl-imām . . . ibn kathīr*, ed. Ḫalīl al-Mayis, Beirut n.d.
Ibn Ḫallikān, *Wafayāt al-aʿyān* = Aḥmad b. Muḥammad Ibn Ḫallikān, *Wafayāt al-aʿyān wa-anbāʾ abnāʾ al-zamān*, ed. Iḥsān ʿAbbās, Beirut 1968–77.
Ibn al-Jawzī, *Muntaẓam* = Ibn al-Jawzī, *al-Muntaẓam fī taʾrīḫ al-mulūk waʾl-umam*, ed. Muḥammad ʿAbd al-Qādir ʿAṭā and Muṣṭafā ʿAbd al-Qādir ʿAṭā, Beirut 1992, vol. XIII.
——, *Talbīs* = ——, *Talbīs Iblīs*, Cairo 1368H.
Ibn al-Murtaḍā, *Ṭabaqāt al-Muʿtazila* = Ibn al-Murtaḍā, *Ṭabaqāt al-Muʿtazila*, ed. S. Diwald Wilzer, Beirut 1961.
Ibn al-Qifṭī = Jamāl al-Dīn Ibn al-Qifṭī, *Taʾrīḫ al-ḥukamāʾ*, ed. J. Lippert, Leipzig 1903.
Ibn Ṣāʿid, *Ṭabaqāt* = Ibn Ṣāʿid al-andalusī, *Ṭabaqāt al-umam—Les catégories des nations*, ed. L. Cheicho, Beirut 1912.
Jāḥiẓ, *Ḥujaj al-nubuwwa* = *Kitāb ḥujaj al-nubuwwa*, in *Rasāʾil al-Jāḥiẓ*, ed. Ḥasan al-Sandūbī, Cairo 1933, 117–54.
Ḫayyāṭ, *Intiṣār* = Abū Ḥusayn . . . b. ʿUthmān al-Ḫayyāṭ, *Kitāb al-Intiṣār waʾl-radd ʿalā Ibn al-Rāwandī al-mulḥid (Le livre du triomphe et de la réfutation d'Ibn al-Rawandi l'hérétique)*, ed. A. N. Nader, Beirut 1957 (ed. H. S. Nyberg, Cairo 1925).
Kanz al-ʿummāl = al-Muttaqī al-Hindī, *Kanz al-ʿummāl fī sunan al-aqwāl waʾl-afʿāl*, Beirut 1981.
Kirmānī, *al-Aqwāl al-dhahabiyya* = Ḥamīd al-Dīn Kirmānī, *al-Aqwāl al-dhahabiyya* (Golden Sayings), ed. S. al-Sawy, Tehran 1977.
Kindī, *Risāla* = G. Tartar, *Dialogue Islamo-chrétien sous le calife al-Maʾmūn (813-834): Les épîtres d'Al-Hāšimī et d'Al-Kindī, Thèse pour le doctorat de 3e cycle*, Strasbourg 1977.
Kuzari = *Kitāb al-Radd wa-ʾl-Dalīl fīʾl-Dīn al-Dhalīl (al-Kitāb al-Khazārī)*, ed. D. H. Baneth and H. Ben-Shammai, Jerusalem 1977.
Lisān al-ʿarab = Ibn Manẓūr, *Lisān al-ʿarab* Beirut n.d.
Maʿāhid = ʿAbd al-Raḥīm al-ʿAbbāsī, *Maʿāhid al-tanṣīṣ ʿalā šawāhid al-talḫīṣ*, ed. Muḥammad Muḥyī al-Dīn ʿAbd al-Ḥamīd, Cairo 1947.
Maimonides, *Commentary on the Mishah* = Moses Maimonides, *Commentary on the Mishnah: Mishnah ʿim Perush Rabbenu Moshe ben Maimon*, ed. Y. Qafiḥ, Jerusalem 1963–68 (7 vols.).
——, *Dalāla* = Mūsā bn. Maymūn, *Dalālat al-Ḥāʾirīn*, ed. S. Munk and Y. Joel, Jerusalem 1931.
——, *Guide* = Maimonides, *The Guide of the Perplexed*, tr. S. Pines, Chicago 1963.
Mānkdīm, *Šarḥ al-uṣūl al-ḫamsa* = Mānkdīm Šešdīv, *[Taʿlīq] Šarḥ al-uṣūl al-ḫamsa*, ed. ʿAbd al-Karīm ʿUthmān, Cairo 1965.
Masʿūdī, *Murūj* = Abū al-Ḥasan ʿAlī b. al-Husayn al-Masʿūdī, *Murūj al-dhahab wa-maʿādin al-jawhar—Les Prairies d'or*, ed. Barbier de Meynard et Pavet de Courteille, rev. Ch. Pellat, Beirut 1974.
——, *Tanbīh* = ——, *al-Tanbīh waʾl-išrāf*, ed. ʿAbdallāh Ismāʿīl al-Ṣādiq, Cairo 1938.
Mawardī, *Aʿlām al-nubuwwa* = al-Mawardī, *Aʿlām al-nubuwwa*, ed. ʿAbd al-Raʾūf Saʿd, Cairo 1971.

Māturīdī, *Ta'wīlāt ahl al-Sunna* = Abū Manṣūr al-Māturīdī, *Ta'wīlāt ahl al-Sunna*, ed. Muḥammad Mustafīḍ al-Raḥmān, Baghdād 1983.

——, *Tawḥīd* = ——, *Kitāb al-tawḥīd*, ed. F. Kholeif, Beirut 1970.

Mujarrad maqālāt al-Aš'arī = Abū Bakr b. Fūrak, *Muğarrad maqālāt al-Aš'arī (Exposé de la doctrine d'al-Aš'arī)*, ed. D. Gimaret, Beirut 1986.

Muqammiṣ—see Muqammaṣ (on the name, see above, Ch. 1, n. 77).

Muqammaṣ, *Twenty Chapters* = S. Stroumsa, ed. and tr., *Dāwūd ibn Marwān al-Muqammiṣ's Twenty Chapters ('Ishrūn Maqāla)*, Leiden 1989.

Murtaḍā, *Šāfī* = al-Šarīf al-Murtaḍā, *al-Šāfī fī'l-imāma*, ed. 'Abd al-Zahrā' al-Ḥusaynī al-Ḫaṭīb and Fāḍil al-Mīlānī, Tehran 1986.

Naqd al-nathr = (Pseudo) Qudāma b. Ja'far, *Naqd al-nathr*, ed. Ṭaha Ḥusayn and 'Abd al-Ḥamīd al-'Abbadī, Cairo 1938.

Nasafī, *Radd* = M. Bernand, "Le *Kitāb al-radd 'alā l'Bida'* d'Abū Muṭī' Makḥūl al-Nasafī," *Annales Islamologiques* 16 (1980), 8–126.

Nāṣir-i Ḫusraw, *Jāmi' al-ḥikmatayn* = Nasir e Khosraw, *Kitab-e Jāmi' al-hikmatain—Le livre réunissant les deux sagesses*, ed. H. Corbin and M. Mo'in, Tehran and Paris 1953; Arabic tr. by Ibrāhīm al-Dāsūti Shatā, Cairo 1974.

Qiṣṣa = D. J. Lasker and S. Stroumsa, ed. and tr., *The Polemic of Nestor the Priest: Qiṣṣat Mujādalat al-Usquf and Sefer Nestor ha-Komer*, Jerusalem 1996 (2 vols.).

Qāsim b. Ibrāhīm, *Rādd 'alā al-mulḥid* = *Al-rādd 'alā al-mulḥid li'l-imām al-Qāsim b. Ibrāhīm*, ed. Muḥammad Yaḥyā Sālim 'Azzān, Ṣan'ā' 1992.

Qirqisānī, *Anwār* = Abū Ya'qūb al-Qirqisānī, *Kitāb al-anwār wa'l-marāqib, Code of Karaite Law*, ed. L. Nemoy, New York 1939–43.

Rāzī, *Mafātīḥ al-ghayb* = *Tafsīr al-Faḫr al-Rāzī al-muštahir bi'l-tafsīr al-kabīr wa-Mafātīḥ al-ghayb*, Cairo 1985.

Rāzī, *Rasā'il* = Abū Bakr Muḥammad b. Zakariyyā' al-Rāzī, *Rasā'il falsafiyya*, ed. P. Kraus, Cairo 1939.

——, *Šukūk* = ——, *Kitāb al-Šukūk 'alā Jālīnūs*, ed. M. Mohaghegh, Tehran 1993.

Sa'adya, *Amānāt* = Sa'adya ben Yosef Fayyūmī, *Kitāb al-Muḫtār fī al-amānāt wa'l-i'tiqādāt*, ed. Y. Qāfiḥ, Jerusalem and New York 1970 (English translation in *The Book of Beliefs and Opinions*, tr. S. Rosenblatt, New Haven 1948).

——, *Sefer ha-Galuy* = ——, *Sefer ha-Galuy*, in A. E. Harkavy, *Zikkaron la-Rishonim*, (in Hebrew), St. Petersburg 1881, 133–235.

Sijistānī, *Ithbāt al-nubu'āt* = Abū Ya'qūb al-Sijistānī's, *Kitāb ithbāt al-nubu'āt*, ed. 'Ārif Tāmir, Beirut 1966.

Šadharāt al-dhahab = Ibn al-'Imād al-Ḫanbalī, *Šadharāt al-dhahab fī aḫbār man dhahab*, Cairo 1931.

Šahrastānī, *Milal* = Muḥammad al-Šahrastānī, *K. al-Milal wa'l-niḥal—Book of Religious and Philosophical Sects*, ed. W. Cureton, London 1842.

Shahrastani, *Livre des religions* = Shahrastani, *Livre des religions et des sectes*, tr. D. Gimaret and G. Monnot, Paris 1986.

Tawḥīdī, *Imtā'* = Abū Ḥayyān al-Tawḥīdī, *Kitāb al-Imtā' wa'l-Mu'ānasa*, ed. Aḥmad Amīn and Aḥmad al-Zayn, Cairo 1944.

Ṭabarī, *al-Dīn wa-l'dawla* = 'Alī b. Rabbān al-Ṭabarī, *K. al-Dīn wa'l-dawla fī ithbāt nubuwwat al-nabī Muḥammad* (= *The book of Religion and Empire*), ed. A. Mingana, Manchester 1923.

Zamaḫšarī, *Kaššāf* = Abū al-Qāsim Jār-Allāh Maḥmūd b. 'Umar al-Zamaḫšarī, *al-Kaššāf 'an ḥaqā'iq al-tanzīl wa-'uyūn al-aqāwīl fī wujūh al-ta'wīl*, Dār al-fikr 1977.

Modern scholarship

A'sam, *History* = 'Abd al-Amīr al-A'sam, *Ta'rīḫ Ibn al-Rīwandī al-mulḥid; nuṣuṣ wa-wathā'iq min al-maṣādir al-'arabiyya ḫilāl alf 'ām* [= A. A. al-A'asam (sic!), *History of Ibn ar-Rīwandī, The Heretic*], Beirut 1975.

——, *References* = ——, *Ibn al-Rīwandī fī al-marājiʿ al-ʿarabiyya al-ḥadītha* [= *Ibn ar-Rīwandī in the Modern-Arabic References*], Beirut 1978, 2 vols.
Abrahamov, "Barāhima" = B. Abrahamov, "The Barāhima's Enigma: A Search for a New Solution," *WO* 18 (1987), 72–91.
Bist Guftar = M. Mohaghegh, *Bist Guftar: Twenty Treatises on Islamic Philosophy, Theology, Sects and History of Medicine, with an English introduction by J. van Ess* (Wisdom of Persia Series XVII, McGill University, Tehran 1976).
Bar-Asher, "Abū Bakr al-Rāzī" = M. M. Bar-Asher, "Abū Bakr al-Rāzī (865–925)," *Klassiker der Religionsphilosophie: von Platon bis Kierkegaard*, ed. F. Niewöhner, Munich 1995, 99–111.
——, "L'éthique de Rāzī" = ——, "Quelques aspects de l'éthique d'Abū-Bakr al-Rāzī et ses origines dans l'œuvre de Galien," *SI* 69 (1988), 5–83, 70 (1989), 119–47.
Brion, "Philosophie et révélation" = F. Brion, "Philosophie et révélation: Traduction annotée de six extraits du *Kitāb Aʿlām al-Nubuwwa* d'Abū Ḥātim al-Rāzī," *Bulletin de Philosophie Médiévale* 28 (1986), 134–62.
——, "Le temps" = ——, "Le temps, l'espace et la genèse du monde selon Abū Ḥātim al-Rāzī; Présentation et traduction des chapitres 1, 3–4 du *Kitāb aʿlām al-nubuwwa* d'Abū Ḥātim al-Rāzī," *Revue Philosophique de Louvain*, 87 (1989), 139–64.
Calder, "Literary Construct" = N. Calder, "The Barāhima: Literary Construct and Historical Reality," *BSOAS* 57 (1994), 40–51.
Chokr, *Zandaqa et zindīqs* = M. Chokr, *Zandaqa et zindīqs en Islām au second siècle de l'hégire*, Damascus 1993.
Cohen-Somekh = M. R. Cohen and S. Somekh, "In the Court of Yaʿqūb ibn Killis: A Fragment from the Cairo Genizah," *JQR* 80 (1990), 283–314.
Cook, "ʿAnan and Islam" = M. A. Cook, "ʿAnan and Islam: The Origins of Karaite Scripturalism," *JSAI* 9 (1987), 161–82.
Daiber, "The Ismaili Background" = H. Daiber, "The Ismaili Background of Fārābī's Political Philosophy—Abū Ḥātim ar-Rāzī as a Forerunner of Fārābī," in U. Tworuschka, ed., *Gottes ist der Orient, Gottes ist der Okzident: Festschrift für Abdoldjavad Falaturi, zum 65. Geburtstag*, Köln 1991, 143–50.
——, "Political Philosophy" = ——, "Political Philosophy," *History of Islamic Philosophy*, ed. O. Leaman and S. H. Nasr, London and New York 1996, II, 841–85.
——, "Unity and Diversity" = ——, "Abū Ḥātim ar-Rāzī (10th century A.D.) on the Unity and Diversity of Religions," *Dialogue and Syncretism: An Interdisciplinary Approach*, ed. J. D. Gort, Amsterdam 1989, 87–104.
Davidson, *Hiwi* = I. Davidson, *Saadia's Polemic Against Ḥiwi al-Balkhi*, New York 1915.
De Smet, "polémique ismaélienne" = D. De Smet, "Al-Mu'ayyad fī d-Dīn Šīrāzī et la polémique ismaélienne contre les 'Brahmanes' d'Ibn al-Rāwandī," in U. Vermeulen and D. De Smet (eds.), *Egypt and Syria in the Fatimid, Ayyubid and Mamluk Eras; Proceedings of the 1st, 2nd and 3rd International Colloquium organized at the Katholieke Universiteit Leuven in May 1992, 1993 and 1994*, Leuven 1995, 85–97.
——, *Quiétude* = ——, *La quiétude de l'intellect: Néoplatonisme et gnose ismaélienne dans l'œuvre de Ḥamīd ad-dīn al-Kirmānī* (X^e/XIe s.), Leuven 1995.
Dhanani, *Physical Theory* = A. Dhanani, *The Physical Theory of Kalām: Atoms, Space, and Void in Basrian Muʿtazilī Cosmology*, Leiden 1994.
Druart, "Rāzī's Conception of the Soul" = T.-A. Druart, "Al-Rāzī's Conception of the Soul: a Psychological Background to his Ethics," *Medieval Philosophy and Theology* 5 (1996), 245–69.
——, "Ethics" = ——, "The Ethics of al-Rāzī (865–925?)," *Medieval Philosophy and Theology* 6 (1997), 47–71.
EI = *The Encyclopedia of Islam* 1st ed., Leiden 1913–38.

*EI*² = *The Encyclopedia of Islam*, 2nd ed., Leiden 1960ff.

Emerson, "Deism" = R. L. Emerson, "Deism," in *Dictionary of the History of Ideas: Studies of Selected Pivotal Ideas*, New York 1973, I, 647–52.

Fabro, "Atheism" = C. Fabro, "Atheism," *New Encyclopædia Britannica* (1973), I, 479–480.

Fakhry, *Islamic Philosophy* = M. Fakhry, *A History of Islamic Philosophy*, New York 1983².

Fierro, "Ibn Ḥazm" = M. Fierro, "Ibn Ḥazm et le *zindīq* juif," *Revue de l'Occident Musulman et de la Méditerranée* 63–64 (1992), 81–89.

Fück = J. Fück, "Some Hitherto Unpublished Texts on the Muʿtazilite Movement from Ibn al-Nadīm's *Kitāb al-Fihrist*," in *Professor Muḥammad Shafi' Presentation Volume*, Lahore 1955.

Gabrieli, "Zandaqa" = F. Gabrieli, "La 'Zandaqa' au 1er siècle abbasside," *l'Elaboration de l'Islam – colloque de Strasbourg, 12–13–14 Juin 1954*, Paris 1961, 23–38.

Gardet-Anawati, *Introduction* = L. Gardet and M.-M. Anawati, *Introduction à la théologie musulmane*, Paris 1970.

García Gómez, "Polémiqua," = E. García Gómez, "Polémiqua Religiosa entre Ibn Ḥazm e Ibn al-Naghrīla," *Al-Andalus* 4 (1936), 1–28.

Gimaret, "Bibliographie" = D. Gimaret, "Bilbliographie d'Ašʿarī: un réexamen," *JA* 273 (1985), 223–92.

——, "document" = ——, "Un document majeur pour l'histoire du kalām: Le Muğarrad Maqālāt al-Ašʿarī d'Ibn Fūrak," *Arabica* 32 (1985), 185–218.

Griffith, "Comparative Religion" = S. H. Griffith, "Comparative Religion in the Apologetics of the First Christian Arabic Theologians," *Proceedings of the Patristic, Medieval and Renaissance Conference*, Villanova, Penn. 1979.

Goodman, "The Epicurean Ethic" = L. E. Goodman, "The Epicurean Ethic of Muḥammad ibn Zakariyâ' ar-Rāzī," *SI* 34 (1971), 5–26.

——, "Myth" = ——, "Rāzī's Myth of the Fall of Soul: Its Function in His Philosophy," in G. F. Hourani, ed., *Essays on Islamic Philosophy and Science*, Albany 1975, 25–40.

——, "Rāzī" = ——, art. "al-Rāzī," *EI*², VIII, 474–77.

——, "Rāzī's psychology" = ——, "Rāzī's psychology," *Philosophical Forum* 4 (1972), 26–48.

——, "Muḥammad ibn Zakariyyā' al-Rāzī" = ——, "Muḥammad ibn Zakariyyā' al-Rāzī," *History of Islamic Philosophy*, ed. O. Leaman and S. H. Nasr, London and New York 1996, I, 198–215.

Guidi, *La Lotta* = M. Guidi, *La lotta tra l'Islam e il Manicheismo*, Rome 1927.

Hamdānī, "Unknown Ismāʿīlī Authors" = Ḥ. F. al-Hamdānī, "Some Unknown Ismāʿīlī Authors and their Works," *JRAS* (1933), 359–78.

Iskandar, "Bibliography" = A. Z. Iskandar, "The Medical Bibliography of al-Rāzī," in G. F. Hourani, ed., *Essays on Islamic Philosophy and Science*, Albany 1975, 41–46.

——, "Rāzī" = ——, "al-Rāzī," *The Cambridge History of Arabic Literature: Religion, Learning and Science in the ʿAbbasid Period*, ed. M. J. L. Young et al., Cambridge 1990, 370–73.

James, "Atheism" = G. A. James, "Atheism," *Encyclopedia of Religion*, ed. M. Eliade, New York and London 1987, 479–80.

Kraemer, "Heresy" = J. L. Kraemer, "Heresy versus the State in Medieval Islam," in S. R. Brunswick, ed., *Studies in Judaica, Karaitica and Islamica Presented to Leon Nemoy on his Eightieth Birthday*, Ramat Gan 1982, 167–80.

Kraus, *Gesammelte Aufsätze* = P. Kraus, *Alchemie, Ketzerei, Apokryphen im frühen Islam; Gesammelte Aufsätze*, ed. R. Brague, Hildesheim 1994.

——, "Ketzergeschichte," = ——, "Beiträge zur islamischen Ketzergeschichte: das *kitab az-zumurrud* des Ibn ar-Rāwandī," *RSO* XIV (1933), 93–129, XIV (1934), 335–79 (reprinted in Kraus, *Gesammelte Aufsätze*, 109–90).

——, "Raziana," = ——, "Raziana," *Orientalia*, N.S., 4 (1935), 300–34; 5 (1936), 35–56, 358–78 (reprinted in Kraus, *Gesammelte Aufsätze*, 221–98).
——, *Rasā'il*—see Rāzī.
Kraus and Pines, *EI* = "al-Rāzī," *EI*, 3.II, 1134–1136 (P. Kraus and S. Pines).
Laoust, "Classification" = H. Laoust, "La classification des sectes dans le *Farq* d'al-Bagdādī," *REI* 29 (1961), 19–59.
Lawrence, *Shahrastānī* = B. B. Lawrence, *Shahrastānī on Indian Religions*, The Hague and Paris 1976.
Lazarus-Yafeh, *Intertwined Worlds* = H. Lazarus-Yafeh, *Intertwined Worlds: Medieval Islam and Bible Criticism*, Princeton 1992.
Le Boulluec, *Controverse religieuse* = A. Le Boulluec, ed., *La controverse religieuse et ses formes*, Paris 1995.
Lewis, "Heresy" = B. Lewis, "Some Observations on the Significance of Heresy in the History of Islam," *SI* 1 (1953), 43–63.
Liebes, *Elisha's Sin* = Y. Liebes, *Elisha's Sin: The Four who entered Pardes and the Nature of Tamudic Mysticism* (in Hebrew), Jerusalem 1990.
Madelung, "Abū 'Īsā" = W. Madelung, "Abū 'Īsā al-Warrāq über die Bardesaniten, Marcioniten und Kantäer," in Hans R. Roemer and Albrecht Noth, eds., *Studien zur Geschichte und Kultur des Vorderen Orients, Festschrift für Bertold Spuler zum siebzigsten Geburtstag*, Leiden 1981, 210–24 (reprinted in idem, *Religious Schools and Sects in Medieval Islam*, London 1985, XX).
——, *Qāsim ibn Ibrāhīm* = ——, *Der Imam al-Qāsim ibn Ibrāhīm und die Glaubenslehre der Zaiditen*, Berlin 1965.
Marcuzzo, *Dialogue* = G. B. Marcuzzo, *Le Dialogue d'Abraham de Tibériade avec 'Abd al-Raḥmān al-Hāšimī à Jérusalem vers 820*, Rome 1986.
Massignon, *Recueil* = L. Massignon, *Recueil de textes inédits concernant l'histoire de la mystique en pays d'Islam*, Paris 1929.
Mawsū'at aṭrāf al-ḥadīth = Abū Hājir Zaghlūl, *Mawsū'at aṭrāf al-ḥadīth al-nabawī al-šarīf*, Beirut 1989.
Meier, "Der 'Urknall'" = F. Meier, "Der 'Urknall' eine Idee des Abū Bakr ar-Rāzī," *Oriens* 33 (1992), 1–21.
Miller "Al-Fârâbî's Dispute" = L. Miller, "Al-Fârâbî's Dispute about the *Adab al-Jadal*," *Acts of the International Symposium of Ibn Turk, Khwârezmî, Fârâbî, Beyrûnî and Ibn Sînâ (Ankara, 9–12 September 1985)*, Ankara 1990, 185–88.
Mohaghegh, "*Kitāb al-'ilm al-ilāhī*" = M. Mohaghegh, "Rāzī's *Kitāb al-'ilm al-ilāhī* and the Five Eternals," *Abr Nahrain* 13 (1972), 16–23.
——, *Filsūf-i Rayy* = ——, *Filsūf-ī-Rayy*, Tehran 1970.
Müller-Lauter, "Atheismus" = W. Müller-Lauter, "Atheismus II: Systematische Darstellung," *Theologische Realenzyklopädie*, Berlin and New York, 1979, 378–436.
Niewöhner, "Are the Founders of Religions Impostors?" = F. Niewöhner, "Are the Founders of Religions Impostors?" in S. Pines and Y. Yovel, eds., *Maimonides and Philosophy – Papers Presented at the Sixth Jerusalem Philosophical Encounter, May 1985*, Dodrecht 1986, 233–45.
——, *Veritas* = ——, *Veritas Sive Varietas: Lessings Toleranzparabel und das Buch Von den Drei Betrügern*, Heidelberg 1988.
Nomanul Haq, "The Indian and Persian background" = Syed Nomanul Haq, "The Indian and Persian background," *History of Islamic Philosophy*, ed. S. H. Nasr and O. Leaman, London and New York 1995, I, 52–63.
Nyberg "Deux réprouvés" = H. S. Nyberg, "'Amr ibn 'Ubaid et Ibn al-Rawendi: deux réprouvés," in R. Brunschvig and G. E. Von Grunebaum, *Classicisme et déclin culturel dans l'histoire de l'Islam*, Paris 1957, 125–39.
Pines, *Atomism* = S. Pines, *Studies in Islamic Atomism*, tr. M. Schwartz, ed. T. Langermann, Jerusalem 1997.
——, *Beiträge* = ——, *Beiträge zur islamischen Atomenlehre*, Berlin 1936.

——, "Nouvelles Etudes" = ——, "Nouvelles Etudes sur Awḥad al-Zamān Abū'l-Barakāt al-Baghdādi; Appendice IV: Notes sur Abū Bakr Rāzī," *Mémoires de la Société des Etudes Juives* I, Paris 1955, 55–60 (reprinted in S. Pines, *Collected Works*, I, Jerusalem and Leiden 1979, 142–48).

——, "Philosophy" = ——, "Philosophy," *The Cambridge History of Islam*, ed. P. M. Holt et al., vol. 2B, 780–823.

——, "Rāzī" = ——, "Rāzī critique de Galien", *Actes du VIIIe Congrès International d'Histoire des Sciences*, Paris 1953, 480–87 (reprinted in S. Pines, *Collected Works*, II, Jerusalem and Leiden 1986, 256–63).

——, "Shī'ite Terms" = ——, "Shī'ite Terms and Conceptions in Judah Halevi's *Kuzari*," *JSAI* 2 (1980), 165–251 (reprinted in S. Pines, *Collected Works*, V, Jerusalem 1997, 219–51).

Plessner, "Heresy" = M. Plessner, "Heresy and Rationalism in the First Centuries of Islam" in G. Baer, ed., *The 'Ulamā' and Problems of Religion in the Muslim World: Studies in Memory of Professor Uriel Heyd*, Jerusalem 1971, 3–10.

Poonawala = I. K. Poonawala, *Biobibliography of Ismā'īlī Literature*, Malibu 1977.

Powers, "Ibn Ḥazm" = D. S. Powers, "Reading/Misreading One Another's Scriptures: Ibn Ḥazm's Refutation of Ibn Naghrella al-Yahūdī," *Studies in Islamic and Judaic Traditions*, ed. W. M. Brinner and S. D. Ricks, I, Atlanta 1986, 109–21.

Poznanski, "Miszellen" = S. Poznanski, "Miszellen über Saadja, V," *Monatsschrift für Geschichte und Wissenschaft des Judentums* 51, 718–32.

Rahman, *Prophecy in Islam* = F. Rahman, *Prophecy in Islam: Philosophy and Orthodoxy*, London 1958.

Ritter, "Philologica" = H. Ritter, "Philologica VI: Ibn al-Ǵauzīs Bericht über Ibn ar-Rēwendi," *Der Islam* 19 (1931), 1–17.

Rosenthal, *Saraḫsī* = F. Rosenthal, *Aḥmad b. aṭ-Ṭayyib as-Saraḫsī*, New Haven 1943.

Rosenthal, "Ḥiwi" = J. Rosenthal, "Ḥiwi al-Balkhi – A Comparative Study," *JQR* 38 (1947/8), 317–42, 419–30; 39 (1948/49), 79–94.

Rudolph, *Māturīdī* = U. Rudolph, *Al-Māturīdī und die sunnitische Theologie in Samarkand*, Leiden 1997.

Ruska, "Alchemie" = J. Ruska, "Die Alchemie ar-Rāzī's," *Der Islam* 22 (1935), 281–319.

——, "Al-Bīrūnī" = ——, "Al-Bīrūnī als Quelle für das Leben und die Schriften al-Rāzī's," *Isis* 5 (1923), 26–50.

Schacht, "New Sources" = J. Schacht, "New Sources for the History of Muhammadan Theology," *SI* 1 (1953), 23–42.

——, *Origins* = ——, *The Origins of Islamic Jurisprudence*, Oxford 1979.

Sezgin, *GAS* = F. Sezgin, *Geschichte des arabischen Schriftums*, Leiden 1967–1984, 9 vols.

Spink, *French Free-Thought* = J. S. Spink, *French Free-Thought from Gassendi to Voltaire*, New York 1969.

Stroumsa, "Barāhima" = S. Stroumsa, "The Barāhima in Early Kalām," *JSAI* 6 (1985), 229–41.

——, "The Blinding Emerald" = ——, "The Blinding Emerald: Ibn al-Rāwandī's *Kitāb al-Zumurrud*," *JAOS* 114 (1994), 163–85.

——, "Ecritures alternatives" = ——, "Ecritures alternatives? Tradition et autorité chez les libres penseurs en Islam médiéval," in E. Patlagean and A. Le Boulluec, eds., *Les Retours aux Écritures: fondamentalismes présents et passés* (Bibliothèque de l'Ecole Pratique des Hautes Etudes, section des Sciences Religieuses) Louvain and Paris 1994, 270–93.

——, "Elisha" = ——, "Elisha Ben Abuya and Muslim Heretics in Maimonides' Writings," *Maimonidean Studies* 3 (1995), 173–93.

——, "From Muslim Heresy" = ——, "From Muslim Heresy to Jewish-Muslim Polemics: Ibn al-Rāwandī's *Kitāb al-Dāmigh*", *JAOS* 107 (1987), 767–72.

——, "*Kalām* juif" = ——, "Le *kalām* juif avant et après Maïmonide" *Genos*, Lausanne, forthcoming
——, "Sabéens de Ḥarrān et Sabéens de Maïmonide" = ——, "Sabéens de Ḥarrān et Sabéens de Maïmonide," *Actes du colloque du CNRS sur Maïmonide: Traditions philosophiques et scientifiques médiévales arabes, hébraïques, latines, Paris-Villejuif 17–20 Juin 1997*, ed. T. Lévy (forthcoming).
——, "Signs of Prophecy" = ——, "The Signs of Prophecy: The Emergence and Early Development of a Theme in Arabic Theological Literature," *Harvard Theological Review* 78 (1985), 101–14.
Stroumsa, "Manichaeism" = S. and G. G. Stroumsa, "Aspects of Anti-Manichaean Polemics in Late Antiquity and Under Early Islam," *Harvard Theological Review* 81 (1988), 37–58.
Tartar, *Dialogue* = G. Tartar, *Dialogue Islamo-chrétien sous le calife al-Maʾmūn (813–834): Les épîtres d'Al-Hashimī et d'Al-Kindī*, Paris 1985.
TG = see van Ess.
Thomas, *Anti-Christian Polemic* = D. Thomas, *Anti-Christian Polemic in Early Islam: Abū ʿĪsā al-Warrāq's Against the Trinity*, Cambridge 1992.
Urbach, *The Sages* = E. E. Urbach, *The Sages: Their Concepts and Beliefs* (in Hebrew) Jerusalem 1971.
Urvoy, *Penseurs libres* = D. Urvoy, *Les penseurs libres dans l'Islam classique: L'interrogation sur la religion chez les penseurs arabes indépendants*, Paris 1996.
Vajda, "Ibn al-Rāwandī," = G. Vajda, s.v. "Ibn al-Rāwandī," *EI*², 905–906.
——, "Finalité" = ——, "La finalité de la création de l'homme selon un théologien juif du IXᵉ siècle," *Oriens* 15 (1962), 61–85.
——, "Prophétologie" = ——, "La prophétologie de Dāwūd ibn Marwān al-Raqqī *al-Muqammiṣ*, théologien juif arabophone du IXᵉ siècle," *JA* 265 (1977), 227–35.
——, "Qirqisani" = "Etudes sur Qirqisānī V: Les règles de la controverse dialectique," *REJ* 122 (1963), 7–74.
——, "Zindīqs" = "Les Zindīqs en pays d'Islam au début de la période abbaside," *RSO* 17 (1938), 173–229.
Van Ess, "L'autorité de la tradition prophétique" = J. van Ess, "L'autorité de la tradition prophétique dans la théologie muʿtazilite," *La notion d'autorité au Moyen Age: Islam, Byzance, Occident*, Paris 1980, 211–26.
——, "Disputationspraxis" = ——, "Disputationspraxis in der islamischen Theologie, eine vorläufige Skizze," *REI* 44 (1976), 23–60.
——, "Fārābī" = ——, "Al-Fārābī wa-ibn al-Rīwandī," in *Al-Fārābī wa'l-ḥaḍāra al-insāniyya*, Baghdād 1975–76, 389–98.
——, *Īcī* = ——, *Die Erkenntnislehre des ʿAḍudaddīn al-Īcī: Übersetzung und Kommentar des ersten Buches seiner Mawāqif*, Wiesbaden 1966.
——, "Image" = ——, "Ibn ar-Rēwandi, or the Making of an Image," *al-Abḥāth* 27 (1978/9), 5–26.
——, "Logical Structure" = ——, "The Logical Structure of Islamic Theology," in G. E. von Grunebaum, ed., *Logic in Classical Islamic Culture*, Wiesbaden 1970, 21–50.
——, "Skepticism" = ——, "Skepticism in Islamic Religious Thought," in C. Malik, ed., *God and Man in Contemporary Islamic Thought*, Beirut 1972, 83–98 (= *al-Abḥāth* 21 [1968], 1–18).
——, *TG* = ——, *Theologie und Gesellschaft im 2. und 3. Jahrhundert Hidschra: Eine Geschichte des religiösen Denkens im frühen Islam*, Berlin and New York, I–III, 1991–1992; IV, 1997; VI, 1995.
Vollers, "Religionsgespräch" = K. Vollers, "Das Religionsgespräch von Jerusalem (um 800 D)," *Zeitschrift für Kirchengeschichte* 29 (1908), 29–71.
Walker, *Early Philosophical Shiism* = P. E. Walker, *Early Philosophical Shiism; The Ismaili Neoplatonism of Abū Yaʿqūb al-Sijistānī*, Cambridge 1993.
Walker, "Political Implications" = ——, "The Political Implications of al-Rāzī's

Philosophy," *The Political Aspects of Islamic Philosophy: Essays in Honor of Muhsin E. Mahdi*, ed. C. E. Butterworh, Cambridge Mass. 1992, 61–94.

Walzer, *Greek into Arabic* = R. Walzer, *Greek into Arabic: Essays on Islamic Philosophy*, Oxford 1962.

Wasserstein, *The Party-Kings* = D. J. Wasserstein, *The Rise and Fall of the Party-Kings; Politics and Society in Islamic Spain 1002–1086*, Princeton 1985.

——, "Samuel ibn Naghrīla" = ——, "Samuel ibn Naghrīla ha-Nagid and Islamic Historiography in al-Andalus," *Al-Qanṭara* 14 (1993), 109–25.

Wasserstrom, *Between Muslim and Jew* = S. M. Wasserstrom, *Between Muslim and Jew: The Problem of Symbiosis under Early Islam*, Princeton 1995.

Wensinck, *Concordance* = Wensinck, *Concordance et indices de la tradition musulmane*, Leiden 1965.

Wright, "Against the Galileans" = "Against the Galilaeans," in W. C. Wright, *The Works of the Emperor Julian*, Cambridge, Mass. and London, 1923, III, 318–427.

Wolfson, *Philo* = H. A. Wolfson, *Philo: Foundations of Religious Philosophy in Judaism, Christianity and Islam*, Cambridge 1947.

——, "Veracity of Scripture," = ——, "The Veracity of Scripture in Philo, Halevi, Maimonides and Spinoza," *Alexander Marx Jubilee Volume*, New York 1950, 603–30.

GENERAL INDEX

INDEX OF SOURCES

www.ingramcontent.com/pod-product-compliance
Lightning Source LLC
LaVergne TN
LVHW102227300926
849160LV00005B/47

* 9 7 8 9 0 0 4 3 1 5 4 7 1 *